INTERPRETING SILENT ARTEFACTS
Petrographic Approaches to Archaeological Ceramics

Edited by
Patrick S. Quinn

Archaeopress
Gordon House
276 Banbury Road
Oxford
OX2 7ED

www.archaeopress.com

ISBN 978 1 905739 29 5

© Archaeopress and the individual authors 2009

All rights reserved. No part of this book may be reproduced, stored in retrieval system, or transmitted, in any form or by any means, electronic, mechanical, photocopying or otherwise, without the prior written permission of the copyright owners.

Printed and bound in Great Britain by
Marston Book Services Ltd, Oxfordshire

Cover images

Thin sections of Late Prehistoric ceramics from southern California seen under the polarizing microscope

CONTENTS

Foreword by Ian Whitbread   v

HENRY CLIFTON SORBY (1826-1908) & THE DEVELOPMENT OF THIN SECTION PETROGRAPHY IN SHEFFIELD   1
Noel Worley

THE PROVENANCE POTENTIAL OF IGNEOUS GLACIAL ERRATICS IN ANGLO-SAXON CERAMICS FROM NORTHERN ENGLAND   11
Rob Ixer & Alan Vince

TECHNOLOGICAL INSIGHTS INTO BELL-BEAKERS: A CASE STUDY FROM THE MONDEGO PLATEAU, PORTUGAL   25
Ana Jorge

INDIGENOUS TABLEWARE PRODUCTION DURING THE ARCHAIC PERIOD IN WESTERN SICILY: NEW RESULTS FROM PETROGRAPHIC ANALYSIS   47
Giuseppe Montana, Anna Maria Polito & Ioannis Iliopoulos

PETROGRAPHIC & MICROSTRATIGRAPHIC ANALYSIS OF MORTAR-BASED BUILDING MATERIALS FROM THE TEMPLE OF VENUS, POMPEII   65
Rebecca Piovesan, Emmanuele Curti, Celestino Grifa, Lara Maritan & Claudio Mazzoli

PROVENANCE & PRODUCTION TECHNOLOGY OF EARLY BRONZE AGE POTTERY FROM A LAKE-DWELLING SETTLEMENT AT ARQUÀ PETRARCA, PADOVA, ITALY   81
Lara Maritan, Claudio Mazzoli, Marta Tenconi, Giovanni Leonardi & Stefano Boaro

CERAMIC TECHNOLOGY & SOCIAL PROCESS IN LATE NEOLITHIC HUNGARY   101
Attila Kreiter, György Szakmány & Miklós Kázmér

EARLY POTTERY TECHNOLOGY & THE FORMATION OF A TECHNOLOGICAL TRADITION: THE CASE OF THEOPETRA CAVE, THESSALY, GREECE   121
Areti Pentedeka & Anastasia Dimoula

Contents

FINE-GRAINED MIDDLE BRONZE AGE POLYCHROME WARE — 139
FROM CRETE: COMBINING PETROGRAPHIC &
MICROSTRUCTURAL ANALYSIS
Edward W. Faber, Peter M. Day & Vassilis Kilikoglou

POTTERY TECHNOLOGY & REGIONAL EXCHANGE — 157
IN EARLY IRON AGE CRETE
Marie-Claude Boileau, Anna Lucia D'Agata
 & James Whitley

THE MOVEMENT OF MIDDLE BRONZE AGE — 173
TRANSPORT JARS A PROVENANCE STUDY BASED ON
PETROGRAPHIC AND CHEMICAL ANALYSIS OF CANAANITE
JARS FROM MEMPHIS, EGYPT
Mary Ownby & Janine Bourriau

PETROGRAPHIC ANALYSIS OF EB III CERAMICS FROM — 189
TALL AL-'UMAYRI, JORDAN: A RE-EVALUATION
OF LEVELS OF PRODUCTION
Stanley Klassen

COMPARISON OF VOLCANICLASTIC-TEMPERED — 211
INCA IMPERIAL CERAMICS FROM PARIA, BOLIVIA
WITH POTENTIAL SOURCES
Veronika Szilágyi & György Szakmány

MULTI-VILLAGE SPECIALIZED CRAFT PRODUCTION — 227
& THE DISTRIBUTION OF HOKOHAM SEDENTARY PERIOD
POTTERY, TUSCON, ARIZONA
James M. Heidke

A PRELIMINARY EVALUATION OF THE VERDE — 245
CONFEDERACY MODEL: TESTING EXPECTATIONS
OF POTTERY EXCHANGE IN THE CENTRAL ARIZONA
HIGHLANDS
Sophia E. Kelly, David R. Abbott, Gordon Moore, Christopher Watkins
& Caitlin Wichlacz

CERAMIC PETROGRAPHY & THE RECONSTRUCTION — 267
OF HUNTER-GATHERER CRAFT TECHNOLOGY IN
LATE PREHISTORIC SOUTHERN CALIFORNIA
Patrick Quinn & Margie Burton

Acknowledgments — 297

FOREWORD

This volume presents a range of petrographic case studies as applied to archaeological problems, primarily in the field of pottery analysis, i.e. ceramic petrography. Petrographic analysis involves using polarising optical microscopy to examine microstructures and the compositions of rock and mineral inclusions in thin section, and has become a widely used technique within archaeological science. The results of these analyses are commonly embedded in regionally specific reports and research papers. In this volume, however, the analytical method takes centre stage and the common theme is its application in different archaeological contexts.

The volume was inspired by the meeting on *Petrography of Archaeological Materials*, co-hosted by the Department of Archaeology, University of Sheffield, UK and the Ceramic Petrology Group (www.ceramicpetrology.com), on 15-17 February 2008. It is a natural successor to two earlier volumes on petrographic studies produced by the British Museum. The Sheffield conference was attended by around 60 participants from Britain and other regions in Europe and from North America. As might be anticipated the participants, including specialists on pottery, plaster, mortar, mudbrick and other materials, also attended a microscope workshop in the Materials Science Laboratory in the Department of Archaeology. This maintains a tradition long established at Ceramic Petrology Group meetings of fostering 'hands-on' shared experience in materials identification and interpretation.

Petrographic analysis can be relatively easy to learn at a basic level because it is visual and descriptive, and relates to hand specimen properties and geological materials. Nevertheless, identification and interpretation of unusual inclusions and varied micromorphologies arising from choices in raw materials, processing, forming and firing can be demanding even for specialists. Shared experience is therefore a key resource for addressing these issues, but the major problem faced by all petrographers, is how to communicate complex visual and analytical information in a concise and effective manner. A photomicrograph helps, but it represents only a fraction of the sample area and poorly conveys microstructural diversity. Nor can a photomicrograph incorporate the range of optical properties necessary to identify inclusions accurately. This information needs to be communicated in reports via comprehensive description and interpretation to aid other petrographers in recognising technological properties and comparing results. Publication on the Internet offers opportunities to expand beyond the limitations of traditional paper reports, but face-to-face discussion over a polarising microscope remains an essential means of exchanging information.

Petrography pioneer Henry Clifton Sorby had begun to address archaeological questions as early as the 1860s, only a few years after his remarkable advances in developing the thin section analysis of rocks (Worley, this volume). It is not at all surprising that petrography was quickly recognised within archaeological research as a valuable tool for answering questions. Procedures for examining mineral and rock inclusions in composite materials were subsequently developed in the earth sciences, especially in the field of sedimentary petrology, but properties specific to the anthropogenic manipulation of these materials also need to be addressed, and this can only be done within archaeology. Prime examples are the pioneering works of Anna Shepard in the USA and David Peacock in the UK, who established ceramic petrography as a specialist archaeological field.

Foremost amongst the archaeological issues to have exploited petrography is the study of ancient trade and exchange, which is the subject of several papers in this volume (Ixer and Vince; Montana *et al.*; Boileau *et al.*; Ownby and Bourriau; Heidke; Kelly *et al.*). The most basic question raised by many excavators is whether pottery at their site was locally produced or imported from nearby or remote sources. Ceramic petrography lends itself to such issues because inclusions in pottery can be compared with local and regional geology, as well as with fabrics of pottery from known sources. It may be suspected that excavators like to find that their site was integrated into regional exchange networks, since this indicates a wider social significance for the settlement under investigation. On the other hand, sites with limited evidence for such engagement have their own stories to interpret (Jorge; Maritan *et al.*; Pentedeka and Dimoula; Szilágyi and Szakmány; Kelly *et al.*) and are just as critical for gaining insight into social organisation at the regional scale. Dealing with large quantities of imported fabrics is as challenging as it is exciting because it takes considerable effort to identify remote source areas, especially where different regions of similar geological character present viable alternatives. This is where petrography benefits from the ability to build on previous research through re-analyse of thin sections from earlier studies (Ownby and Bourriau). We can see here that well archived (Worley) petrography projects can remain an active research resource long after the initial studies have been completed.

Analysis of composition and microstructure lends itself to investigations of ancient technology. Recent research has highlighted the socially embedded nature of technology and its role in identifying social identities and boundaries. Technology is therefore not just of interest to materials specialists, but is relevant to a wide range of archaeological research on social issues. Several papers in this volume focus on technological and social aspects of pottery production (Jorge; Piovesan *et al.*; Kreiter *et al.*; Klassen; Quinn and Burton). Raw materials prospection, sampling and processing are frequently used when interpreting local pottery fabrics. Moreover, this type of analysis can reveal something of the interaction between ancient craftspeople and their natural environment, as much in identifying materials that were not selected as those that might have been used. Several contributions to this volume employ this approach to assist in identifying processing technologies such as refining (e.g. sieving and levigation), intentional addition of materials as temper and clay mixing, which may characterise particular technological traditions. One type of temper in particular that excites petrographers is grog, or crushed pottery (Jorge; Maritan *et al.*; Piovesan et al.; Klassen; Quinn and Burton). Grog is material that clearly was added intentionally, but the reasons for incorporating fragments of old pottery within new products are varied and rarely identifiable with surety. There is potential for developing the identification of vessel forming techniques from ceramic microstructures (Pentedeka and Dimoula; Quinn and Burton), though challenges are encountered in the small areas covered by thin sections and the disruption of traces during clay working. Finally, firing conditions can be estimated from the optical properties of fired clay and the condition of certain inclusions (Pentedeka and Dimoula; Quinn and Burton), but SEM studies have proved to be especially successful in this respect (Faber *et al.*).

Naturally, petrographic studies encounter problems where the technique is pushed to its limits. Taking on these problems is fundamental to advancing the role of the technique and to opening new and alternative avenues for addressing archaeological questions. Contributions to this volume illustrate some of these challenges. Ceramic petrography works best where there is regional diversity in rock types. While too much diversity compounds investigations with alternative potential sources (Kelly *et al.*), detailed analysis is necessary to discriminate sources in regions of overwhelmingly similar geology (Ixer and Vince; Jorge; Szilágyi and Szakmány; Kelly *et al.*; Quinn and Burton). Fabrics with relatively large and compositionally distinctive inclusions are especially sensitive to petrographic analysis, but chemical analysis is usually more effective for finer fabrics or those dominated by common minerals such as quartz. There are circumstances where petrography can be usefully applied to finer material, however, especially with the aid of associated techniques such as SEM (Faber *et al.*) and clustering routines to aid fabric classification (Montana *et al.*). Combining petrography and chemical analysis is especially productive (Maritan *et al.*; Kelly *et al.*). Familiar methods can also benefit from modification and application with well-targeted aims. Point counting, for example, has been successfully applied to pottery in Arizona (Heidke), showing that enhancements in technique and appropriate application can bring significant rewards.

Finally, there are some areas of future research not included in this volume that might be considered, such as digital image analysis, experimental petrography and the *chaîne opératoire*. The former has potential to generate valuable quantified petrographic data, but so far its application has been limited. This may be due in part to the complexity of optical properties and microstructures encountered in many thin sections, the relatively small area of many samples and the time needed to capture, process and interpret the data for large numbers of samples. Experimental petrography is an extension of the raw materials analysis referred to above, with the aim of generating specific microstructural properties in modern samples in order to better understand those found in ancient materials. This applies particularly to questions of clay mixing and the generation of distinctive void structures. The *chaîne opératoire*, or sequence of actions and choices in a technological process, is widely recognised as a means investigating technological traditions. Several papers in this volume touch on this area (Jorge; Kreiter *at al.*; Pentedeka and Dimoula; Klassen; Quinn and Burton). There is clearly an avenue of research in integrating more closely raw materials studies with those of forming, decorating and firing to better distinguish technological traditions and integrate them into broader studies of identity and social relations.

Such opportunities to develop petrographic methods and their application are encouraging. They encapsulate a flexibility that allows the technique to address wide-ranging problems and diverse archaeological assemblages, not just textbook case studies, and further develop the work initiated in the mid-nineteenth century by pioneers such as Henry Clifton Sorby.

Dr. Ian K. Whitbread
School of Archaeology and Ancient History
University of Leicester, UK

HENRY CLIFTON SORBY (1826-1908) & THE DEVELOPMENT OF THIN SECTION PETROGRAPHY IN SHEFFIELD

Noel Worley

British Gypsum, Geological & Mining Services, Loughborough, UK
(noel.worley@saint-gobain.com)

Introduction

Henry Clifton Sorby, the son of a 19th century Sheffield factory owner, inherited a private fortune, which he used to pursue a career in science. Sorby was a pioneer in many fields of study and one of his first breakthroughs in 1851 was the development of polarised light microscopy and the production of the first thin sections of rock specimens. Sorby is acknowledged as the founder of microscopical petrography, an essential technique that is applied in the investigation of archaeological materials such as pottery, lithics and plaster. The methods that he established remain fundamentally unaltered and continue to be standard practise in geology and archaeological petrography. Sorby applied his petrographic knowledge of rocks to the identification of the provenance of archaeological materials and developed further interests in other areas of archaeology, art and anthropology.

Sorby's skills as a microscopist and scientist also led him to develop other revolutionary ideas. He made groundbreaking contributions to sedimentology, cosmology, metallurgy, spectroscopy, and biology. The scientific community was slow to recognise his achievements, particularly in Britain. However, his status as a groundbreaking Victorian scientist has since been re-established, especially outside the Britain, where his name has become associated with several prestigious academic awards.

This paper, which draws on evidence from a variety of sources including some fresh documentary evidence, describes Sorby's scientific developments, how he came to develop the technique of thin section petrography, and the equipment he used.

Henry Clifton Sorby

Henry Clifton Sorby was born in 1826 in Woodbourne in the eastern part of Sheffield as the only son of a wealthy factory owner Henry Sorby and his wife Amelia. Sheffield, then a town, grew into an industrial city and became the leading centre for the manufacture of steel, engineering goods and cutlery. Sorby's background provided him with a private fortune that supported him throughout his life and enabled him to pursue a scientific career (Higham, 1963, p. 2).

Sorby's initial interests were in chemistry. However, he soon discovered a passion for geology and made his first scientific discovery in 1851, developing the use of polarised light microscopy and producing the first thin sections of rocks. The implications of these developments were not initially recognised in Britain. Elsewhere in Europe, however, their potential was quickly recognised, particularly by Ferdinand Zirkel in Germany and Achille Delesse in France, who used thin sections and polarising light microscopy to establish the fundamentals of igneous and metamorphic petrology and optical mineralogy.

This pattern of pioneering scientific inventiveness, later to be taken up by others, characterised much of Sorby's career. After developing petrography, he made equally important breakthroughs in the study of meteorites, developing the techniques of reflected light microscopy for the investigation of opaque materials. This method was subsequently applied by Sorby to the examination of metals. He was the first to establish that solid metals had a crystalline structure and is therefore credited as the Father of Metallography.

Sorby's research in the field of sedimentology was equally fundamental. He developed the first apparatus to observe and measure the formation of sediments and he devised mathematical equations to explain their development. He also introduced a petrological classification of sedimentary rocks, which forms the basis of all contemporary systems used at the present day (Sorby, 1879b).

The Development of Thin Section Petrography

The process of manufacturing thin slices of solid materials was demonstrated to Sorby by his friend William Crawford Williamson, the Professor of Zoology at Owens College Manchester. Crawford Williamson had developed a technique to permit the examination of the structure of teeth, bones, wood and fossil foraminifera by grinding them on a glass slide until they became transparent to light (Sollas, 1909; Higham, 1963, p. 36). The exact date of this exchange is not known, although Wilcockson (1947) argues that it took place between 1842 and 1849. Sorby appears to have perfected the technique at least by 1849 and realised that it could be applied to the preparation of slices of rocks, thereby revealing the relations between the mineral aggregates and surrounding cements of which they were composed. In 1849 Sorby produced his first thin sections of Pre-Cambrian rocks (c. 542 Ma BP - million years before present) collected from the Malvern Hills in England. The findings of this investigation were reported in December of the same year to the Sheffield Literary and Philosophical Society and represent the first ever attempt to draw geological conclusions from thin sections (Wilcockson, 1947).

Sorby had been educated in the study of light and optics by his private science tutor Reverend Walter Mitchell. Since the development of the Nicol prism in 1829, the use of polarised light for microscopy was well established and would have been generally available. Sorby realised its potential and began applying it to the investigation of the thin sections of rocks that he was producing (Nuttall, 1981). He published the first results from thin sections of the Jurassic (c. 158 MA BP) Lower Calcareous Grit from the coast of Yorkshire, England (Figure 1A,B) in the Journal of the Geological Society and the Proceedings of the Yorkshire Geological Society, following the reading of the paper in Sheffield (Sorby, 1851). Whilst this landmark study attracted little attention amongst the geological community, it is now universally recognised as the first published petrographic account (Judd, 1908).

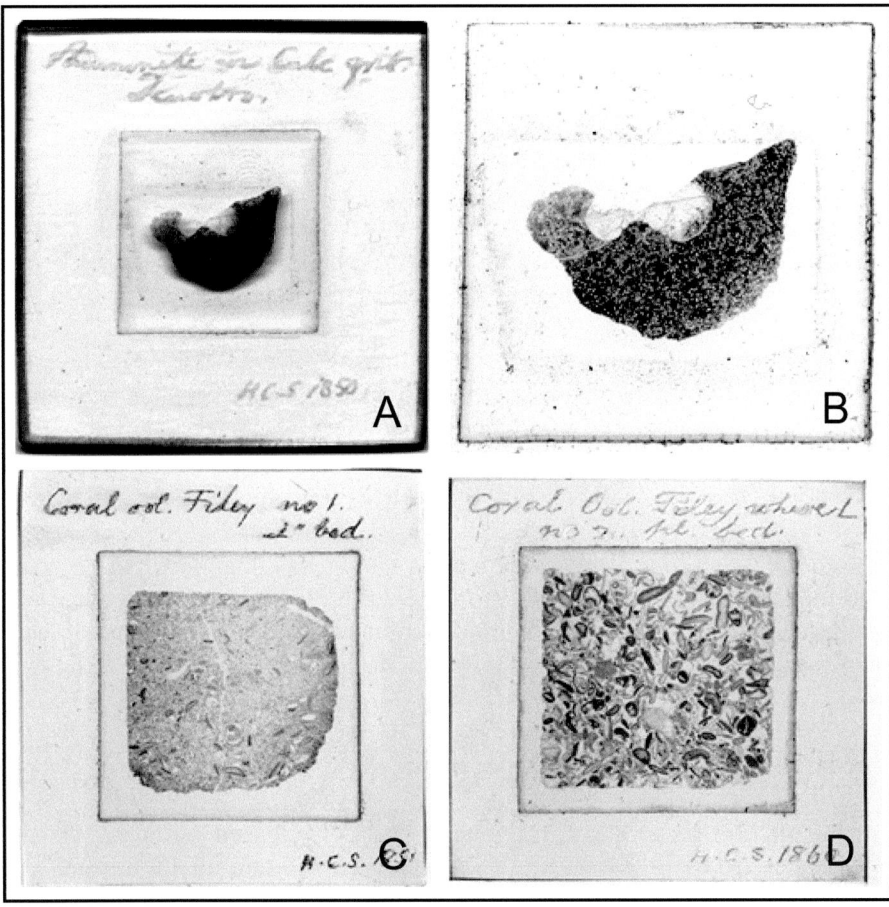

Figure 1. Examples of Sorby's early thin sections. a) Ammonite shell filled with calcareous sandstone sediment from the Cretaceous Lower Calcareous Grit Formation near Scarborough Castle, North Yorkshire, UK. This specimen, the findings of which were published in Sorby (1851), is amongst the first petrographic thin sections ever made. The square slide measures 40 x 40 mm, b) Enlarged image of the previous slide, with visible sand-sized quartz grains. The balsam used to adhere the specimen to the slide is highlighted by a yellow oxidation rim, c) Thin section of biosparite Jurassic Coral Oolite Formation, from Filey, North Yorkshire, UK, dated 1851, d) Thin section of the Coral Oolite Formation, dated 1860. In this later section the sample has been trimmed square and represents an improvement in quality over the earlier sections.

Sorby's Petrographic Microscopes

Sorby was fortunate that during the 19th century Britain led the world in microscope technology. It was during this period that the achromatic lens was invented by Joseph Jackson Lister (Nuttall, 1973). This development overcame the chromatic and spherical aberrations associated with earlier lens designs, thereby permitting observations at far higher resolutions than were previously possible. Sorby had the financial resources to obtain the best available equipment and on 28 April 1848, he purchased his first microscope, 'The best smaller model', made by Smith and Beck of London (Nuttall, 1981). A record of the purchase is contained in the Smith and Beck archive and in the Sorby collection housed in the Sheffield Records Office (SLP. 51–180) and a photograph, reproduced in Figure 2A, shows Sorby with his instrument. The increased resolution of this achromatic microscope must have greatly aided Sorby in his research and he skilfully exploited the potential of the new instrument. It was this microscope that he used for his early work, developing the technique of thin section preparation, investigating slatey cleavage in metamorphic rocks (Sorby, 1851; 1856a,b) and the nature of inclusions within mineral crystals (Sorby, 1858).

There is no description of the adaptation of this microscope for use with polarised light, although it is known that a polarising apparatus was supplied. Unfortunately Sorby's earliest microscope can no longer be traced. Nevertheless, similar instruments such as that illustrated in Figure 2B are housed in the Royal Microscopical Society Collection at the Museum of the History of Science, Oxford University, England.

In 1861 Sorby purchased a second microscope, which was manufactured especially for him by Smith and Beck (Humphries, 1967) (Figure 2C). This microscope featured a polarising apparatus, a stage that could accommodate his square format slides, and a binocular eyepiece that used the Wenham prism design.

Sorby's Method of Thin Section Preparation

Modern thin section preparation techniques are largely based on those, which were originally developed by Sorby in the 19th century. It was not until 1868, some 16 years after his ground breaking 1851 paper, that he provided a brief outline of his method (Sorby, 1868, 1882). Sorby collected 25 mm rock specimens in the field using either a hammer or a type of handsaw referred to as a 'slitting' saw. One side of the specimen was then ground flat on a slab using emery powder and fixed to a 1 5/8" (40 mm) square glass microscope slide with Canada Balsam adhesive. The mounted sample was then ground progressively thinner using finer grades of emery paper and lastly 'Water of Ayr' stone, until they had reached 30 µm thick. As today, the thickness of the slide was determined using the interference colour of quartz. Sorby left more fragile materials including limestone somewhat thicker and ground very fine grained rocks to less than beyond 30 µm in thickness. To avoid scratching the glass slide, he attached thin 'shims' made from zinc sheeting of suitable thickness to the corners of the slides. Sorby labelled each slide with its location and the date that it was manufactured, using a diamond stylus (Figure 1).

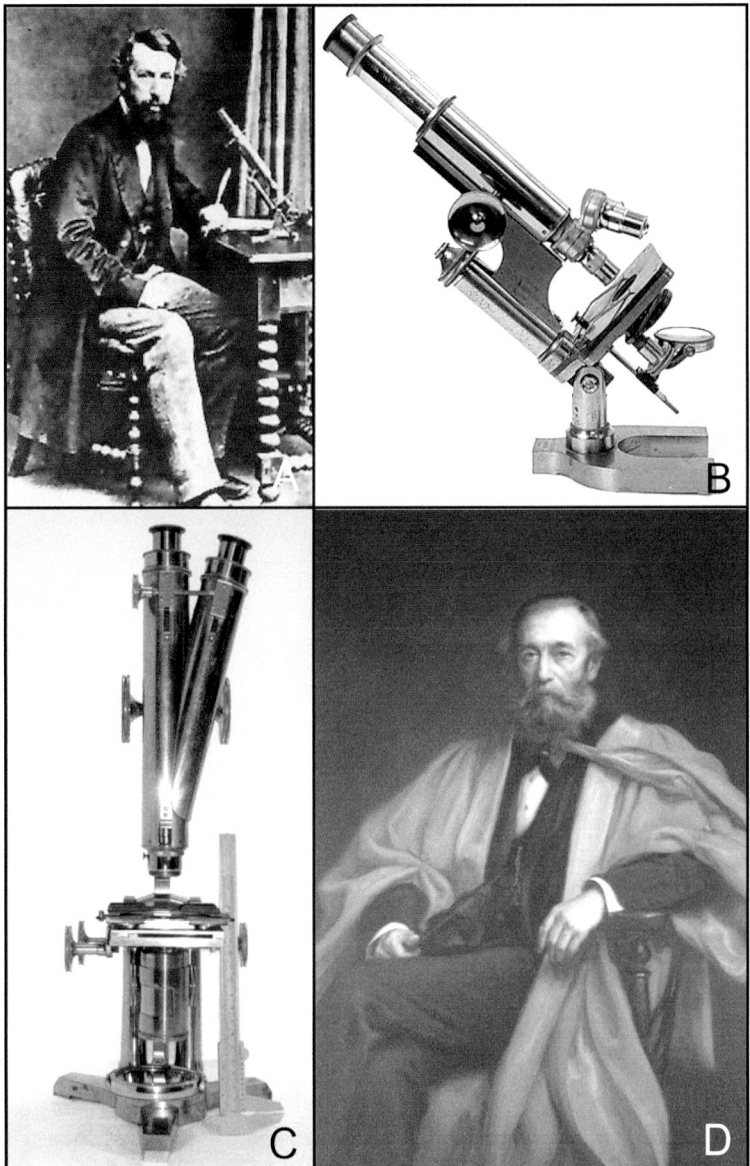

Figure 2. Sorby and his petrographic microscopes. a) Sorby with his Smith & Beck 'Best Smaller Model' achromatic microscope purchased in 1848. This instrument, numbered 201, no longer survives but a description in Nuttall (1981) states that this was the single pillar design with graduated sliding tube. b) A similar but not identical microscope to Sorby's first instrument. The microscope is 58 cm in height, c) A similar instrument to Sorby's Smith & Beck, 'Large Best Microscope', purchased at a cost of £87 2s 6d in 1861 and made binocular with a Wenham type adaptation. A polarising extra large prism apparatus was fitted. The objectives provided a maximum magnification of 400x and also contained concave reflectors for opaque subjects (Humphries, 1967). This microscope is in the possession of Museums Sheffield, d) A painting of Henry Clifton Sorby by M.L.Walker in 1898, commissioned by the Sheffield Literary and Philosophical Society. Image courtesy University of Sheffield.

Some examples of Sorby's thin sections from the Sorby Collection of Rocks, Minerals, Fossils and Shells held at the University of Sheffield, which contains over 1000 slides, are shown in Figure 1. The quality of his slides is remarkable, although in some the Canada Balsam mounting medium has deteriorated around the margins of the glass cover slips. Sorby's wisdom of writing directly onto the glass slide had overcome the problem of using fragile paper labels, which would have no doubt have disintegrated over the intervening 160 years since the thin sections were made.

Sorby's achievements in petrography were not confined to thin section preparation. He also made several other important contributions to the use and development of the optical microscope, the use of polarised light and spectroscopy. In 1877, Sorby first described the use of the quartz wedge accessory, which is now regarded as an essential tool in petrography for studying birefringence colours. In addition, he devised a method for the determination of the refractive index of minerals and he recognised the value of this property in determinative study (Humphries, 1967).

Sorby and Archaeology

By the 1860s Sorby's reputation as a leading scientist was well established and his petrographic expertise was of world renown. George Rolleston, Professor of Anatomy and Physiology at Oxford University and also an amateur archaeologist, was at the forefront of the application of scientific methods in archaeology. The Ashmolean Museum Records (Rolleston, 032) show that he sent rock specimens from excavation of a cemetery at Frilford, Oxfordshire to Sorby for identification. In a letter dated 26th March 1869 to Rolleston, Sorby describes how he used his extensive reference collection of rocks and identified the provenance of the specimen, a volcanic rock probably from Nieder Mendig in the Eiffel Mountains, Germany.

After the death of his mother in 1878, Sorby developed an interest in sailing and spent the last part of his active life navigating his yacht along the southeast coast of England. The geological exposure of rocks in this area is somewhat limited and as a consequence Sorby became interested in marine biology, as well as architecture and archaeology. He applied his scientific mind with typical rigour in an investigation of the provenance of Roman and Medieval bricks and tiles in East Anglia. This research culminated in his 1887 lecture in Sheffield entitled 'On the Character of bricks made at various periods as means for estimating the Date of the Erection or Repair of Ancient Buildings' (Higham, 1963).

Sorby's interest in archaeology flourished in the latter years of his life, yet uncharacteristically he did not publish the results of his studies. However, he presented a series of popular annual lectures between 1883 and 1900 at the Firth College, Sheffield, which later became the University of Sheffield. A list of these presentations obtained from the former Geology Department of the University (Table 1) illustrates Sorby's eclectic archaeological interests (Fearnsides, 1914).

Date	Title of Lecture
1887	The Rise and Decline of some Great English Ports of early Times – A Study in History and physical geography
1890	Pre Norman Pictorial Art
1892	The development of Architecture in Norman Times
1893	Natural History symbolism in Medieval Times
1895	Medieval Geography
1897	Egyptian Art and Literature before Moses

Table 1. A selected list of lectures presented by Sorby at the Firth College, Sheffield.

Sorby's Legacy

Throughout his life, Henry Clifton Sorby lived in Victorian Sheffield and pursued a scientific career in relative isolation. It was argued by his biographer, Norman Higham that this was because the opportunity to conduct scientific research had yet to become an established feature within the universities. This aspect of Sorby's development offers an explanation about why he pioneered many new and widespread areas of investigation but then left it to others to follow them up. His most important contribution to science is undoubtedly the development of microscopy as a basic technique for the investigation of materials, archaeological, biological, geological, metallurgical or mineralogical, He also was able to demonstrate how microscopic observations could be used to discover entirely new areas of science and was directly responsible for establishing the processes involved in the formation of sedimentary, igneous and metamorphic rocks, as well as metals. The basic techniques that Sorby devised remain in use today and are testimony to his genius.

Sorby received many accolades and awards, including the Royal Society Gold Medal in 1874 and Hon LLD University of Cambridge in 1879. He was elected as the 18th foreign members of the Academy of Lynall in Rome, the worlds' oldest scientific Society. He was a Member of the Imperial Mineralogical Society of St Petersburg, awarded the Great Boerhaave Medal by the Dutch Society of Science, elected a Member of the Academy of Natural Science, Phildelphia and a corresponding member of the Lyceum of Natural History, New York

Sorby devoted a large amount of his time and resources in civic society for he had become a scientist of international renown but firmly attached to his home city. He did much to establish scientific education within Sheffield, firstly with his leadership of the Literary and Philosophical Society, and secondly by bringing to Sheffield in 1879 the British Association for the Advancement of Science Meeting. Afterwards he concentrated on developing university education in the city and the first success in this area was the formation of Firth College, where Sorby was appointed President. This was followed by formation of the Technical School in 1886 where he played a seminal role. The institutes were amalgamated with the Medical School in 1896 to become the University College of Sheffield, which later became the University of Sheffield in 1905.

Sorby died at home in Broomhall, Sheffield, near the university in 1908 and is buried adjacent to his mother Amelia in Ecclesall Parish Church. For the last five years of his life he had become disabled as a result of fractured leg following a fall. He died from a respiratory infection.

In his will Sorby was extremely generous to the University of Sheffield and bequeathed all his research materials including his archaeological library. His financial generosity to scientific research in Sheffield was significant and took the form of the endowment of the Royal Society Fellowship and Professorship of Geology. The influence of this philanthropy has helped shape many of the later research achievements from the University of Sheffield.

Sorby is undoubtedly the most important scientific figure that Sheffield has produced and his reputation has diminished with the passage of time. Higham (1963, p. 2) wrote that "Outside Sheffield and outside geology and metallurgy, the name of Henry Clifton Sorby is little known, and even within these circles all that is known by many is the name". It is however satisfying to know that following publication of Higham's biographical account, Sorby's international reputation has grown, largely in North America, where his achievements are revered. His name is associated with awards made by the International Association of Sedimentologists, the International Metallographic Society, and the University of Warsaw. Important research laboratories in microscopy and fluid dynamics at the Universities of Sheffield, Leeds and Reading in England also bear his name. More recently a measure of his importance to Science has been recognised by NASA in the naming of a Lunar feature Dorsa Sorby, alongside other great natural scientists of the world such as Darwin and Lyell.

Acknowledgements

The author would like to thank Martin Whyte for access to the Sorby Collection and other documents held at the University of Sheffield, Alistair Mclean and Museums Sheffield for assistance with the Smith & Beck microscope in Figure 2C, Ken Dorning, President of the Sorby Natural History Society 2006-2008 for the images in Figure 1 and for helpful comments, the Sheffield Record Office and Early Technology Edinburgh for the image used in Figure 3B, and finally Patrick Quinn for his encouragement and invaluable assistance in writing this article.

References

Beck, J. 1848. In: *Sorby Collected letters*, SLP. 51–180. City Library, Sheffield.

Fearnsides, W.G. 1914. *The First Sorby Lecture*. Unpublished manuscript, Geography Department, University of Sheffield.

Judd, J.W. 1908. Henry Clifton Sorby and the Birth of Microscopical Petrology. *Geological Magazine,* N S decade 5, V: 193.

Higham, N.H. 1963. *A Very Scientific Gentleman. The major achievements of Henry Clifton Sorby*. Pergamon Press, Oxford.

Humphries D.W. 1992. *The preparation of thin sections of Rocks, Minerals and Ceramics*. Royal Microscopical Handbook, Oxford University Press.

Humphries D.W. 1967. The Contributions of Henry Clifton Sorby (1826–1908) to Microscopy. *The Microscope and Crystal Front*: 15: 351-362.

Nuttall, R.H, 1981. The First Microscope of Henry Clifton Sorby. *Technology and Culture*, 22: 275-281.

Nuthall, R.H. 1973. C.R. Goring, J.J. Lister and the Achromatic Microscope. *Microscopy*, 32: 253-261.

Rolleston, G, 1869. Letter from Mr. H. C. Sorby at Sydenham Villa, Tunbridge Wells. George Rolleston Archaeological Archive, 32. Ashmolean Museum Records.

Sollas W.J.S. 1909. Anniversary address to the Geological Society. *Proceedings of the Geological Society, London*, 65: 21-27.

Sorby, H.C. 1851. On the microscopical structure of the Calcareous Grit of the Yorkshire coast. *Quarterly Journal of the Geological Society of London*, 7: 1–6.

Sorby, H.C. 1856a. On Slatey-cleavage, as Exhibited in the Devonian Limestones of Devonshire. *Philosophical Magazine*, 9: 20-37.

Sorby, H.C. 1856b. On the Theory of the Origin of Slatey Cleavage. *Philosophical Magazine,* 9: 127-129.

Sorby, H.C. 1868. In: Beale, L.S. *How to work with the microscope*. Harrison, London: 170-183.

Sorby, H.C. 1879a. On the colouring matter found in Human Hair. Congress of Anthroplogy and Pre-Historic Archaeology. *Journal of the Anthropological Institute of Great Britain and Ireland*, 8: 402-425.

Sorby, H.C. 1879b. The structure and origin of Limestones. *Proceedings of the Geological Society, London*, 35: 56-95.

Sorby, H.C. 1882. Preparation of transparent sections of rocks and minerals. *The Northern Microscopist,* 2: 101-104.

Wilcockson, W.H. 1947. The geological work of Henry Clifton Sorby. *Proceedings of the Yorkshire Geological Society*, 27: 1-22.

THE PROVENANCE POTENTIAL OF IGNEOUS GLACIAL ERRATICS IN ANGLO-SAXON CERAMICS FROM NORTHERN ENGLAND

Rob Ixer

Department of Geology, University of Leicester, UK
(r.ixer@btinternet.com)

Alan Vince

Alan Vince Archaeological Consultancy, Lincoln, UK

Introduction

A notable feature of prehistoric and Anglo-Saxon pottery in northern England and the English midlands is the presence of fabrics with angular, mainly igneous rock clasts, in a relatively fine-grained paste. These fabrics are dominant at sites in East Yorkshire and the Vale of Pickering and have attracted the attention of researchers for many years. The first petrographic analysis of this distinctive type of pottery was undertaken by Ian Stead as part of excavations of Yorkshire Iron Age settlements and cemeteries. Later studies, by members of the Department of Scientific Research of the British Museum (BM) (e.g. Rigby, 2004), attempted to put Stead's excavations, into a regional context.

The publication of Freestone and Middleton (1991) summarises the BM interpretation of these wares. In thin section their non-plastic inclusions are dominated by angular clasts of a limited number of rock types, usually igneous in origin. Both the restricted range of lithologies and their angular shape suggests that the inclusions represent crushed rock temper. These wares are so abundant in parts of East and North Yorkshire, that they must have been produced in the region and are not, for example, imports from Scandinavia, northeast England or Scotland. Given the lack of primary outcrops of igneous rocks in this area, a possible source of the temper could be erratic clasts from mixed glacial drift deposits left by ice sheets that covered much of northern England during the Pleistocene. Freestone and Middleton (1991) suggested that potters selectively extracted igneous erratics of specific composition from these glacial gravels, based upon their dark-coloured, coarse-grained appearance, and crushed them to make temper.

In his study of earlier prehistoric pottery from North Yorkshire, Wardle (1991) also encountered igneous-tempered fabrics and reported that the use of glacial erratic temper started as early as the Bronze Age in this part of Britain. He modified the interpretation of Freestone and Middleton (1991) by demonstrating experimentally that crushing fresh igneous rocks was extremely labour intensive and that heating rocks in a fire and then rapidly cooling them with water (i.e. 'fire-cracking'), was a more effective way of producing temper.

Following on from these important early studies, the present paper examines the nature of Anglo-Saxon igneous-tempered ceramics from several sites in Yorkshire in order to further investigate the origin and technology of this long-lived tradition of pottery manufacture. Thin sections and polished thin sections of ceramics have been used to readdress the 'BM/Wardle' model of northern English igneous-tempered ceramics. A comparison of the igneous inclusions in the ceramics with the analysis of primary sources of igneous rocks, a knowledge of the distribution of glacial deposits and the movement of ice during the Pleistocene is then used to interpret the source of this distinctive type of pottery.

Materials and Methods

Standard thin sections and polished thin sections were prepared from 47 igneous-tempered Anglo-Saxon ceramics from 13 sites in East Yorkshire, North Yorkshire and the southern borders of County Durham (Figure 1). The sections were investigated using transmitted and reflected light microscopy. For comparison, geological samples of a range of primary igneous rock sources in northern England (Shap adamellite and Whin Sill dolerite), the Midlands (Leicestershire Mountsorrel granodiorite), Scotland (Aberdeen granite, Criffel-Dalbeattie granodiorite and Cairnsmore of Fleet granite) and Scandinavia (laurvikite from the Oslo Graben) were also studied. Several thin sections of each type of igneous source were examined in order to establish its composite petrography and account for natural lithological variation that is inherent in most large igneous bodies.

In analysing the ceramics and geological samples, particular attention was given to key primary minerals such as microcline, amphibole and muscovite, the textures and mineral intergrowths, as well as the presence/absence and type of secondary/alteration minerals especially those of the epidote group. The inclusions in the igneous-tempered Anglo-Saxon ceramic samples were matched where possible to a their primary source and the ceramics were then grouped according to the igneous sources that they contain. Finally, the geographic distribution of these igneous-tempered fabric groups and a consideration of the direction of Pleistocene ice movements in northern Britain was used to determine the provenance of the ceramics.

Distribution of Primary Igneous Sources

Coarse-grained acid to intermediate rocks, such as granites, granophyres and adamellites crop out in Cumbria in northwestern England. These include the Skiddaw granite and the Eskdale granite, the Ennerdale granophyre and the small Shap adamellite. Of these sources, only the Shap adamellite is found as glacial erratics in northeastern England. In Dumfries and Galloway, southwestern Scotland, contemporaneous granite and granodiorite outcrops include the Cairnsmore of Fleet Granite and south of this, the Criffel-Dalbeattie Granite-Granodiorite complex. The Cheviots of Northumberland contain small, poorly exposed granite and granodiorite bodies among the more volcanic rocks that comprise the bulk of these rolling hills.

Nearby, the Whin Sill of County Durham and Northumberland is a medium-grained doleritic sill that runs east west across northeast England and crops out in several places.

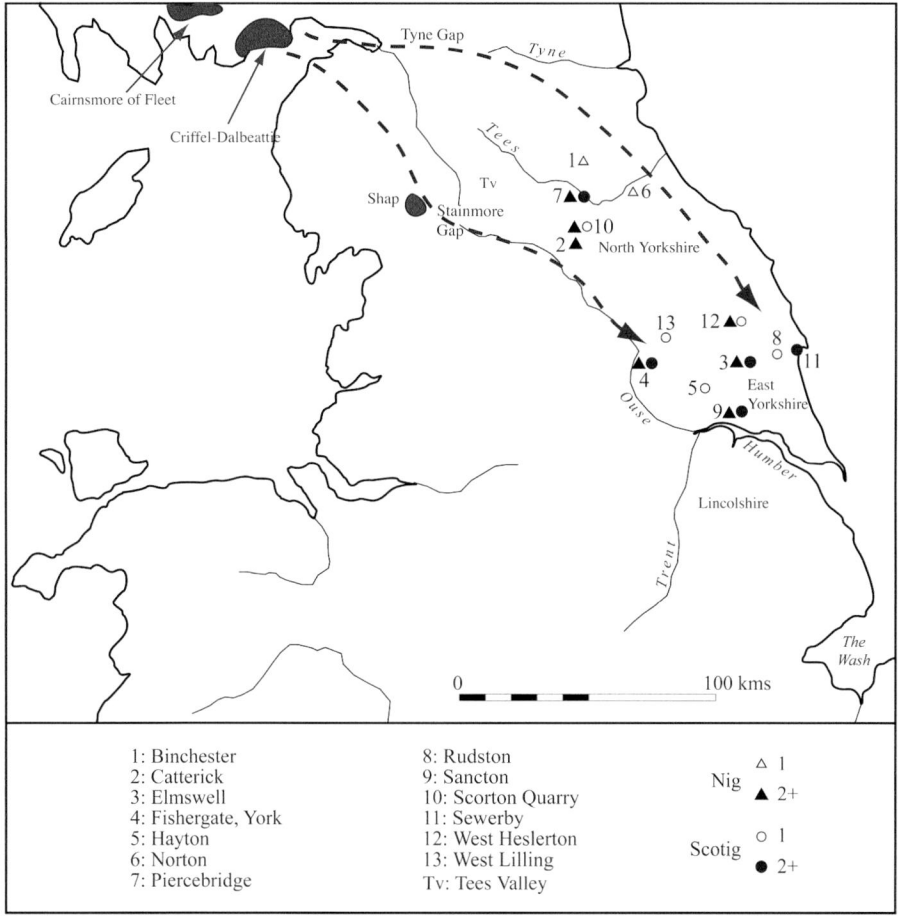

Figure 1. Location of Anglo-Saxon igneous-tempered ceramics analysed in this study, with occurrence of source related fabric groups, primary igneous sources and direction of Pleistocene ice flow. Nig = Shap adamellite-tempered ceramics, Scotig = Scottish igneous-tempered ceramics.

Although the Cumbrian and southern Scottish igneous sources are related geochemically, they can be distinguished from one another petrographically. All sources contain quartz, sodic plagioclase, potassium feldspar and biotite, though the relative amounts of these phases vary. None exhibit significant epidotisation of their feldspars. The Shap adamellite is characterised by biotite with abundant apatite and zircon, that alters to green chlorite and is associated with sphene. In this rock, zoned plagioclases are commonly altered to fine-grained white mica and amphibole, microcline and primary muscovite are all absent. The Cairnsmore of Fleet two-mica granite has primary coarse-grained muscovite and biotite and some of its potassium feldspar occurs as microcline. The Criffel-Dalbeattie granodiorite contains green amphibole, altered plagioclase, altered biotite and locally abundant sphene. The Cheviot Hills 'granite' is highly distinctive and consists of pyroxene-bearing (diopside and augite) granodiorite and pink granophyre. Finally, the intrusion that forms the Whin Sill is a mainly an unaltered quartz dolerite with abundant plagioclase and pyroxene. Potassium feldspars, biotite and amphiboles are essentially absent in this rock.

Although a very wide range of plutonic igneous rocks crop out in Leicestershire, the 'granites' of this area are in fact mainly granodiorites with green amphibole, altered and unaltered biotite and altered plagioclase. Symplectitc intergrowths of quartz and feldspars such as myrmekite are locally present in the groundmass as are epidotised feldspars. Microcline potassium feldspar is absent in this rock. The Oslo Graben Laurvikite is a sodic syenite with very coarse-grained feldspar, intergrown with titanaugite, olivine and opaque minerals.

Glacial Drift and Sources of Igneous Erratics

The oldest glacial drift deposits north of the Humber estuary are exposed at the surface along the east coast of Yorkshire (e.g. the Skipsea Till, Neale and Catt, 1994), where rivers have removed younger covering material. Elsewhere, these deposits are not exposed and probably never covered the chalk, or extended into the Vale of York. Erratic igneous rocks of Scandinavian (Oslo Graben laurvikite), southeastern Scottish (Aberdeen Granite) and northeastern English origin (Cheviot Hills igneous rocks and Whin Sill) occur in this drift. Whilst clasts of biotite granite are present in some places, they are not a prominent feature of the tills and beach sands. Comparative glacial material also occurs in the Lindsey Marshes of Lincolnshire (Wilson, 1963) and the northern and northeastern Norfolk coast (Chatwin, 1961).

Inland, this older drift is covered by more recent glacial deposits that were transported by ice that traveled from the north and northwest along three major routes (Figure 1). Material from the igneous rocks of the Cheviot Hills and Whin Sill was carried by ice flowing southwards along the east coast of England and deposited in the Withernsea Till, which overlies the older Skipsea Till. Ice traveling southeastwards from southwest Scotland, through the Tyne gap and merging with the ice flowing down the east coast of England, carried material from the southwest Scottish granites and granodiorites, but not the Shap adamellite. Erratics from this latter source was transported by ice traveling southeastwards from southwest Scotland along the Eden Valley, through the Stainmore Gap and into the lower Tees valley, along with southwest Scottish granite and granodiorite and Whin Sill. Lastly, a branch of ice from this later route flowed south along the Vale of York and Trent valley (King, 1976).

In each of the three routes of glacial ice flow a different mix of Scottish and northern English igneous rocks, mainly of granite and granodiorite, were entrained in their corresponding drift, allowing them to be distinguished from each other. With this in mind, ceramics produced from the glacial material left in East and North Yorkshire by the different Pleistocene ice movements should also be petrographically distinct.

Igneous Temper in Ceramics from Anglo-Saxon Yorkshire

Vessels containing igneous rock fragments form a distinctive and common group of early-mid Anglo-Saxon pottery fabrics in northern England and the English midlands. In many cases, it is possible to distinguish these from similar prehistoric igneous-tempered fabrics through the form or decoration of the vessels in which they occur. In some parts of Yorkshire and the neighboring counties there is evidence for the survival of pre-Roman igneous-tempering potting traditions through the Roman period and into the early Anglo-Saxon period. This is clearest at Yeavering, where a series of smashed vessels from the occupation levels associated with Anglo-Saxon halls has all the characteristics expected of pre-Roman Iron Age vessels in Northern England, including large angular igneous rock fragments (Hope-Taylor, 1977). Similarly, a female inhumation in an Anglo-Saxon context at Binchester was accompanied by a flat-based jar, analysed in this study, which has a fabric containing large angular rock fragments of sphene-bearing biotite granite. With these exceptions, the majority of the early-mid Anglo-Saxon igneous-tempered pottery found north of the Humber is finer textured than comparative prehistoric material. Common inclusions that are visible to the naked eye include sheaf-like aggregates of biotite and euhedral crystals of feldspar.

Results

Based upon the petrography of their inclusions, the 47 igneous-tempered Anglo-Saxon ceramics can be ascribed to one of three groups that relate to specific primary igneous sources and the glacial deposits derived from them (Tables 1-3; Figures 2-4).

Site	Samples	Shap adamellite-tempered ceramics Interpretation
Binchester	1	Possibly Shap adamellite
Catterick	7	Apatite in biotite in 3 sections (Figure 2C). Sphene noted in 3 sections - possibly Shap adamellite
Piercebridge	5	One sample has biotite granite clasts together with coarse-grained sandstone fragments - possibly Shap adamellite
Scorton Quarry	3	Brown amphibole present in one section - possibly Sharp adamellite (Figure 2B)
Norton	1	Contains inclusions of sphene-bearing, fine-grained biotite granite with a little myrmekite - possibly Shap adamellite
Sancton	1	Contains sphene-bearing biotite granite - possibly Shap adamellite or Criffel-Dalbeattie granodiorite

Table 1. Interpretation of Anglo-Saxon Shap adamellite-tempered ceramics analysed in this study.

Figure 2. Shap adamellite-tempered ceramics. a) Typical rod perthite in sample V1457, b) Simply twinned, zoned plagioclase altering to fine-grained white mica and biotite altering to chlorite along cleavage planes in sample V1197, c) Single biotite lath with many hexagonal apatite sections and a few grey non-basal sections in sample V1429, d) Zoned plagioclase altering to very fine-grained clay in sample V1457). Images taken in crossed polars.

Shap adamellite-tempered ceramics

This group of ceramics includes clasts of altered biotite granite with quartz-potassium feldspar intergrowths (mainly rod and bead perthites) (Figure 2A), plagioclase altering to fine-grained white mica (Figure 2B,D) and biotite altering to chlorite (Figure 2B). In some clasts, the biotite contains abundant apatite inclusions (Figure 2C). The ceramics in this group usually contain little or no amphibole, primary muscovite or microcline and the feldspars have not been altered to epidote. Based upon the mineralogy of their inclusions, the source of igneous material is suspected to be the Shap adamellite. Although it is conceivable that clasts deriving from the southwestern Scottish granites and granodiorites, but without their characteristic minerals, could have a composition that matches the inclusions in the ceramics, they can quite confidently be interpreted as wholly or partly derived from the Shap adamellite.

With the primary origin of the erratic clasts as the Shap adamellite of the Cumbria, it is likely that the drift that was used as temper for these ceramics was transported through the Stainmore Gap. A more specific provenance assignment is not possible. Samples assigned to this group include the pot from Binchester and 17 early-mid Anglo-Saxon vessels.

Site	Samples	Scottish igneous-tempered ceramics Interpretation
Catterick	5	Four sections contain granodiorite (biotite plus green amphibole, Figure 3A,B). Biotite is very extensively altered to chlorite. In one of these amphibole is only present in trace amounts but the alteration of plagioclase is similar to the other three – possible Scottish source. A fifth section contains fragments of a two mica/microcline granite (Figure 4A). Some granodiorite is also present. Scottish source.
Fishergate, York	3	All three sections contain coarse-grained single microcline clasts (Fig 3c) and muscovite flakes and rarer granite clasts and no biotite. In each case these inclusions are less frequent than coarse-grained sandstone fragments (more frequent in one section than in the other two) - Scottish
Hayton	1	This section is granodiorite tempered with biotite and green amphibole. Possible Scottish source.
Piercebridge	3	Three sections contain granodiorite fragments with green amphibole - possible Scottish source. The fourth section contains a two mica granite plus possible gneiss or high grade metamorphic psammite. Quartz with mica pressure shadows is widespread. Re-examination suggests that some of the metamorphic material may be highly foliated granite. Although probably from SW Scotland the amount of metamorphic material is unusual.
Rudston	1	This section is tempered with microcline-two mica granite - possible Scottish source
Sancton	4	One section contains microcline-muscovite granite with some microporphyritic lava from the Cheviots - possible Scottish source. One section is a granodiorite tempered pot with biotite-green amphibole - possible Scottish source. One section has a mixed range of non-plastics including some Whin Sill dolerite, possible biotite granite and medium grade metamorphics including quartz-garnet and quartz-muscovite phyllite - possible Scottish source. One section has biotite granite plus fresh Whin Sill dolerite and a little phyllite - possible Scottish source
Sewerby	3	One section contains granodiorite with green amphibole and biotite - possible Scottish source. One section contains a two mica and microcline granite and its constituents - possible Scottish source. One section contains sandstone with kaolinite in its pore spaces and trace amounts of granodiorite - possible Scottish source
West Heslerton	1	This section is tempered with a microcline-bearing granite. Muscovite was not recognised but the biotite is pale-coloured. Re-examination confirms that the temper includes microcline perthite-biotite granite - possible Scottish source
West Lilling	1	The section looks to be biotite granite tempered although some green amphibole is present as discrete grains - possible Scottish source
West Heslerton	2	Both sections contain biotite granite. The first contains a sphene-bearing, biotite granite. possibly Shap adamellite or Criffel-Dalbeattie granodiorite. The second contains an apatite-biotite-rich granite with some microcline - possible Scottish source

Table 2. Interpretation of Anglo-Saxon Scottish igneous-tempered ceramics analysed in this study.

Scottish igneous-tempered ceramics

A total of 24 of the Anglo-Saxon ceramic thin sections contained igneous inclusions that could be assigned to the Scottish Criffel-Dalbeattie granodiorite or the Cairnsmore of Fleet granite. Inclusions from these two sources can be easily distinguished from those of the Shap adamellite in the previous group of ceramics. They include granodiorite composed of green amphibole (Figure 3A,B), altered biotite, plagioclase, quartz and potassium feldspar (mainly perthite) and microcline-bearing, two-mica granite with quartz, plagioclase, potassium feldspar (including coarse-grained microcline) (Figure 3C), biotite and primary, often coarse-grained, muscovite (Figure 3D).

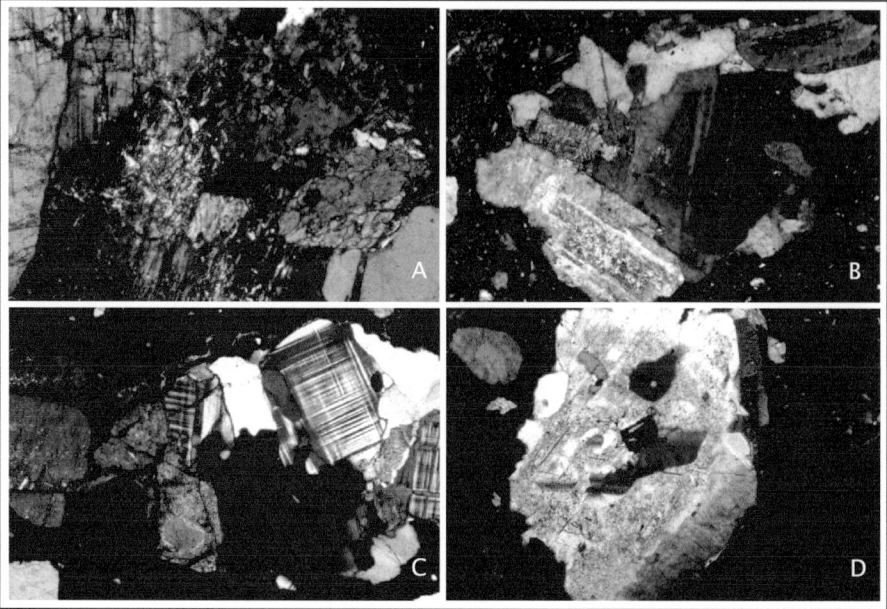

Figure 3. Scottish igneous-tempered ceramics. a) Simply twinned amphibole, quartz and plagioclase, altering to fine-grained white mica in sample V1432, b) Quartz and zoned plagioclase, showing cores altering to fine-grained white mica and less altered margins, plus biotite laths and a single amphibole lath in sample V1433, c) Quartz-microcline with characteristic crossed-hatched twinning and plagioclase altering to fine-grained clay minerals in sample V1547, d) Clear quartz, cloudy plagioclase, muscovite and biotite in sample V1846. Images taken in crossed polars.

Leicestershire granite-tempered ceramics

In five of the Anglo-Saxon ceramics analysed, the igneous temper cannot be matched to either the Shap adamellite or any of the southwestern Scottish sources. Samples from Piercebridge, Scorton Quarry, Sancton and Elmswell all contain minor green amphibole, significant amounts of biotite and plagioclase that is altered to epidote (Figure 4A). The two sections from the site of Elmswell show cuneiform intergrowths between quartz and feldspar (Figure 4B). The alteration of plagioclase to epidote and symplectite-like intergrowths between quartz and feldspar are not features of the Shap adamellite and are very rare, in the southwest Scottish granites. However, both are characteristic features of some igneous rocks from Leicestershire, including the Mountsorrel granodiorite.

Site	Samples	Leicestershire granite-tempered ceramics Interpretation
Elmswell	2	One section has a granodiorite with biotite altering to chlorite and plagioclase to epidote (Figure 4A) - possibly Leicestershire source. One section has an atypical altered biotite granite with cuneiform intergrowths between quartz and feldspar (Figure 4b,C). Plagioclase is altered to epidote- possibly Leicestershire source
Piercebridge	1	In one section the plagioclase has partially altered to epidote (Figure 4D) - possibly Leicestershire source
Scorton Quarry	1	The section contains biotite-amphibole granodiorite fragments with biotite extensively altering to chlorite. Minor amounts of epidote are also present - possibly Leicestershire source
Sancton	1	One section contains a mixed temper including coarse-grained microcline (Figure 4C,D), plagioclase altering to epidote and sediments. Some muscovite is present as discrete grains - possible Scottish or Leicestershire source

Table 3. Interpretation of Anglo-Saxon Leicestershire granite-tempered ceramics analysed in this study.

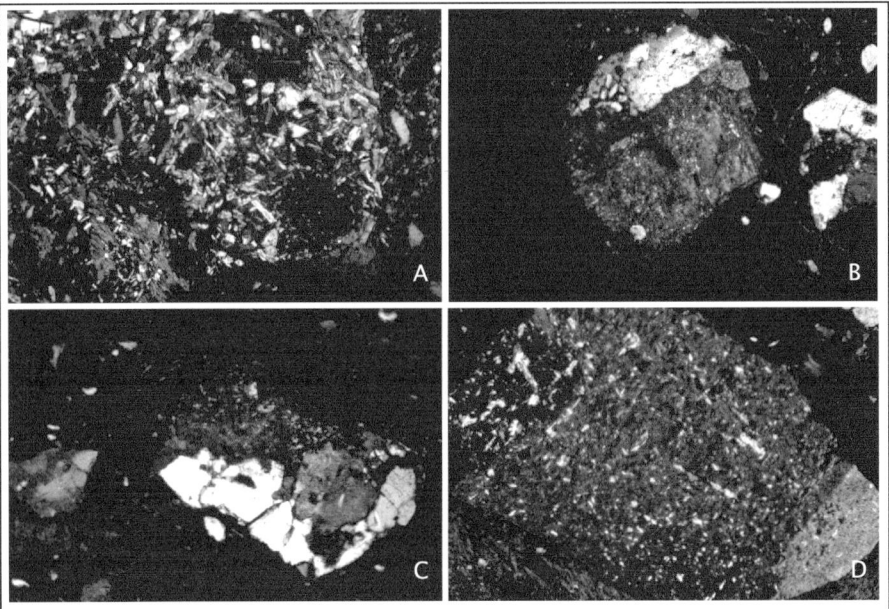

Figure 4. Leicestershire granite-tempered ceramics. a) Twinned plagioclase extensively altered to tabular epidote and lath-shaped biotite altered along cleavage planes to chlorite in sample V668. b) Plagioclase altering to fine-grained clay minerals with cuneiform intergrowth with quartz in sample V669, c) Plagioclase altering to fine-grained clay minerals and epidote with sub-graphic intergrowth with quartz in sample V669, d) Plagioclase altered to fine-grained white mica and clinozoisite in sample V1481. Images taken in crossed polars.

Discussion

Anglo-Saxon igneous-tempered ceramics with inclusions deriving from the Shap adamellite, the southwestern Scottish sources and the Mountsorrel granodiorite occur at a range of sites in East and North Yorkshire and the southern Durham borders. Based upon the samples analysed, several important patterns emerge. Those sites studied in the Stainmore Gap/Tees Valley either exclusively contain ceramics of the Shap group (Binchester, and Norton) or are dominated by igneous-tempered ceramics of this composition (Piercebridge, Catterick and Scorton Quarry). Glacial till composed of material derived from the northwest, including Cumbria would have been available in this area and may be the source of the igneous material used to make these ceramics.

Ceramics dominated by inclusions that could be ascribed to the southwestern Scottish igneous sources occur at sites to the east of the Vale of York. In some samples, these inclusions are accompanied by igneous material from northeastern sources such as the Cheviot andesites and Whin Sill dolerite and coarse-grained sandstone fragments, which are likely to be of Carboniferous origin. Carboniferous rocks are the dominant type of clast in glacial gravels in the Vale of York and elsewhere in the north of England. It is likely that these deposits, such as the Withernsea Till, left by ice flowing southwards along the east coast of England was used to produce the Scottish igneous-tempered ceramics.

In most of the Anglo-Saxon ceramics analysed in this study the inclusions appear to be of a single igneous rock type. Tills or gravels containing such high frequencies of igneous and igneous-derived minerals have not so far been detected in previous petrographic studies of ceramics from the region, nor does the geological literature suggest that such deposits exist (King, 1976). It is therefore possible that potters selected specific, easily recognisable erratic clasts to use as temper. The Shap adamellite is a highly distinctive rock and stands out when it occurs within glacial deposits. Indeed its presence was used in early Quaternary studies to delineate past ice movements. Glacial erratic rocks suitable for the preparation of this type of temper though not common, have a widespread distribution throughout the north of England. The co-occurrence in some Scottish-igneous tempered ceramics of igneous and sandstone inclusions may suggest that both types of inclusions were naturally occurring in the clay used to produce these sample or perhaps that mixed gravel temper was used as temper.

The small numbers of Anglo-Saxon ceramics that contain clasts with strong petrographic similarities to igneous sources in Leicestershire, are not likely to have been produced from glacial material in East or North Yorkshire. Instead they appear to have been tempered with Mountsorrel granodiorite and anthropogenically transported northwards into the region.

Conclusions

Ceramic vessels made using a similar procedure to that employed in northern England from the Bronze Age to the Roman conquest, continued to be made in the post-Roman period. Definite examples of this type occur at Yeavering and Binchester in contexts where a post-Roman date is undeniable. The analysis in this study of the Binchester vessel indicates that it contains coarse angular fragments of biotite granite that could have originated in the Lake District or southwestern Scotland. Glacial ice flowed into northern England from these two areas during the Pleistocene, through the Tyne Gap and Stainmore Gap and southwards into the Vale of York and East Yorkshire. It is therefore possible that the igneous inclusions in this vessel originated from glacial deposits left when the ice sheets retreated.

The majority of igneous-tempered pottery found in Northern England in post-Roman contexts is culturally of Anglo-Saxon character and the igneous material is finer and more evenly distributed than in prehistoric and Roman examples. Given the small size of the inclusions, it is not easy to determine whether they represent temper or a natural component of the clay used to make the pottery. However, the frequency of igneous fragments and associated minerals in some of the vessels analysed in this study appears to be higher than that of boulder clay or glacial gravels in the region. This suggests that igneous erratics were selectively extracted by potters from glacial deposits and intentionally added as temper.

Detailed petrographic analysis of the igneous inclusions within 47 igneous-tempered Anglo-Saxon ceramics from Yorkshire and Durham and their comparison with thin sections of primary igneous bodies in northern England, Scotland and Scandinavia, indicates that several distinct lithologies are present in the pottery. These can be classified into three groups based upon their composition and probable source(s). Igneous inclusions with no original muscovite or microcline, with biotite altered to chlorite and sphene as a common accessory mineral may have originated from the Shap adamellite in Cumbria or possibly from the Criffel-Dalbeattie granodiorite or the Cairnsmore of Fleet granite in southwestern Scotland. Inclusions of green amphibole-bearing granodiorite and two-mica granite with original coarse-grained muscovite and microcline are good matches for the Criffel-Dalbeattie granodiorite and the Cairnsmore of Fleet granite respectively. A third, less common group is characterized by igneous inclusions with secondary epidote and/or cuneiform intergrowths between quartz and feldspar. Based upon the geological samples analysed, this rock did not originate from northern England. A probable match can be found in the Mountsorrel granodiorite of Leicestershire. Pottery tempered with this type of igneous rock is very common in the East Midlands (Williams and Vince, 1997) and is also found in Lincolnshire, Nottinghamshire and South Yorkshire.

Ceramics with igneous inclusions identified as the Shap adamellite occur at sites at the northern end of the Vale of York, where boulder clay containing material of Cumbrian origin would have been most common. Those with inclusions attributed to the southwestern Scottish sources occur at sites to the east of the Vale of York, where ceramics with Shap adamellite inclusions are absent. Thus, the igneous temper within the Anglo-Saxon ceramics analysed reflect variation in the Pleistocene sediments of the region and the routes of the different ice movements during the last glacial period.

The data recorded in this study suggest that Anglo-Saxon igneous-tempered ceramics were produced in several different locations in northeastern England. Differences between the igneous-tempered ceramics of some adjacent sites may indicate the local production of pottery in the early Anglo-Saxon period. Whether pottery was produced in the household for individual consumption, or was traded between neighboring communities, however is not yet clear. The occurrence at sites in East and North

Yorkshire of ceramics tempered with the Mountsorrel granodiorite in this study provides evidence for the long-distance movement of ceramics or raw materials.

Acknowledgements

The analysis presented in this paper was undertaken as part of the English Heritage-funded Northumbrian Kingdom Anglo-Saxon Pottery Project and also supported by the Constantine XI Palaeologos Research Fund. Thin sections were prepared by Steve Caldwell of University of Manchester. Sadly, Alan Vince passed away during the preparation of this manuscript. His illuminating mind and highly stimulating 'quick questions', which invariably took a few hours to a few days or even months to work out, will be missed. Patrick Quinn is thanked for his encouragement and unstinting assistance in publishing this work despite Alan's untimely death.

References

Chatwin, C.P. 1961. *East Anglia and Adjoining Areas. British Regional Geology*. Her Majesty's Stationery Office. London

Freestone, I.C. and Middleton, A.P. 1991. Report on the petrology of pottery from Iron Age cemeteries at Rudston and Burton Fleming. In: Stead, I.M. (Ed.) *Iron Age cemeteries in East Yorkshire: Excavations at Burton Fleming, Rudstone, Garton-on-the-Wolds, and Kirkburn*. English Heritage Archaeological Report 22. English Heritage in association with the British Museum, London: 162-164.

Hope-Taylor, B. 1977. *Yeavering. An Anglo-British centre of early Northumbria*. Her Majesty's Stationery Office. London.

King, C.A.M. 1976. *Northern England. The Geomorphology of the British Isles*. Methuen and Co. London.

Neale, J. and Catt, J. 1994. Jurassic, Cretaceous and Quaternary rocks of Filey Bay and Speeton. In: Scruton, C. (Ed.) *Yorkshire Rocks and Landscape, A field guide*. Ellenbank Press, Maryport Cumbria: 183-191.

Rigby, V. 2004. Pots in Pits. The British Museum Settlement Project 1988-92. *East Riding Archaeologist*, 11.

Taylor, B.J., Burgess, I.C., Land, D.H., Mills, D.A.C., Smith, D.B. and Warren, P.T. 1971. *Northern England. British Regional Geology. (Fourth Edition)*. Her Majesty's Stationery Office. London

Wardle, P. 1992. *Earlier Prehistoric Pottery Production and Ceramic Petrology in Britain*. BAR British Series, 225, Tempus Reparatum, Oxford.

Wilson, V. 1963. *East Yorkshire and Lincolnshire. British Regional Geology.* Her Majesty's Stationery Office. London.

Williams, D. and Vince, A. 1997. *The Characterization and Interpretation of Early to Middle Saxon Granitic Tempered Pottery in England.* Medieval Archaeology, 41: 214-219.

TECHNOLOGICAL INSIGHTS INTO BELL-BEAKERS: A CASE STUDY FROM THE MONDEGO PLATEAU, PORTUGAL

Ana Jorge

University of Sheffield, Department of Archaeology, University of Sheffield, UK
(anasjorge@hotmail.com)

Introduction

The second half of the 3rd millennium BC saw significant changes in material culture in most of Western Europe and is generally viewed as a period of great social transformation. One of its emblematic and controversial elements is the Bell-Beaker vessel, found in a variety of forms, decorative styles and contexts, within a wide temporal frame. However, Beakers are but one component of much broader, stylistically varied ceramic assemblages, where they are largely outnumbered by other contemporary pottery types in many regions. Singling out Beakers from their 'accompanying pottery' risks abstracting these vessels from the contextual relationships in which they are entangled. Such a separation tends to essentialise Beakers, conceiving them as intrinsically different. The social function of Beakers can be understood only when the production and use of pottery assemblages *as a whole* is investigated. A technological approach provides a means to explore ceramic variability beyond stylistic distinctions and opens up the possibility for local diversity to be discussed past the narrow focus on provenance and systemic exchange.

Ceramic analysis has shown repeatedly that the large majority of Beakers in Western Europe were produced locally (e.g. Querré, 1992; Convertini and Querré, 1998; Dias *et al.*, 2000; Clop, 2007). Nevertheless, some results reveal complex and ambiguous patterns of chemical and petrographic variation, suggesting that at a regional (Querré, 1992) and even site level (Dias *et al.*, 2000) Beakers were not produced in standardised ways. The focus on Beakers and their assumed participation in long-distance exchange has meant that this local heterogeneity has mostly gone unexplored. Equally, patterns of variability among non-Beaker pots generally receive little attention.

The case study presented in this paper is a comparative analysis of pottery from three sites (Fraga da Pena, Malhada and Carapito III) located in the Upper Mondego Plateau, Portugal (Figure 1), using thin-section petrography. The aim is two-fold. First, to characterise practices of pottery production in the late 3rd millennium BC in the region by considering technological, stylistic and contextual evidence, and second, to explore the ways in which technological choices informing Beaker production refer to the interplay between local ceramic traditions and more wide ranging regional and 'international' concepts about vessel form and function.

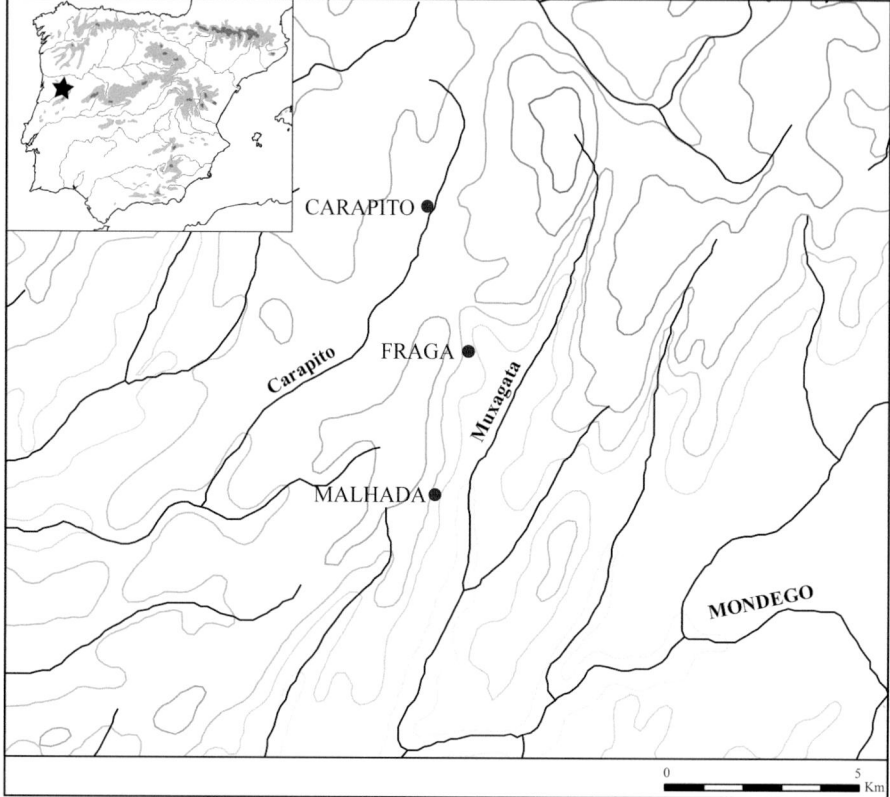

Figure 1. Location of the Late Chalcolithic and EBA sites of Malhada, Fraga and Carapito III in the Upper Mondego Plateau, in Northern-Central Portugal.

Archaeological Context

In Iberia, Beakers were produced and used for almost a millennium and incorporated in both Chalcolithic (c. 2700-2200 BC) and Early Bronze Age (EBA) (c. 2200-1630 BC) contexts (Díaz-del-Río, 2006, p. 96). As argued by Díaz-del-Río (2006), this pattern prevents the definition of chronologically meaningful 'Beaker cultural phases' within regional sequences and shows that Beakers had relevance within different socio-historical conditions both across the Peninsula and throughout the millennium. The long lifespan of this vessel type challenges the idea that they played a uniform, stable role among the societies of this period. Their participation in diverse and changing practices, both domestic and funerary, implies that, even at a regional scale, the use of Beakers could not have been connected to fixed or singular associations.

In most of northwestern Iberia, Beakers have been collected mainly from burials and are very scarce in almost all non-funerary sites (Garrido Pena, 1995; Cruz, 2001; Sanches, 2000/2001; Prieto Martínez, 2001; Jorge, 2002, 2003; Díaz-del-Río, 2006; Valera, 2006). A few sherds were recovered from the later phases of the Chalcolithic stone-built monuments that dot the western regions and, more rarely, small open-air sites and rock shelters. At the onset of the EBA, Beakers were also deposited in ceremonial places, specifically small stone enclosures that integrate natural rock formations and occupy commanding positions in the landscape. In every case, Beakers are found associated with a stylistically varied pottery, both 'common' and 'fine'.

Figure 2. Ceramic stylistic categories of the late 3rd millennium BC in the Upper Mondego Plateau. a) Examples of shapes of the Late Neolithic-Chalcolithic tradition (bowl and cups), b) Beakers (linear impressed, banded 'puntilhado', nail-impressed), c) Novel types from the Late Chalcolithic-EBA transition (inverted conical vessel and parabolic vessels). Scale in cm.

An example of these sites in the Upper Mondego Plateau is Fraga da Pena, located close to settlements such as Malhada, with roughly contemporary, if slightly earlier occupations, which date from the second half of the 3rd millennium BC (Valera, 2006, p. 199) (Figure 3). Excavations at Fraga have revealed a large number of Beakers. These consist of Maritime and Impressed Bell-Beakers and Beaker-shaped vessels produced with nail-impressions that have been considered to represent a 'local style' (Valera, 2000, 2006) (Figure 2B). This motif is unknown on any other vessel type in Central Portugal. Dated to 2282-1922 BC (Valera, 2006, p. 244), the assemblage from

Fraga includes a range of ceramic shapes and decorations characteristic of the transition to the EBA in the region, which combines long-lived and novel styles (Figure 2). Some ceramic styles, such as comb-incised pottery, characteristic of the Late Chalcolithic, have wide regional distributions that have been interpreted as the participation of local groups in regional social networks (Jorge, 1986; Valera, 2006). Therefore, Beakers were not the only vessels with potential for evoking associations beyond the local.

Fraga typifies the diversification of the ceramic repertoire at the end of the 3rd millennium BC, with the appearance of S-profiled pots, inverted conical vessels and large parabolic storage vessels (Figure 2C). These new types are present in small numbers at the settlement of Malhada, a cluster of natural platforms on a steep slope, of which some show more than one phase of occupation (Valera, 2006). Inverted conical vessels were the main offerings in nearby Carapito III, a dolmen re-used at the onset of the EBA and from where Beakers are absent (Leisner and Ribeiro, 1968; Senna-Martinez, 1989) (Figure 3).

Geological Setting

The Upper Mondego Plateau consists largely of the Beira Hercynian granitic batholith complex and smaller outcrops of metamorphic rocks (Figure 4). With the exception of minor bodies of gabbros, monzodiorites and granodiorites cropping out west of Real (Valle Aguado *et al.*, 2005, p. 174), the igneous lithologies are fairly homogeneous in terms of their mineralogical composition.

The archaeological sites discussed in this paper are set on an extensive outcrop of coarse two-mica granites, although, their immediate surroundings differ (Figure 4). Fraga sits on the top of the plateau's ridge, overlooking the valley of the Muxagata River, whereas Carapito III is located on the undulating ground cut by the Carapito River. Further south along the ridge, Malhada is situated in an area of contact between two-mica granite, biotite granite and the highly deformed gneiss-granite of Maceira (Valle Aguado *et al.*, 2005, p. 178-179). The nearest outcrops of metamorphic rocks belong to the Matela-Matança formation, composed by Ordovician quartzites and low-grade metapelites extensively transformed into hornfels and spotted slate by the intrusion of granite (Azevedo and Nolan, 1998, p. 3). Further west, the metasediments of the Real formation belong to the Pre-Ordovician Beira Schist-Greywacke Complex and consist of interbedded sequences of metapelitic and psammitic rocks with thin calc-silicate layers (Silva, 2005, p. 8-9). Here, regional metamorphism reached sillimanite grade before being subjected to retrograde metamorphism (Azevedo and Nolan, 1998, p. 3).

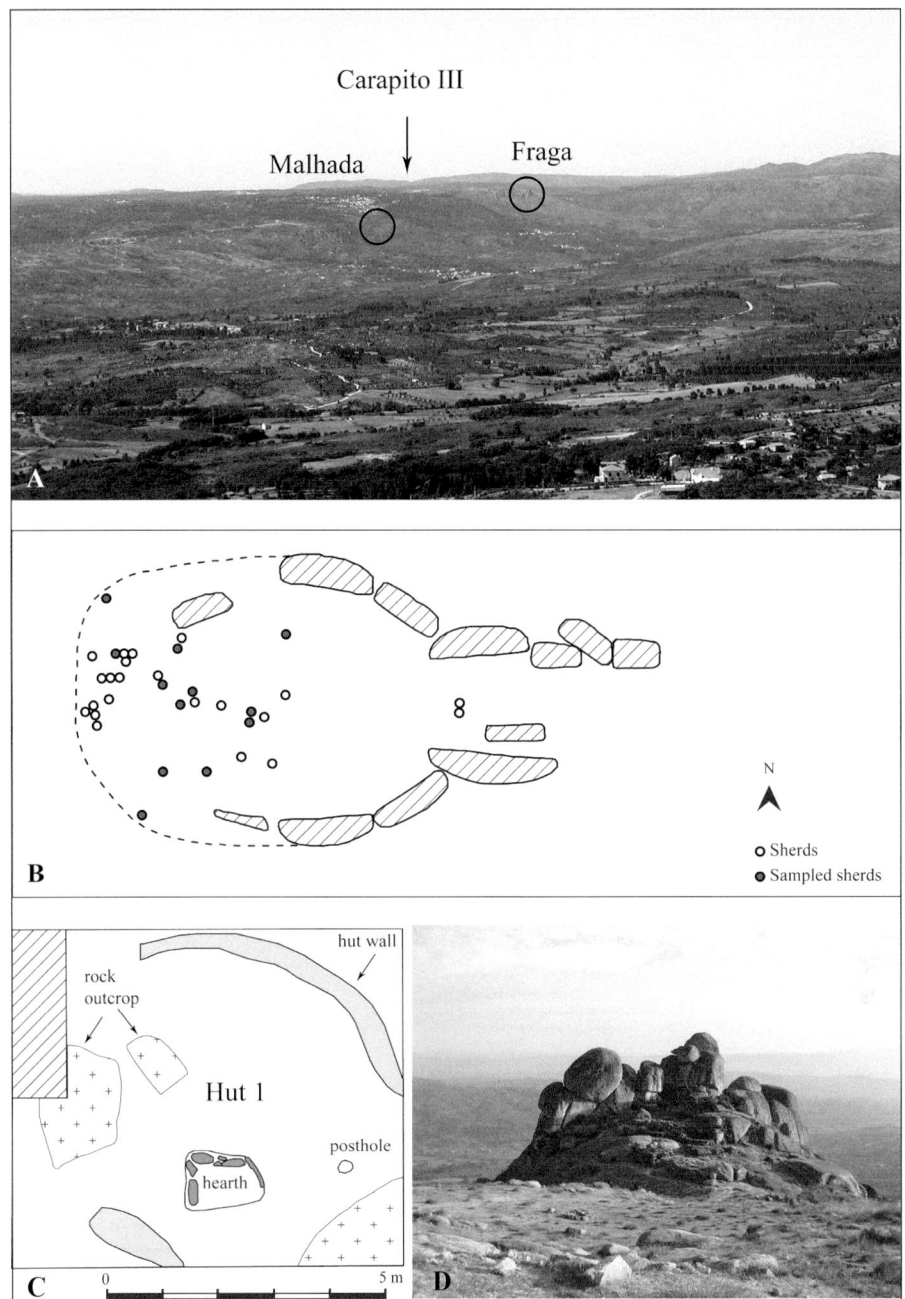

Figure 3. Archaeological sites discussed in this paper. a) View of the plateau's ridge from the left bank of the Mondego, with location of the archaeological sites, b) Plan of Carapito III with distribution of sherds (redrawn from Leisner and Ribeiro, 1968), c) Schematic plan of Hut 1, Sector B of Malhada (adapted from Valera, 2006, plate 39, p. 879), d) photograph of Fraga da Pena taken from the top of the ridge.

Figure 4. Simplified geological map of the Upper Mondego Plateau with location of the archaeological sites studied (adapted from Boorder 1965; Azevelo and Nolan, 1998; Silva, 2005).

Aplites, pegmatites, microgranite and basic dykes are widespread throughout the Upper Plateau. Dolerite dykes are commonly altered into fine clayey materials that were exploited for tile and brick production in the region during the first half of the twentieth century (Teixeira and Assunção, 1958, p. 82). Prehistoric communities also used these clays of doleritic origin for their pottery, which represents around 14% of the samples analysed (Table 1).

Clay origin	Fabric groups (FG)	Fabric loners (sample no.)	N	%
Metamorphic	FG 1, 2, 3, 6, 7, 24	A5/92, FP-5, FP-25, FP-53, Mal-16, Mal-133/96, Mal-223/96	42	24.9%
Ambiguous igneous/metamorphic	FG 10, 11, 12		19	11.2%
Granitic (residual)	FG 13, 14, 15	FP-31, FP-39, FP-45, FP-65, FP-67, Mal-41, Mal-52, Mal-61, Mal-74, Mal-654/98	43	25.4%
Granitic (transported)	FG 4, 29, 30		23	13.6%
Basic igneous (doleritic)	FG 21, 22	FP-21, Mal-194/96	23	13.6%
Intermediate igneous (granodioritic)	FG 19, 20		9	5.3%
Undetermined	FG 23, 27	A5/86, FP-51	10	5.9%
Total N of samples analysed			169	100.0%

Table 1. Geological origin of the ceramic samples analysed.

Petrographic Analysis: Clay Sources and Paste Recipes

<u>Materials and approach</u>

The study presented here is based on a total of 169 samples, 91 from Malhada, 63 from Fraga and 14 from Carapito III, representing 11%, 26% and 74% of the total number of classified vessels at each site, respectively. Among these, 21 of the 34 Beakers from Fraga were sampled. Previous analytical work undertaken in the Upper Mondego Plateau has sought to explore the relationship between different ceramic categories by sampling across the typological spectrum (Dias *et al.*, 2000, 2005). However, the methods employed were unable to detect much of the variability subsequently revealed by systematic petrographic examination (Jorge *et al.*, 2009). By comparison, thin-section petrography has proven particularly suitable for the identification of clay sources and ceramic recipes in the region.

Fabrics were defined on the basis of both mineralogy and texture, following Whitbread (1995). Whenever possible, evidence for differential weathering processes was also considered, a strategy that proved useful when dealing with very coarse ceramics and a regional hard-rock geology composed of relatively homogeneous mineralogical suites.

<u>Analytical results</u>

One of the most striking characteristics of the assemblages is their high compositional variability. Indeed, the results show the selection of a wide range of clay sources, including acid, intermediate and basic igneous rocks as well as low-, medium- and high-grade metamorphic rocks. Several distinct methods of paste preparation such as tempering and clay mixing were also employed. This diversity translates into 22 fabric groups (FG), often divisible into sub-groups, and 22 loners (i.e. fabrics represented by a single sample) (Table 1).

A noteworthy feature of the pottery analysed is the prevalent use of what appear to be residual clay or at least material that suffered minimal transport. This is indicated by the high angularity of the inclusions in the ceramics and the frequent chemical alteration of large, angular feldspars that, in some samples, can be seen to have disaggregated during ceramic manufacture. It is also common for quartz crystals to show fresh breaks along grain boundaries consistent with processes of leaching, which again supports the likelihood of clays having been collected in areas of *in situ* weathering (Delvigne, 1998). In accordance with this hypothesis, the majority of fabrics are composed of related inclusions (i.e. derived from a single parent rock) or of mineralogically distinct inclusions likely to result from the addition of temper rather than natural processes, as indicated by differential sphericity and strongly bimodal grain-size distributions. There are exceptions to this pattern, since very few fabrics point towards the use of deposits of mixed origin and finer sediments. As a result, the pottery from this area is poorly sorted and often very coarse, with fabric coarseness relating directly to the grain-size of the parent rock.

Although around half of the 169 samples analysed were produced with clays derived from granite and related metamorphosed rocks, prehistoric potters seem to have exploited widely the geological diversity of the Upper Mondego Plateau during this period. If we consider the probable geological origin of the pottery studied, determined on the basis of the mineral and rock fragments naturally occurring in the clay, it can be seen that a wide variety of lithologies are represented, most of which are available close to the sites (Table 2; Figure 4). Only 7% of the samples analysed appear to have been made with clays derived from rock types found over 10 km from the sites or at even greater distances of up to 20 km, as in the case of hornblende-bearing igneous rocks. Some fabrics cannot be easily related to a specific geological formation despite being broadly compatible with the regional geology. For instance, the clay used to make FG 5 derives from tremolite-bearing schist (Figure 6G,H), however, neither the tremolite-schist of the Satão formation (Silva, 2005, p. 8-9) nor the metabasites of greenschist facies formed by metamorphism of doleritic dykes (Teixeira and Assunção, 1958, p. 113-114) can be excluded as potential sources.

In a region dominated by single extensive lithologies, pinpointing geographically restricted sources is not possible for most ceramic raw materials, with the exception of the hornblende-bearing granodiorites and quartz monzodiorites, limited to the Trancozelos area (Azevedo and Nolan, 1998, p. 3) and, to a lesser extent, the metasediments of the Real formation (Figure 4). Nonetheless, there are other aspects of variability that can offer further insights into clay selection. The textural characteristics of the clays depend on the weathering conditions as much as parent rock composition; therefore, textural differences between fabrics made with clays undistinguishable in terms of lithological origin can be revealing of discrete sourcing areas (Harrad, 1998). The ceramics made with clays derived from intermediate igneous rocks, for example, belong to two very distinct fabric groups (Figure 5). The first (FG 19) has the characteristics of a freshly weathered product, with large inclusions composed exclusively of related rock fragments and minerals. The second (FG 20) is much finer, with highly altered inclusions and rare surviving rock fragments. These textural differences suggest that the clays used in FG 19 and FG 20 are likely to have been collected at different locations at or around the same outcrops. The fact that the two fabric groups belong to different sites further highlights the potential significance of this kind of petrographic variation.

Fabric Group	Clay origin	Clay recipe	Typological category	Decoration	Site	N samples
1	phyllite	grog temper	beakers	decorated & plain	Fraga	6
2	phyllite		beakers	decorated & plain	Fraga	2
3	phyllite	granite temper	LN-Chalc	mostly decorated	Malhada & Fraga	5
4	transported phyllite & granite		LN-Chalc	plain	Fraga	2
5	tremolite schist	granite temper	LN-Chalc	all decorated	Malhada & Fraga	2
6	muscovite schist		LN-Chalc & LChalc-EBA	all plain	Carapito	4
7	cataclasic rock		LN-Chalc & LChalc-EBA	decorated & plain	Carapito	3
10	mixed granitic & metamorphic		LN-Chalc & LChalc-EBA	mostly plain	Malhada	4
11	igneous/metamorphic		LN-Chalc	all plain	Malhada	3
12	igneous/metamorphic (muscovite)		LN-Chalc & LChalc-EBA	mostly plain	Malhada & Fraga	12
13	two-mica granite		beakers, LN-Chalc & LChalc-EBA	decorated & plain	Malhada, Fraga & Carapito	16
14	muscovite granite	clay mixing	LN-Chalc & LChalc-EBA	all plain	Carapito	3
15	biotite granite		LN-Chalc & LChalc-EBA	mostly plain	Malhada	7
15 A	biotite granite		beakers, LN-Chalc & LChalc-EBA	mostly decorated	Fraga	8
19	hornblende-granodiorite/ quartz monazite		LN-Chalc	all plain	Malhada	5
20	hornblende-granodiorite/ quartz monazite		LChalc-EBA	mostly decorated	Fraga & Carapito	4
21 A	dolerite		LN-Chalc	all plain	Malhada	8
21 B	dolerite	granite temper	beakers, LN-Chalc & LChalc-EBA	only the beaker is decorated	Malhada & Fraga	6
22	mixed dolerite & granite		beakers & LN-Chalc	all plain	Malhada & Fraga	7
23	undetermined		LN-Chalc & LChalc-EBA	mostly decorated	Malhada & Fraga	5
24	fine biotite schist/ metasandstone?		mostly LN-Chalc	mostly plain	Malhada & Fraga	12
27	undetermined	tempered?	LN-Chalc	all plain	Malhada	3
29	transported (granitic)		beakers, LN-Chalc & LChalc-EBA	decorated & plain	Malhada, Fraga & Carapito	18
30	granite (transported?)		LN-Chalc	plain	Malhada	2
Loners	various	various	beakers, LN-Chalc & LChalc-EBA	decorated & plain	Malhada, Fraga & Carapito	22

Table 2. Summary of the fabric groups and stylistic categories (shape and decoration) discussed in the text. LN-Chal = Late Neolithic-Chalcolithic tradition; LChal-EBA = Late Chalcolithic-EBA transition. Fabric group numbering as published in Jorge et al. (2009). Differences in numbering result from their integration in a regional database.

Similarly, most of the 22 loners differ from 'local' fabric groups on the basis of textural differences or technological features indicative of distinct clay 'recipes', as in instances where the same clay is selected but prepared differently. Only two samples do not seem to relate directly to the geology of the Upper Mondego Plateau, and their possible provenance is currently under investigation. They correspond to a Beaker (sample FP-5) and a large parabolic vessel (FP-25), both from Fraga.

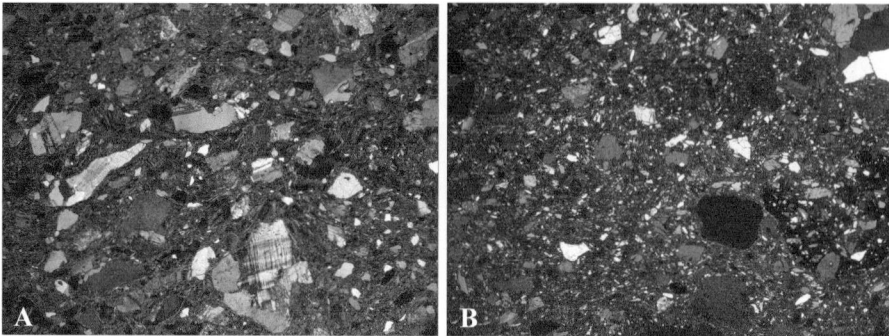

Figure 5. Thin-section micrographs of fabrics made with clay derived from intermediate igneous rocks. a) FG 19 (sample Mal-64), b) FG 20 (sample FP-57). Images taken in crossed polars. Image width = 5.5 mm.

Another characteristic of the pottery analysed is the flexible use of particular kinds of clays. For instance, clays of phyllitic origin were employed in the production of ceramic pastes as different as the grog-tempered FG 1, the un-tempered FG 2 and the granite-tempered FG 3 (Figure 7A-D). Equally, dolerite-derived clays are more commonly tempered with granite or used without temper, but grog was added to loner FP-21 (Figure 7E-H). Granitic clays have also been occasionally tempered with grog (two samples) and mixed with other clays, as in FG 14, where material derived from a muscovite granite was mixed with an altered argillaceous rock (Figure 6C). Other tempers were employed rarely, for example, siltstone in a small dish made with the clay used in FG 24 (Figure 6B) and sand in one Beaker. This kind of petrographic variability reflects deliberate decisions made early in the manufacturing process that cannot be explained exclusively by evoking technical constraints imposed by the physical properties of the clays. Indeed, the same clay was sometimes used in several different recipes and grog was added to clays that already contained abundant naturally occurring non-plastic inclusions.

When the ceramics of the three sites are compared, it becomes apparent that these technological choices are often site-specific (Table 2). Half of the 22 fabric groups consist of samples from a single site. This pattern is significant, since the sites are separated by only short distances. It shows that raw material availability is only one factor influencing clay selection. Differences in the range of ceramic raw materials employed at Carapito, on one hand, and Malhada and Fraga, on the other, could arise partially from the location of the dolmen on a different watershed and its potential use by a different community; however, the fact that Carapito is a burial site must be taken into consideration when investigating such differences.

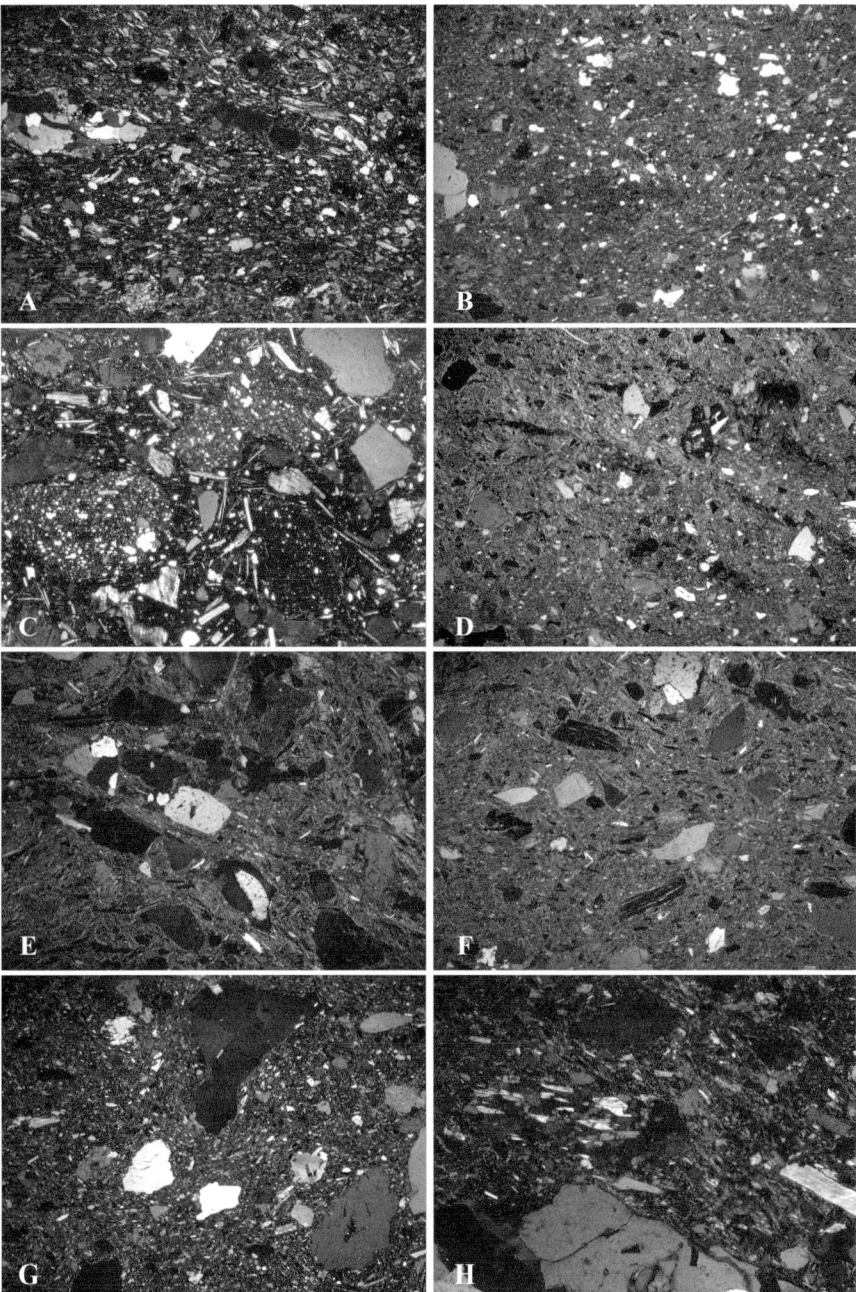

Figure 6. Thin-section micrographs of selected fabric groups defined in this study. a) Mica-schist fabric group (FG 6, sample FP-56), b) Biotite schist/metasandstone fabric group (FG 24B, sample Mal-214/96), c) Muscovite granite and argillaceous rock mixed fabric group (FG 14, sample A5/89), d) Granitic fabric group (FG 29, sample FP-30), e) Biotite granite fabric group (FG 15, sample Mal-20), f) Biotite granite fabric subgroup (FSG 15A, sample FP-22), g) tremolite-schist fabric group (FG 5, sample FP-28), h) Detail of biotite granite fabric subgroup (FSG 15A, sample FP-22) with diagnostic rock fragment. All images taken in crossed polars. Image width = 5.5 mm, except h = 1.4 mm.

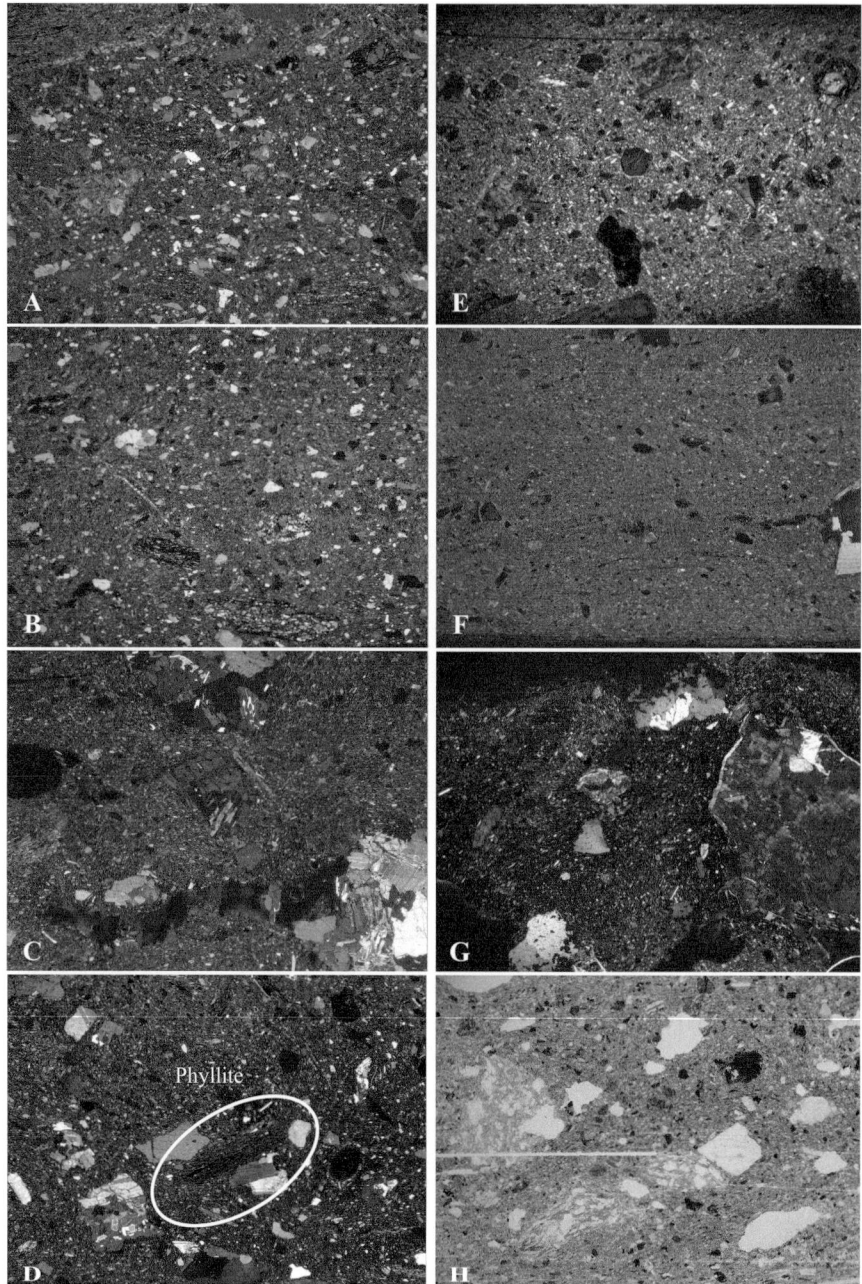

Figure 7. Thin-section micrographs illustrating technological differences between fabrics produced with similar clays. a-d) Phyllitic clay, tempered with: a) Grog (sample FP-9, FG 1), b) Un-tempered (sample FP-10, FG 2), c) Tempered with granite (sample Mal-12, FG 3), d) Tempered with granite (sample Mal-170/96, FG 3). e-g) Doleritic clay: e) Tempered with grog (sample FP-21, loner), f) Un- tempered (sample Mal-69, FG 21A), g) Tempered with granite (sample Mal-47, FG 21B), h) detail of grog in sample FP-21. All images taken in crossed polars except h – PPL. Image width = 5.5 mm, except c, d = 5.9 mm, h = 2.8 mm.

Contextualising Technological Variability: Style and Place

Fabric diversity results from technological choices that are informed by particular cultural logics, the practice of which encompasses both improvisation and the following of less flexible rules. These practices are linked to considerations about the social contexts in which the pots would have been used. This would have involved assessing which pots were appropriate for particular situations, such as burial, the birth of a child, the welcoming of kin, the serving of specific foodstuffs, and who would participate in their production and handling. Such considerations are equally important at times when the ceramic repertoire underwent changes, as vessel shapes were being re-designed. It is from this perspective that the appearance of new pottery types and decorations in the second half of the 3rd millennium BC and the integration of ceramic categories of international resonance such as Bell-Beakers within local contexts like the Mondego Plateau need to be discussed. In order to do so, technological variation must be investigated in light of stylistic and contextual information.

Prehistoric potters would have chosen their clay according to colour, plasticity and coarseness, characteristics that would have guided them in the identification of 'good clay'. However, the selection of distinct clays would have also prompted others in the community to recognise pots with different functions and/or pots made by different people. Vessels made with granite-derived clay tend to be brown and their large micas often give the surface of the vessels a glittering effect, whereas those made with doleritic clays are reddish, 'smooth', and sometimes spotted by red nodules (Figure 2A). What remains to be understood is how meaningful these and other technological differences might have been for the ways in which this diverse pottery was produced and consumed.

The comparison between petrographic and typological categories provides an important first insight into these issues by showing that technological choices do not seem to relate primarily to morphology. Regional types have been defined through a geometric approach to morphological classification that takes into account both form and size (Senna-Martinez, 1989; Valera, 2006) since in northwestern Iberia specific associations between shapes and functions are very difficult to establish until later in the Bronze Age. Cooking or drinking vessels, for example, are hard to differentiate among the Chalcolithic ceramic repertoire, even though variation in volume and profile allow archaeologists to infer the use of certain pots for storage rather than consumption of food or drink. The fabric groups defined in this study crosscut these typologies suggesting that ceramic production was informed by strategies that fall outside the idea of a standardised design. In fact, within each of the three sites studied, almost every fabric group encompasses a range of different shapes and often transcends the divide between earlier Chalcolithic typologies and the new shapes and decorative styles that emerge during the late 3rd millennium BC (Table 2). These new types represent a conceptual change in ceramic design, from spherical and globular to conical and S-profiled shapes, but their production seems to have conformed to existing understandings about forming techniques and paste recipes. The fact that these new shapes do not seem to have involved changes in technological practice is particularly significant as it suggests a process of innovation embedded in the local ceramic tradition.

When the relationship between fabrics and stylistic categories is scrutinised at site level, it emerges that at each site the majority of fabric groups includes samples of both decorated and undecorated vessels of various shapes, size and wall thickness. At Malhada, some fabrics seem to have been used to produce only plain vessels of 'Chalcolithic morphology' but with the exception of FG 27, they nevertheless cover a range of shapes. The assemblage from Fraga, with a high incidence of Late Chalcolithic-EBA types, decorated vessels and a large set of Beakers, shows a similar lack of systematic correlation between technological and morphological attributes. The only exception is, as discussed below, the presence of two type-specific fabrics, both related to Beakers (Table 2).

Also, in general decorated and plain vessels cannot be separated into specific fabric groups (Table 2). Nonetheless, given that decorated pots represent less than 30% of the samples analysed, the fact that both vessels in FG 5, as well as most of those in the FG 1, 2 and 3 are decorated could be meaningful. It is worth noting that FG 5 was made with raw materials occurring possibly at some distance from Malhada and Fraga, whereas FG 1 and 2 refer exclusively to Beakers and FG 3, the only other fabric made using phyllite-derived clays, includes five pots of various Chalcolithic shapes mostly from Malhada, four of which are decorated. These clays were processed differently for the production of Beakers by adding grog rather than granite or by not tempering the clay (Figure 7A-D). Yet, the fact that local Beakers used the same raw materials as other decorated pots from a neighbouring site could suggest that making and using Beakers in the Upper Mondego Plateau was worked into local understandings of appropriateness and quality and into existing routines of practice.

Beakers, as a category, show a similarly varied albeit more complicated picture than the rest of the pottery. They are made in a variety of fabrics and, in half of the cases analysed, cannot be distinguished from the other samples petrographically. This is revealing of the ways in which vessels with 'international resonance' were integrated into local contexts for two main reasons. First, it suggests that the production of Beakers deposited at Fraga was not regulated by rigid norms regarding either raw materials or paste recipes. Clays derived from phyllite, dolerite and granite of different kinds were used to produce Beakers, as for any other ceramic type. Second, it shows that no specific clay type or method of clay preparation can be exclusively associated with the production of Bell-Beakers. Some Beakers are petrographic loners but not in greater percentage than other shapes, and not necessarily indicating more distant sources than other vessels.

Yet, there are also Beaker-specific fabrics (FG 1 and FG 2) that use locally available clays and represent 40% of the 21 Beaker samples analysed. The selection of similar phyllitic clay for both fabrics and the use of grog in FG 1, a very rare tempering agent in the regional prehistoric pottery, reveal specific strategies for making at least some of the local Beakers. Interestingly, these are mostly 'local style Beakers', i.e. decorated with nail impressions, although a single impressed Beaker and a plain Beaker were also made in these two fabrics. In contrast, the three 'international style Beakers' (puntilhado) that were analysed fit into fabric groups that include other kinds of vessels: FP-47 and FP-50 are indistinguishable from other pots from Fraga (FG 15A) (Figure 6F) and FP-38 is identical to an inverted conical vessel from Malhada, among others.

The great technological variation seen not only amongst Beakers but also within the pottery assemblage of Fraga more generally seems to be related to the character of the occupation of the enclosure, a place where people would have congregated possibly from a wider area. Fabric diversity is likely to result both from a long succession of separate events and the participation of various groups with vessels manufactured possibly in a variety of locales within the regional landscape. The fact that two related fabrics were used exclusively for Beakers, however, suggests that more complex social mechanisms were at times implicated in the production of these vessels and their consumption at the site. This fabric homogeneity should not be seen necessarily as a cultural signature for a single group of people, such as the inhabitants of a settlement, since evidence from Malhada shows that pottery can be petrographically very diverse at one single domestic site. It could nevertheless indicate shifts in the social organisation of ceramic production related to either style (Beaker) or place (ceremonial site).

The Local and the International: Technology, Tradition and Beakers in the Upper Mondego Plateau

A non-standardised mode of ceramic production would have created highly variable assemblages only when the routines guiding the manufacturing process were flexible and able to encompass synchronic diversity as well as diachronic change. The fact that so many small, homogenous fabric groups could be defined for all three sites can therefore be seen to reflect recurrent, small-scale production.

In the settlement of Malhada, petrographic diversity seems to result from the activity of various potters rather than compliance to discrete pottery functions. In this context, pots made with different clay sources or recipes might be taken to stand for different potters or 'communities of potters', in the sense that Wenger (1998) gives to communities of practice and learning. These are likely to operate along family lines. The majority of fabrics are found throughout the site and its different phases of occupation but their distribution is not always random. Fabric groups 27 and 24 and related loners, for example, are associated exclusively or almost exclusively with Hut 1, in the earliest level of Sector B, where domestic activities such as roasting and grinding of acorns took place (Valera, 2006, p. 151) (Figure 3C). However, these fabric groups represent just under half of the 34 samples analysed for this particular hut. The fabric variability in Hut 1 supports the idea of flexibility in raw material selection for pottery manufacture during the use of this structure, which could have been either a work-area shared by different families or the 'house' of a single social unit? The first hypothesis could help explain the presence of various fabrics and the second could perhaps more easily account for the technological homogeneity of a significant percentage of this assemblage.

The restricted area excavated and the patchy preservation limits an understanding of the organisation of production on site. Nevertheless, as proposed elsewhere (Jorge, *et al.*, 2009), technological variation at Malhada seems to point towards ceramic production aimed at supplying the range of pots needed for daily life, possibly made in batches encompassing various shapes at any given time. These could have included vessels later selected for decoration, as suggested by the presence of decorated pots within most fabric groups at the site. Hence, there is no evidence for specialist pottery manufacture at Malhada. This does not exclude a scenario where specific tasks, such as decorating, were undertaken by particular individuals. For example, among some rural communities in Maghreb, North Africa, all women in the family participated in the production of the annual supply of hand-made vessels but only the mistress of the house decorated the pots (Balfet, 1966). She alone knew "exactly for what purpose she intend[ed] to use each of the objects" (Balfet, 1966, p. 166). The result is a very homogeneous assemblage both technologically and stylistically, yet characterised by "great diversity of details" (Balfet, 1966, p. 168) arising from the contribution of people with different abilities and the simultaneous production of vessels intended for a variety of purposes and social occasions. Ethnographic examples such as this provide avenues to explore variation in archaeological contexts by showing the interplay between conformity (reproduction of the same shapes over long periods of time) and flexibility in technological practices through which cultural transmission and innovation operate. They illustrate the social considerations underlying domestic production and the way that objects and their contextual relationships are thought of and anticipated during manufacture.

The domestic production of pottery is far from 'neutral' or politically inconsequential. Bowser's (2000) study of women potters in Ecuadorian Amazon shows how subtle variations in the decoration of domestic pots are able to explicitly convey ethnic boundaries and political alliances, including cues to political differences within and between groups in the same village. It is within this framework that the production and use of Beakers in the Mondego Plateau needs to be understood. While in some instances pots could have been made separately for specific events, such as burials, it is likely that, although purposefully conceived for feasts, gatherings or as gifts, many Beakers could have been made alongside more 'mundane' pottery.

The interpretation of technological variation at sites such as Fraga is complicated by the fact that specific prescriptions could be in place for the production of objects intended for ceremonial use. It is therefore feasible that at least some of the specificities identified at Fraga, such as the use of grog temper, relates to these rather than other factors. Still, the evidence reveals an assemblage formed mostly by small sets of pots that are technologically similar but stylistically different, many of which include Beakers. This pattern is consistent with that defined for Malhada and further supports the hypothesis that if social boundaries were being expressed, they might be those between familial units. While congregations at Fraga might have been fundamental arenas in the maintenance of group cohesion and the creation of notions of community beyond settlements (Valera, 2000, 2006), it can be suggested that belonging to smaller scale groupings such as nuclear or extended families was at least as important and could have played a role in inter-community politics. As far as the sites analysed allow, this seems to be the case during domestic (Malhada), burial (Carapito III) and ceremonial practices (Fraga) alike. It may be that other concepts of community, operationalised at larger scales, were brought into relief through stylistic rather than technological regularity. Beakers might have had an important role to play here, and the absence of Ciempozuelos-style Beakers in the region albeit predominant to the east, in the Meseta, is in this sense significant.

The petrographic study of pottery from these three sites reveals a ceramic tradition constituted by fragmented patterns that suggests great flexibility in technological practices. This variation was underlain by specific local conditions and processes of cultural transmission, where family was likely to have had an important role. Such fine-scale, yet intertwined traditions point to the absence of an overarching system of control or power relations able to successfully influence the homogenisation of technology. These results concur with current models for the Upper Mondego Plateau (Valera, 2006) and central-northern Iberia more generally (e.g. Jorge, 1999; Díaz-del-Río, 2006) that see very little evidence of institutionalised social inequalities among the societies of the late 3rd millennium BC. According to these models, Beakers would have been used within strategies of social differentiation possibly involving forms of competitive feasting and unequal participation in regional alliances. Despite diverging emphasis on the means through which such unequal power relations were achieved, authors agree that social status would have been a fluid rather than a hereditary category (Jorge, 1999; Díaz-del-Río, 2006; Valera, 2006).

Whereas the technological variability discussed above seems to be consistent with that expected for societies based on some form of domestic context of production, Beakers do not always conform to the pattern. This situation allows for several possible interpretations. First, the presence of Beaker-specific recipes FG 1 and FG 2 could suggest changes in the status quo over time, with Beakers becoming increasingly either more or less standardised during the occupation of Fraga. Unfortunately, the lack of precise dates does not allow this hypothesis to be tested. Second, the homogeneity of FG 1 and FG 2 could indicate a single event requiring special apparatus. In fact, different kinds of events are likely to have taken place at Fraga (Valera, 2006). The presence of technologically very homogeneous Beakers and highly varied ones could, then, reveal the interplay between different kinds of ceremonies at the site, such as local events (involving 'family pots') and those with broader social dimension (agreed 'clan pots') or events requiring the use of Beakers for discrete functions, for example as pots for feasting and vessels for other ritual purposes. Certain rituals could have demanded more prescribed technological behaviour. These may have been interlinked with the rare paintings in the walls of Fraga, which have led Valera to consider the performance of special ceremonies taking place inside the inner enclosure (Valera, 2000). Finally, Beaker fabric variation could relate to varied rituals nested within a broader ceremonial repertoire, as proposed for prehistoric monuments in Britain (e.g. Barrett, 1994; Brück, 2001). In simplistic terms, there could be pots for the people and pots for the gods. The prevailing pattern is, nevertheless, one of great diversity, even if its temporal scale is difficult to establish.

As the ceramic evidence from the Upper Mondego Plateau demonstrates, it is essential to consider that, at a local level, the production of Beakers and their design is the result of improvisations, re-interpretations, innovations, and changes of fashion, which took place throughout several generations. This organic process of development, change and diversification has long been recognised, although often de-emphasised, during stylistic analysis. Not only do vessels commonly defy clear affiliation with 'classic' stylistic groupings, leading to the creation of 'hybrid' taxa and 'variants', but also, even within apparently unproblematic style classes, decorative techniques are far from normative (e.g. Garrido Pena, 2000). An evocative example from Iberia is the occurrence of combined stylistic forms that bring together Neolithic elements ('symbolic' decoration) and Beaker attributes (shape and decoration), which are seen to materialise syncretistic processes between Neolithic beliefs and Beaker rituals (Garrido Pena and Muñoz López-Astilleros, 2000).

Beakers can be viewed as manifestations of an international world in local contexts. However, as Holtzman (2004) argues, the local is a complete reworking of broader 'global' processes through the lens of the local. It follows that the locally reworked 'global' is, from the onset, a local understanding of that world, not a copy or emulation. People would not have been necessarily aware of the larger scale geographical picture, as Beakers were not part of a systematic, long-distance exchange system as previously thought.

Conclusions

In this paper, the relationship between local ceramic traditions and Beakers has been explored through the lens of technology. Such an approach provides insights into the ways in which Beakers refer to long-standing technological practices and were reworked through existing social conditions of production in the Upper Mondego Plateau. Here, as among communities elsewhere, in the meeting of 'local' and 'global', the physical production of Beakers would have considered existing technological and aesthetic frameworks; their characteristics would have been negotiated within active traditions as they provided the possibility for social boundaries to be reassessed. Consequently, recognising that the Beaker itself "had pluralistic roles within a range of both special and everyday pottery" (Case, 2007, p. 249) is insufficient if the rest of the pottery remains a backdrop for the 'international phenomenon'. Beakers would have been understood relationally within the material world rather than in isolation. In conclusion, rather than concentrating solely on characterising Beakers, it is more fruitful to try to understand whether or not they were made and circulated differently from other regional vessel types of the same period, and whether these patterns shift according to context.

Acknowledgements

This paper draws from my doctoral research, under the supervision of Peter Day at the University of Sheffield, whom I thank for his continuous support. Bob Johnston, Jeff Oliver, Becky Wragg Sykes, John Barrett and Mike Parker Pearson read earlier drafts and provided invaluable comments. I thank Patrick Quinn for discussions during petrographic description. This research was made possible through doctoral funding from the University of Sheffield and the Arts and Humanities Research Council. The American Institute of Archaeology funded fieldwork through the Archaeology of Portugal Fellowship 2006. António Valera provided access to the collections from Malhada and Fraga and his typological analysis of these assemblages. Geological sampling was undertaken in collaboration with M. Isabel Dias, who advised on related issues. The Portuguese National Museum of Archaeology granted permission to sample Carapito. Finally, the Instituto Tecnológico e Nuclear hosted analytical work in Lisbon; and the Fitch Laboratory made its equipment available.

References

Azevelo, M. and Nolan, J. 1998 Hercynian late-post-tectonic granitic rocks from the Fornos de Algodres area (Northern Central Portugal). *Lithos*, 44: 1-20.

Balfet, H. 1966. Ethnographic observations in North Africa and archaeological interpretation: the pottery of the Maghreb. In: Matson, F. (Ed.) *Ceramics and Man*. Methuen & Co., London: 161-177.

Barrett, J.C. 1994. *Fragments from antiquity: an archaeology of social life in Britain, 2900-1200 BC*. Blackwell, Oxford.

Boorder, H.D. 1965. *Petrological investigations in the Aguiar da Beira granite area, Northern Portugal*. Grafisch Centrum Delto, Rotterdam.

Bowser, B.J. 2000. From pottery to politics: an ethnoarchaeological study of political factionalism, ethnicity, and domestic pottery style in the Ecuadorian amazon. *Journal of Archaeological Method and Theory*, 7: 219-248.

Brück, J. 2001. Monuments, power and personhood in the British Neolithic. *The Journal of the Royal Anthropological Institute*, 7: 649-667.

Case, H. 2007. Beakers and the Beaker culture. In: Burgess, Ch., Topping, P. and Lynch, F. (Eds.) *Beyond Stonehenge: essays on the Bronze Age in honour of Colin Burgess*. Oxbow Books, Oxford: 237-254.

Clop, X. 2007. *Materia prima, cerámica y sociedad. La gestión de los recursos minerals para manufacturer cerámicas del 3100 al 1500 ANE en el nordeste de la Península Ibérica*. BAR International Series 1660, Archaeopress, Oxford.

Convertini, F. and Querré, G. 1998. Apport des études céramologiques en laboratoire à la connaissance du Campaniformes: résultats, bilan et perspectives. *Bulletin de la Société Préhistorique Française*, 95: 333-341.

Cruz, D.J. 2001. O *Alto Paiva: megalitismo, diversidade tumular e práticas rituais durante a Pré-História recente*. Unpublished doctoral thesis, University of Coimbra.

Delvigne, J.E. 1998. *Atlas of micromorphology of mineral alteration and weathering*. The Canadian Mineralogist Special Publication 3, Mineralogical Association of Canada, Ottawa.

Dias, M.I., Prudêncio, M.I., Prates, S., Gouveia, M.A., and Valera, A.C. 2000. Tecnologias de produção e proveniência de material-prima das cerâmicas campaniformes da Fraga da Pena (Fornos de Algodres - Portugal). In: Jorge, V.O. (Ed.) *Actas do 3º Congresso de Arqueologia Peninsular, 4 (Pré-História Recente da Península Ibérica)*. ADECAP, Porto: 253-268.

Dias, M.I., Valera, A.C., and Prudêncio, M.I. 2005. Pottery production technology throughout the third millennium BC on a local settlement network in Fornos de Algodres, central Portugal. In: Dias, M.I. Prudêncio, M.I., and Waerenborg, J.C. (Eds.) *Understanding people through their pottery. Proceedings of the 7th European Meeting on Ancient Ceramics (EMAC'03)*. IPA, Lisbon: 41-48.

Díaz-del-Río, P. 2006. An appraisal of social inequalities in Central Iberia (c. 5300-1600 CAL BC). In: Díaz-del-Río, P. and García Sanjuán, L. (Eds.) *Social Inequality in Iberian Late Prehistory*. BAR International Series 1525, Archaeopress, Oxford: 67-79.

Garrido Pena, R. 1995. El campaniforme en la Meseta Sur: nuevos datos y propuesta teóricas. *Complutum*, 6: 123-151.

Garrido Pena, R. 2000. *El campaniforme en la Meseta central de la Peninsula Iberica (c.2500-2000 AC)*. BAR International series 892. J & E.Hedges, Oxford.

Garrido Pena, R. and Muñoz López-Astilleros, K. 2000. Visiones sagradas para los líderes. Cerámicas campaniformes con decoración simbólica en la Península Ibérica. *Complutum*, 11: 285-300.

Harrad, L. 2004. Gabbroic clay sources in Cornwall: a petrographic study of prehistoric pottery and clay samples. *Oxford Journal of Archaeology*, 23: 271-286.

Holtzman, J. 2004. The Local in the Local. Models of Time and Space in Samburu District, Northern Kenya. *Current Anthropology*, 45: 61-84.

Jorge, A., Day, P.M., and Dias, M.I. 2009. Technological choices at the onset of the Iberian Bronze Age: pottery from the Mondego Plateau, Portugal. In: Biró, K.T.,

Szilágyi, V., and Kreiter, A. (Eds.) *Vessels inside and outside. EMAC'07. 9th European Meeting on Ancient Ceramics.* Hungarian National Museum, Budapest: 241-246.

Jorge, S.O. 1986. *Povoados da Pré-história Recente da Região de Chaves -Vila Pouca de Aguiar (Trás-os-Montes Ocidental)*. Instituto de Arqueologia da FLUP, Porto.

Jorge, S.O. 1999. *Domesticar a terra. As primeiras comunidades agrárias em território português.* Gradiva, Lisbon.

Jorge, S.O. 2002. Um vaso campaniforme cordado no Norte de Portugal. Castelo Velho de Freixo de Numão (V. N. de Foz Côa). Breve notícia. *Ciências e Técnicas do Património*, 1: 27-98.

Jorge, S.O. 2003. Revisiting some earlier papers on the late prehistoric walled enclosures of the Iberian Peninsula. *Journal of Iberian Archaeology*, 5: 89-135.

Leisner, V. and Ribeiro, L. 1968. Die Dolmen von Carapito. *Madrider Mitteilungen*. 9: 11-62.

Prieto Martínez, M.P. 2001. *La cultura material cerâmica en la Prehistória Reciente de Galicia: Yacimientos al aire libre. TAPA 20.* Laboratorio de Arqueoloxia e Formas Culturais, Santiago de Compostela.

Querré, G. 1992. Les céramiques campaniformes au Sud-Finistèrre, nature et provenances; premiers résultats. *Antiquités Nationales*, 24: 25-47.

Sanches, M. J. 2000/2001. O Castro de Palheiros (Murça). Do Calcolítico à Idade do Ferro. *Portugalia*, 21/22: 5-39.

Senna-Martinez, J.M. 1989. *Pré-História Recente da Bacia do Médio e Alto Mondego: algumas contribuições para um modelo socio-cultural.* Unpublished doctoral thesis, University of Lisbon.

Silva, A.F. 2005. *As sucessões metasedimentares do Supergrupo Dúrio-Beirão (Complexo Xisto-Grauváquico) e do Arenigiano expostas na área correspondente à folha 17-B (Fornos de Algodres), na escala 1/50 000.* Unpublished report, INETI, Lisbon.

Teixeira, C. and Assunção, C. 1958. Rochas básicas de fácies gabróica e dolerítica intrusivas nos granitos da Beira. *Revista da Faculdade de Ciências de Lisboa* (2ª série), 6: 81-123.

Valera, A.C. 2000. O fenómeno campaniforme no Interior Centro de Portugal: o contexto da Fraga da Pena. In Jorge, V.O. (Ed.) *3º Congresso de Arqueologia Peninsular, 4 Pré-História Recente da Península Ibérica.* ADECAP, Porto: 269-290.

Valera, A.C. 2006. *Calcolítico e Transição para a Idade do Bronze na Bacia do Alto Mondego: Estruturação e Dinâmica de uma Rede Local de Povoamento*. Unpublished doctoral thesis, University of Porto.

Valle Aguado, B., Azevedo, M.R., Schaltegger, U., Martínez Catalán, J.R., and Nolan, J. 2005. U-Pb zircon and monazite geochronology of Variscan magmatism related to syn-convergence extension in Central Northern Portugal. *Lithos*, 82: 169-184.

Wenger, E. 1998. *Communities of practice: learning, meaning, and identity*. Cambridge University Press, Cambridge.

Whitbread, I.K. 1995. *Greek transport amphorae: a petrological and archaeological study*. Fitch Laboratory Occasional Paper 4, British School at Athens.

INDIGENOUS TABLEWARE PRODUCTION DURING THE ARCHAIC PERIOD IN WESTERN SICILY: NEW RESULTS FROM PETROGRAPHIC ANALYSIS

Giuseppe Montana

Dipartimento di Chimica e Fisica della Terra, Università degli Studi di Palermo, Italy
(gmontana@unipa.it)

Anna Maria Polito

Dipartimento di Chimica e Fisica della Terra, Università degli Studi di Palermo, Italy

Ioannis Iliopoulos

Department of Geology, University of Patras, Greece

Introduction

Painted tableware bearing geometric motifs or, less frequently, decorated with incised and impressed patterns was produced in western Sicily during the 7th-5th centuries BC (Gargini, 1995; Spatafora, 1996; Trombi, 1999, 2000; Campisi, 2003). Archaeological studies have characterized several indigenous sites as potential ceramic production centres of this type of pottery due to the discovery of kiln structures (Guglielmino, 2000). However, the substantial recurrence of forms and decorative styles has inhibited archaeological efforts to identify the various workshops that operated in the area based exclusively upon macroscopic criteria. Therefore, in this study petrographic analysis has been called into play as a means of differentiating between the different ceramic pastes of indigenous tableware and assigning them to specific local workshops.

This study highlights the importance of applying petrography to both ceramic artefacts and fired samples of potential raw materials. Such an approach necessitates a very good knowledge of the local field geology, as several ancient ceramic production centres are suspected to have been active in a restricted territory and, therefore, exhibit similar geological features. Ceramics collected from the indigenous settlements of Entella, Adranone, and Saraceno in the central and southern parts of western Sicily (Figure 1), as well as locally sampled clayey raw materials, were analyzed petrographically and compared.

The sites of Entella and Adranone are both located in the area of the Sicani Mountains. The ancient city of Entella is considered to be one of the most important indigenous centres of western Sicily, and, during the Archaic period (7th-5th centuries BC), it occupied a strategic position between the hinterland and coastal areas. The settlement of Adranone is situated upon the mountain of the same name, at around 1000 m, and is located only 15 km from Entella. The site of Saraceno is in a more remote area that was controlled by the Greek colonies of Agrigento and Gela. It is located close to the town of Ravanusa, around 130 km to the southeast of Entella and 50 km east of Agrigento.

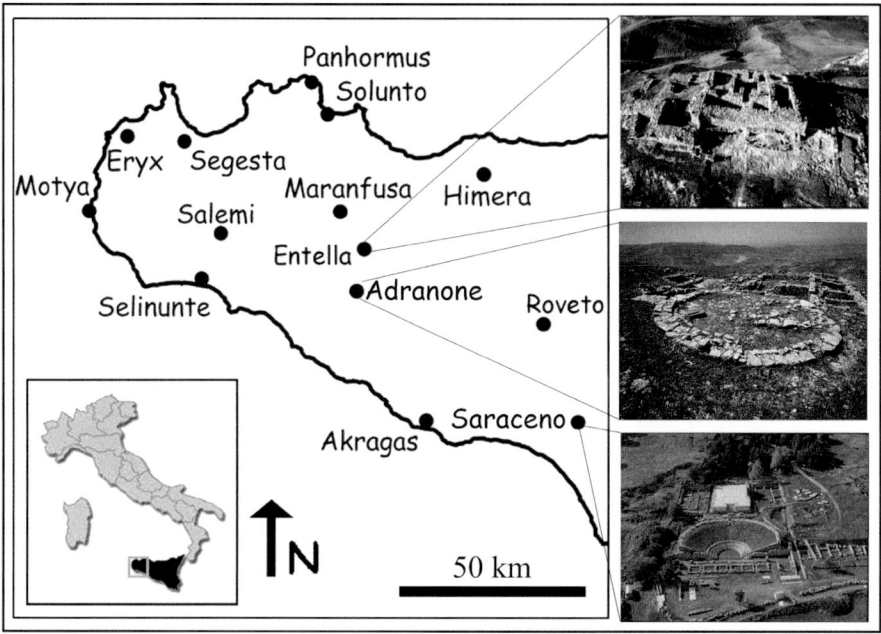

Figure 1. Location of the indigenous settlements of Entella, Adranone and Saraceno in western Sicily. Other main sites reported in the text are also indicated.

The kiln structures discovered in the area of Entella, indicate that intense ceramic production took place there during the sixth century BC (Guglielmino, 2000). Painted tableware produced at this settlement has already been identified in several coeval consumption settlements in the study area, such as Maranfusa, Saraceno, and Adranone (Alaimo *et al.*, 2003; Montana *et al.*, 2007). Moreover, the latter is suspected to have produced it's own pottery contemporaneously with Entella, but employing different clay sources. However, this remains unconfirmed due to the lack of supporting evidence, such as kiln structures. The petrographic assessment of the ceramic material reported upon in this paper aids the identification of the ceramic production centres operating in the study area and, consequently, helps trace the possible trade routes linking the various indigenous settlements. A comparison of the results obtained through petrographic analysis of the ceramic artefacts and the local clay-rich raw materials has permitted the establishment of consistent 'paste groups' at the three settlements, each group defined by compositional and textural features of its aplastic inclusions such as packing, sorting and grain size distribution. The results provide strong evidence for the local production of painted tableware in the settlement of Entella, using local Upper Miocene marine clay. In addition to the interpretation of provenance of the indigenous tableware, petrographic analysis has permitted the reconstruction of several technological steps involved in the production sequence or *chaîne opératoire*, such as the addition of temper and the mixing of raw clays.

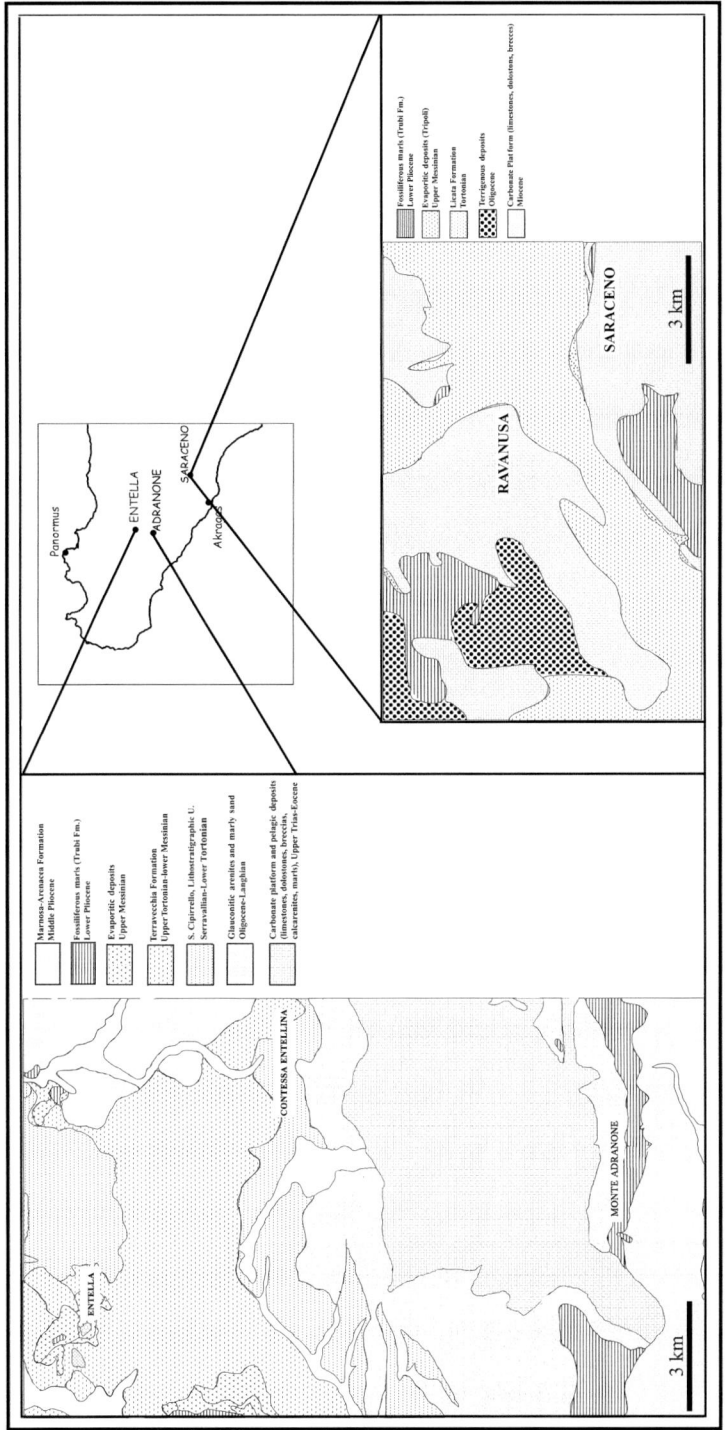

Figure 2. Simplified geological maps of the study area in western Sicily. Modified from Sprovieri *et al.* (1996), Sprovieri *et al.* (2002), Ruggeri & Torre (1974), Hilgen *et al.* (1995) and Krijgsman *et al.* (1995).

Geological Framework

Clay-rich raw material sources appropriate for ceramic manufacture are widespread around the three archaeological sites considered in this study (Figure 2). At Entella two different stratigraphic horizons, dating from the Middle Miocene to the Middle-Upper Pliocene, can be distinguished. San Cipirrello clays are Serravallian-Lower Tortonian grey-blue coloured marls, and are characterized by abundant calcareous microfossils of the foraminifera species *Globigerinoides ruber, Globigerinoides subquadratus, Globorotalia mayeri* and *Globorotalia menardi* (Sprovieri *et al.*, 1996; Sprovieri *et al.*, 2002). The Tortonian-Messinian levels of the Terravecchia clays, which surround the hill of Entella, are not particularly rich in microfossils and are instead characterized by the presence of tiny mica flakes in their coarse silt fraction. The rare foraminifera that occur are typical of the *Globigerinoides obliquus extremus, Globorotalia suterae,* and *Globorotalia conomiozea* Biozones (palaeontological time periods). The species *Orbulina universa, Neogloboquadrina acostaensis,* and *Globorotalia menardii* are also common in this association (Sprovieri *et al.*, 1996; Sprovieri *et al.*, 2002).

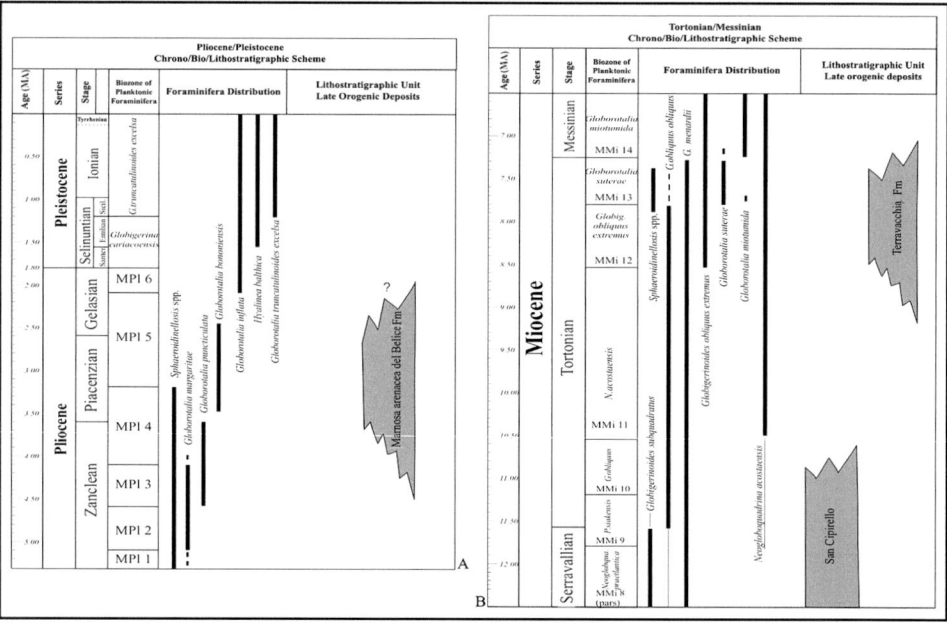

Figure 3. Chronostratigraphy of the main geological units considered in this study. Compiled from Roda (1965), Ruggeri & Torre (1974), Sprovieri *et al.* (1993), Sprovieri *et al.* (1996), Sprovieri *et al.* (2002).

Lower Pliocene clayey raw materials of the Marnoso-Arenacea Formation outcrop approximately two kilometres to the south of Adranone. They comprise grey-blue marls, which become increasingly more clay-rich up sequence. The marls are characterized by abundant foraminifera, as well as calcareous nannofossils including the species *Discoaster tamales, Discoaster pentaradiatus, Discoaster brouweri,* and *Discoaster productus* (Roda, 1965; Ruggeri & Torre, 1974). The Marnoso-Arenacea

Formation is a terrigenous succession, which lies discordantly on older deposits forming a characteristic polyphase thrusting piggy-back sequence.

Clays belonging to the Licata Formation crop out in proximity of the site of Saraceno. This sedimentary formation is continuous with the Lower Pliocene deep-sea marly deposits of the Trubi Formation rich that are in *Globigerinae* foraminifera. It comprises well-stratified grey-blue clays intercalated, at their upper part, with silty layers a few centimetres in thickness, rich in iron oxides (Hilgen *et al.*, 1995; Krijgsman *et al.*, 1995) and planktonic foraminifera. The basal part of the formation is characterized by an Early Langhian foraminifera assemblage of the *Praeorbulina glomerosa* Biozone, while the upper levels contain a Lower Messinian assemblage of the *Globorotalie conomiozee* Biozone (Sprovieri *et al.*, 1993; Sprovieri *et al.*, 1996). The chronostratigraphy of the major lithostratigraphic units in the study area is illustrated schematically in Figure 3.

Materials and Methods

A total of 73 indigenous tableware ceramics were studied, dating mostly from the 6th century BC and including geometrically painted artefacts, with phytomorphic, zoomorphic, and anthropomorphic patterns and achromatic examples. Incised and impressed artefacts are encountered less frequently in the assemblages of the three sites and were therefore not considered during this study. From a morphological and typological point of view the 73 ceramic samples contain globular two-handled vases, jugs, and various kinds of bowls (Figure 4), with an average wall thickness of approximately 8 mm. Macroscopically they are characterized by a fine to very-fine paste with a reddish-brown coloured core and a generally lighter margin. Details of the ceramic artefacts studied are summarized in Table 1.

Settlement	Analysed samples	Typology	Decoration style frequency		Chronology (century B.C.)
Entella	17	two-handled vases (10), bowls (7)	100% painted		6th-5th
Adranone	26	two-handled vases (10), jugs (13), fruit bowls (3)	80% painted	20% achromatic	7th-5th
Saraceno	30	two-handled vases (9), bowls (11), jugs (3), fruit bowls (4), basin (3)	56% painted	44% achromatic	7th-5th

Table 1. Typology, decoration, and chronology of indigenous tableware samples from western Sicily analysed in this study.

Thin sections made from the artefacts were studied under the polarizing microscope, recording the mineralogy and textural characteristics (such as sorting, size distribution, and packing) of their aplastic inclusions, as well as the nature and homogeneity of their clay paste. Samples were classified into several paste groups according to their similarities and differences. Slight variations in the composition and the textural features of the aplastic inclusions led to the establishment of several new paste groups. This approach accommodates the small intra-site and/or intra-deposit variability recorded for the local clay sources in terms of quantity and size distribution of their sand fraction. For samples that were assigned to a group, a similar source of their raw materials could then be asserted. Samples with unique petrographic characteristics that could not be assigned to a paste group were characterized as 'loners'.

Clay briquettes measuring 3.0 x 5.0 x 0.5 cm were moulded from the samples of locally-available raw materials collected from the geological formations described above. These were fired in a muffle furnace at 900°C with for three hours at the maximum temperature. The fired briquettes were then thin sectioned and compared to the archaeological ceramics.

In order to objectively classify the petrographic data and identify possible matches between the ceramics and the local marine clay sources, an automatic grouping procedure was carried out in the manner pioneered by Cau *et al.* (2004). Discriminating characteristics such as the composition, relative abundance, grain-size distribution, packing, size, and mineralogical composition of the aplastic inclusions, the presence or absence of temper and the nature of the clay micromass were transformed into variables in order to generate a data-matrix susceptible to statistical treatment. Codification of the variables was based on the scheme of Cau *et al.* (2004), but with some new variables added in order to better satisfy the present data set (Table 2). The codified petrographic data was converted into binary form and treated statistically using multiple correspondence analysis (MCA) with null categories omitted. MCA was performed using an iterative approach consisting of subsequent runs and removing obvious outliers each time.

Results

Petrography of indigenous tableware ceramics

All but one of the 17 ceramic artefacts from the site of Entella, were assigned to two main paste groups (E1 and E2). These ehibit similar mineralogical characteristics but are differentiated by their packing and grain size distribution (Figure 5A,B). Monocrystalline quartz is the predominant constituent, accompanied by subordinate quantities of polycrystalline quartz, K-feldspar, plagioclase, and white mica. Calcareous bioclasts, partially decomposed due to firing, and pore casts are common, but always less common than the siliceous components. In both pastes the coarser fraction (>0.2 mm) is exclusively of siliceous composition. The most important textural characteristics (i.e., packing, sorting, and grain size of the aplastic inclusions) that led to the distinction of the two paste groups are summarized in Table 3. Sample ENT16 has been treated as a loner due to its textural and compositional peculiarities. Its micromass has a clotted appearance and its packing is around 15%. Calcareous bioclasts predominate over monocrystalline quartz, whilst feldspars and white micas are rare.

	ENTELLA		MONTE ADRANONE			MONTE SARACENO		
Paste Group	E1	E2	A1	A2	A3	S1	S2	S3
Number of individuals	13	3	11	9	3	5	20	2
Groundmass texture								
lumpy appearance					x		x	
homogeneous	x	x	x	x		x		x
Aplastic grains packing								
very low (<3%)								
low (3-10%)			x				x	
medium (10-15%)	x			x				
medium/high (15-25%)		x			x	x		x
very high (>25%)								
Aplastic grains size (*)								
very fine (0.04-0.1 mm)	P	A	P	P	P	P	P	A
fine (0.1-0.2 mm)	C	S	A	C	C	S	C	S
medium (0.2-0.5 mm)	R	C	R	-	S	-	S	C
coarse and very coarse (0.5-2 mm)	-	-	R	-	S	-	R	-
Aplastic grains distribution								
homogeneous	x			x		x		
non homogeneous-serial			x		x		x	
non homogeneous-bimodal		x						x
Calcareous micofossils/micritic clots/decomposition casts								
predominant/abundant			x		x			x
common	x					x	x	
less common/sporadic		x		x				
rare/absent								
Quartz								
predominant/abundant	x	x	x	x		x	x	x
common					x			
less common/sporadic								
rare/absent								
K-Feldspar/plagioclase								
common	x	x						x
less common/sporadic			x	x	x	x		
rare/absent							x	
Micas flakes								
common	x	x		x				
less common/sporadic							x	x
rare/absent			x		x			x
Siliceous lithoclasts								
common								
less common/sporadic	x	x		x				
rare/absent			x		x	x	x	x
Chamotte (grog)								
common								
less common/sporadic			x		x		x	
rare/absent	x	x		x		x		x

Table 2. Compositional and textural features of ceramic paste groups recognised at Entella, Adranone, and Saraceno. P = prevalent, A = abundant, C = common, R = rare, – = not determined.

The majority of the ceramic artefacts from Adranone can be assigned to three paste groups (A1, A2, and A3). As summarized in Table 3, these are differentiated in terms of their texture, as well as by their mineralogical composition. Quartz is the most abundant constituent of paste groups A1 and A2 (Figure 6A, B), whereas in paste group A3 tests of formaminfera predominate (Figure 6C). Crushed ceramic temper or 'grog' was observed in some of the samples assigned to paste groups A1 and A3. Heterogeneity in the colour of the ceramic artefacts from Adranone may indicate either wide variation in the firing conditions, from oxidizing to reducing, or be the result of organic matter in the clay used for their manufacture.

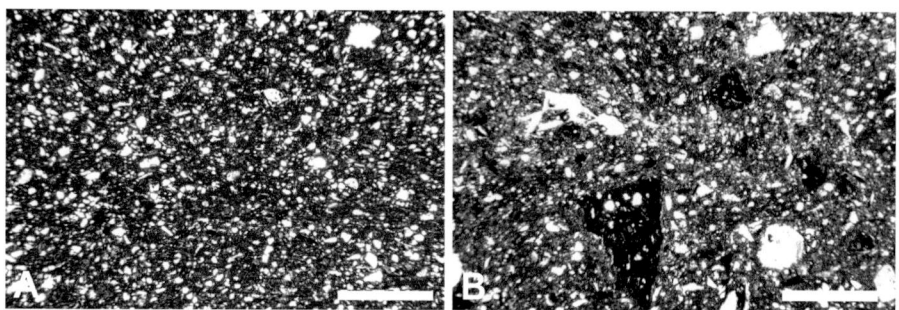

Figure 5. Thin section photomicrographs of indigenous tableware ceramics from Entella analysed in this study. a) Paste group E1, b) Paste group E2. Images taken in crossed polars. Scale bar = 0.5 mm.

Three loner samples from Adranone exhibit paste characteristics that are significantly different from those already described for groups A1-A3. Sample M.A. 2 has a heterogeneous micromass and a packing of 15% (Figure 6D). It is characterized by an abundance of monocrystalline quartz whose size ranges from coarse silt to very fine sand (0.03–0.125 mm). Quartz-rich, over-fired grog is also a common constituent. Calcareous lithoclasts were observed in similar or slightly smaller quantities and exhibit an advanced stage of thermal decomposition. Sample M.A. 23 (Figure 6E) has a homogeneous micromass and very low packing (<5%). Quartz is its predominant constituent with a grain size distribution similar to that recorded for sample M.A. 2. Micritic clots (Cau et al., 2002) and pore casts are very common, whereas mica flakes were only observed rarely. Lastly, sample M.A.25 is characterised by a bimodal grain size distribution of angular to sub-angular quartz, rare chert and polycrystalline quartz, a homogeneous micromass, and a packing of 10–15% (Figure 6F).

Most of the ceramics from Saraceno were assigned to one of three main paste groups (S1, S2, and S3). Among these groups, S2 was the most abundant, accounting for 20 out of the 27 samples analysed. The dominance of this group at Saraceno, suggest that it may have been locally produced. Groups S1 and S3 have a homogeneous micromass (Figure 7A,C), in contrast to the clotted appearance of the groundmass in samples assigned to paste group S2 (Figure 7B). A higher relative abundance of the mineral feldspar in the medium sand fraction is a characteristic of paste group S3 (Figure 7C), whereas the main constituent of paste groups S1 and S2 is monocrystalline quartz. Group S3 is also characterized by abundant foraminifera, which are partially decomposed due to the firing conditions. The main distinguishing characteristics of the three main paste groups from Saraceno are summarized in Table 3.

Of the loners in the Saraceno ceramics analysed, sample Ts4 (Figure 7D) is differentiated by its packing (c. 15%) and the abundance of its calcareous component comprising bioclasts and a few lithoclasts, which were partially thermally decomposed and subsequently recrystallized. Sample Ts8 (Figure 7E) has a heterogeneous micromass and exhibits comparable packing to that reported for sample Ts4. Its most distinguishing characteristic is an abundance of grog and the sporadic presence of metamorphic rock fragments. Lastly, sample Ms4 (Figure 7F) is distinguished by its low packing (5–8%). Its major constituent is angular to sub-angular monocrystalline quartz, which is accompanied by sporadic-to-rare foraminifera that occasionally exhibit evidence of thermal decomposition. Textural concentration features (TCFs) (Whitbread, 1995) rich in iron oxides are sparsely distributed in its groundmass, while chert, feldspar, and polycrystalline quartz are rare constituents.

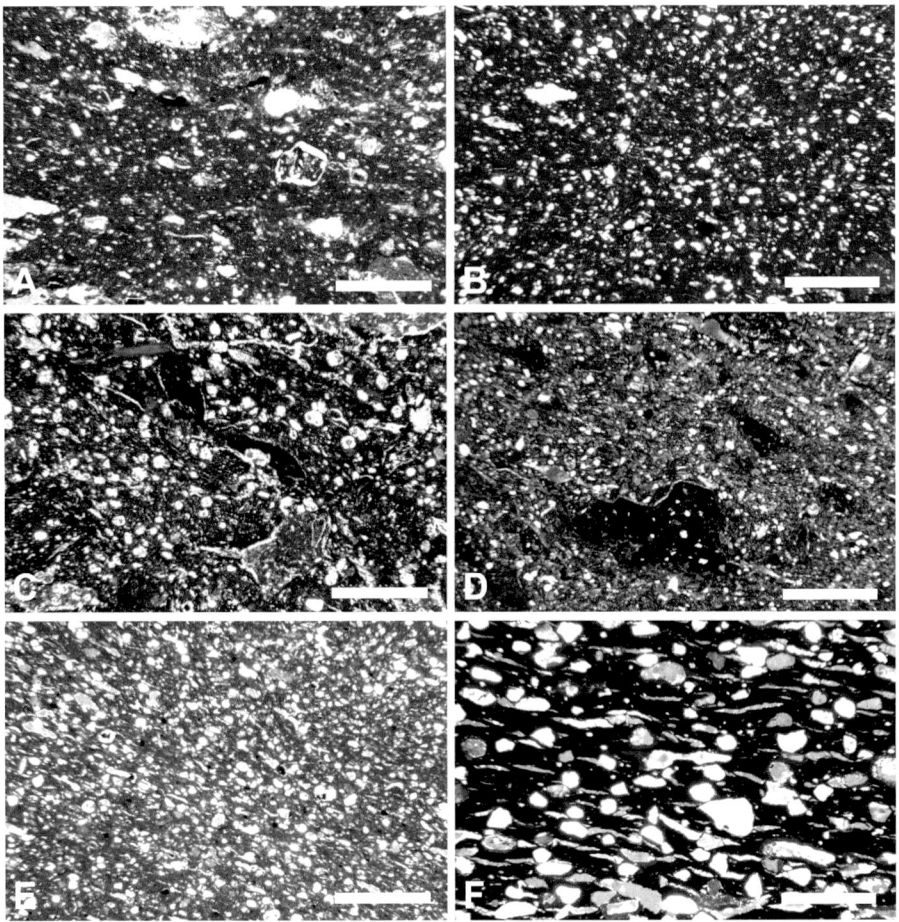

Figure 6: Thin section photomicrographs of indigenous tableware ceramics from Adranone analysed in this study. a) Paste group A1, b) Paste group A2, c) Paste group A3, d) Sample M.A.2, e) Sample M.A.23, (f) Sample M.A.25. Images taken in crossed polars. Scale bar = 0.5 mm.

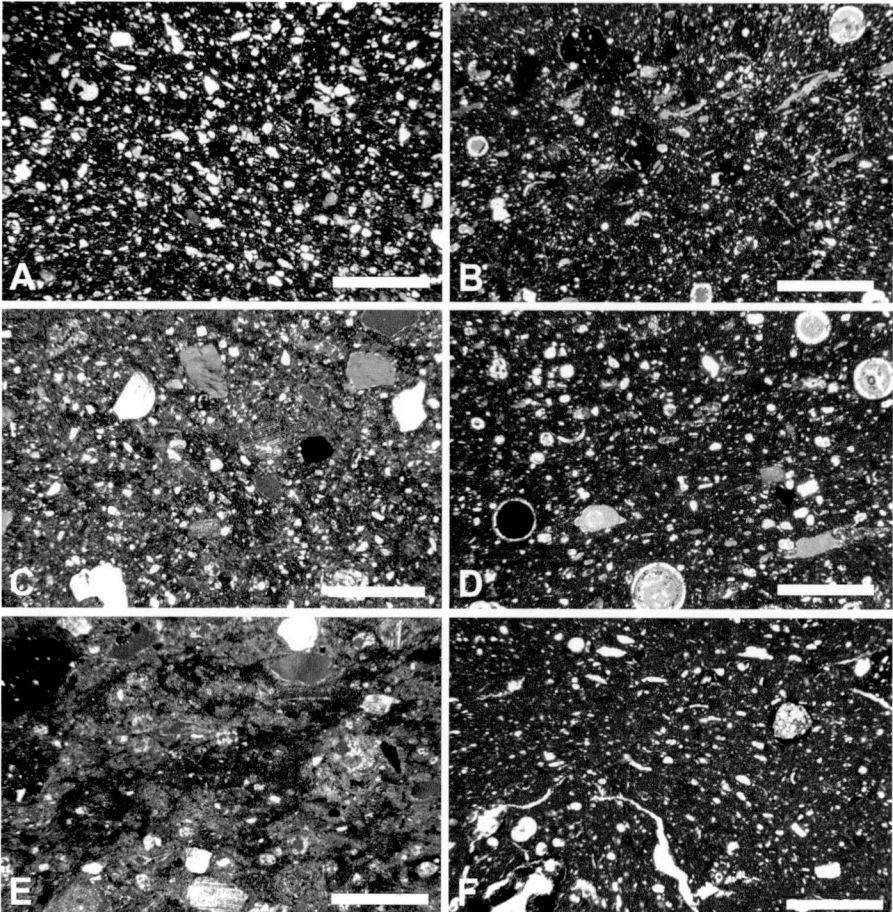

Figure 7. Thin section photomicrographs of indigenous tableware ceramics from Saraceno analysed in this study. a) Paste group S1, b) Paste group S2, c) Paste group S3, d) Sample Ts4, e) Sample Ts8, f) Sample Ms4. Images taken in crossed polars. Scale bar = 0.5 mm.

Petrography of potential local clay sources

Petrographic analysis of the fired experimental briquettes prepared from clay samples of the two main geological formations exposed in the vicinity of Entella revealed substantial differences. Material from the Terravecchia Formation contains a fairly high proportion (10–20% volume) of aplastic inclusions, which consist mainly of abundant quartz grains and common white mica flakes (Figure 8A). By contrast, the briquette from the San Cipirrello Formation exhibits a significantly lower proportion of sand-sized grains (c. 5%), which are is mainly composed of foraminifera that are largely decomposed due firing, while quartz is less abundant (Figure 8B).

Quartz is the predominant constituent of the briquette sample made using the raw material sampled from the area of Adranone (Figure 8C). The calcareous component is abundant but always less frequent than quartz, and is mainly composed of foraminifera. Mica flakes were only encountered sporadically. In general, the proportion of aplastic inclusions has been estimated at around 10–20% by volume. The clay from the Saraceno area (Figure 8D) has a very low proportion (<3%) of aplastic inclusions. Monocrystalline quartz is its predominant constituent, whilst fine mica flakes are common, but not particularly abundant. K-feldspar, plagioclase, chert, and opaque minerals were only encountered only sporadically.

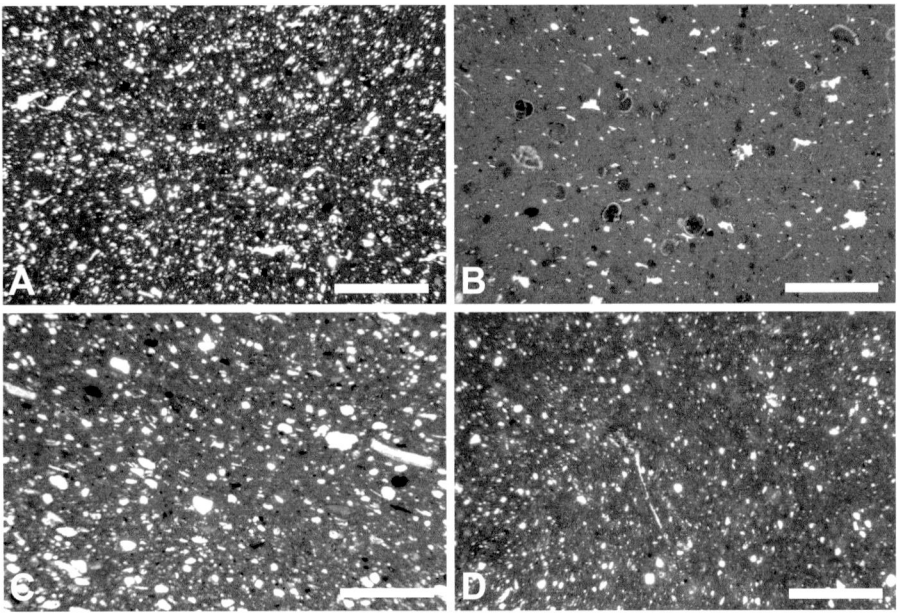

Figure 8. Thin section photomicrographs of experimental briquettes made from clay samples collected in this study, a) Terravecchia Formation, b) San Cipirrello Formation, c) Marnoso-Arenacea Formation, d) Licata Formation. Images taken in crossed polars. Scale bar = 0.5 mm.

Automatic grouping of ceramics and clay samples

Based on the automatic grouping procedure of the codified petrographic data by MCA, several interesting correspondences were found between some of the paste groups, the loners, and the raw material samples analysed in thin section (Figure 9). The clay material sampled from the Terravecchia Formation, which crops out in proximity to the archaeological site of Entella is classified within a cluster that consists of group A2 samples from Adranone and E1 samples from Entella. These samples are characterised petrographically by their abundant siliceous component, represented by very fine monocrystalline quartz, tiny mica flakes, and subordinate content of amounts of foraminifera. Some of their textural aspects such as their packing, size, and the distribution of the aplastic inclusions may also contribute to their classification in the MCA.

VAR 7	VAR 10
Metamorphic rocks	Feldspar
1. Absent	1. Absent
2. Slate-Phyllite few	2. Kfs few
3. Slate-Phyllite frequent	3. Kfs frequent
4. Schist few	4. San./Anort. few
5. Schist frequent	5. San./Anort. frequent
6. Gneiss few	
7. Gneiss frequent	**VAR 11**
8. Amphibolite few	Plagioclase
9. Amphibolite frequent	1. Absent
10. Blueschist/Eclogite few	2. Few
11. Blueschist/Eclogite frequent	3. Frequent
12. Granulites few	
13. Granulites frequent	**VAR 12**
14. Hornfels few	Pyroxenes
15. Hornfels frequent	1. Absent
16. 5 + 9	2. Orthpx. few
17. 5 + 15	3. Orthpx. frequent
18. Serpentinite	4. Clinopx. few
19. 3 + 4 + 18	5. Clinopx. frequent
20. 4 + 18	6. Na px. few
21. 3 + 4	7. Na px. Frequent
22. 3 + 18	
23. 6 + 18	**VAR 13**
24. 3 + 4 + 6	Amphiboles
25. 3 + 6 + 18	1. Absent
26. 3 + 4 + 7	2. Few
27. 4 + 6 + 18	3. Frequent
28. 3 + 6	4. Na amph. few
	5. Na amph. freq.
VAR 8	
Sedimentary rocks	**VAR 14**
1. Absent	Micas
2. Mudstone/Siltstone few	1. Absent
3. Mudstone/Siltstone frequent	2. Chlorite few
4. Sandstone few	3. Chlorite frequent
5. Sandstone frequent	4. Muscovite few
6. Limestone few	5. Muscovite frequent
7. Limestone frequent	6. Biotite few
8. Volcaniclastic few	7. Biotite frequent
9. Volcaniclastic frequent	8. 5 + 7
10. Chert few	9. 4 + 6
11. Chert frequent	10. 3 + 5 + 7
12. 5 + 11	11. 2 + 5
13. 4 + 7	12. 3 + 5
14. 7 + 10	13. 3 + 7
15. 5 + 10	14. 4 + 6
16. 6 + 8	
17. 7 + 11	**VAR 15**
18. 4 + 10	Phyllosilicates
19. 5 + 7 + 11	1. Absent
20. 2 + 4 + 10	2. Present
21. 10 + 4	
22. 2 + 4	**VAR 16**
23. 2 + 10	Carbonates
24. 2 + 10 + 6	1. Absent
25. 2 + 4 + 11	2. Dolomite few
26. 2 + 10 + 4 + 6	3. Dolomite frequent
27. 2 + 11	4. Calcite prim. few
28. 3 + 5	5. Calcite prim. frequent
29. 3 + 4	6. Secondary calcite
30. 3 + 5 + 6 + 10 + 8	7. Microfossils / micritic clots/ cast few

Table 3. Definition of the variables used for the automatic classification of the indigenous tableware from Entella, Adranone, and Saraceno (modified after Cau *et al.*, 2004).

Archaic Indigenous Tableware Production in Western Sicily

Sample	paste group	VAR1	VAR2	VAR3	VAR4	VAR8	VAR9	VAR10	VAR11	VAR14	VAR16
MA3	A1	2	1	8	2	18	7	2	1	4	8
MA4	A1	2	1	7	1	4	7	2	1	4	8
MA9	A1	2	3	6	1	4	7	2	1	4	8
MA10	A1	1	3	6	2	4	7	2	1	4	8
MA16	A1	2	1	7	1	4	7	2	1	4	9
MA17	A1	2	3	6	1	4	7	2	1	4	9
MA20	A1	2	3	7	1	18	7	2	1	4	9
MA21	A1	2	3	6	1	18	7	2	1	4	8
MA22	A1	1	3	6	1	4	7	2	1	4	9
MA27	A1	1	3	6	1	4	7	2	1	4	8
MA28	A1	2	3	6	1	4	7	2	1	4	8
MA29	A1	2	1	7	2	18	7	2	1	4	8
MA6	A2	3	4	1	1	10	7	2	2	9	7
MA11	A2	3	3	1	1	18	7	3	2	9	8
MA14	A2	3	4	1	1	10	5	3	3	9	7
MA15	A2	3	4	1	1	18	7	3	2	9	8
MA18	A2	3	3	1	1	18	5	2	2	9	7
MA19	A2	3	4	1	1	10	5	2	2	9	7
MA24	A2	3	4	1	1	18	7	3	3	9	7
MA26	A2	3	4	1	1	18	7	3	3	9	7
MA5	A3	4	1	8	8	4	5	2	1	1	9
MA8	A3	4	1	7	8	1	5	2	1	1	9
MA13	A3	4	3	7	8	1	7	2	1	1	9
ENT1	E1	3	3	1	1	10	7	2	2	14	7
ENT2	E1	3	4	1	1	10	7	2	2	14	7
ENT3	E1	3	4	1	1	18	5	3	3	9	7
ENT4	E1	3	4	1	1	18	7	3	3	9	7
ENT5	E1	3	4	1	1	18	7	3	3	9	8
ENT6	E1	3	4	1	1	10	5	2	2	9	7
ENT7	E1	3	4	1	1	10	5	3	2	9	7
ENT8	E1	3	4	1	1	18	7	3	3	9	7
ENT10	E1	3	4	1	1	10	7	3	2	9	7
ENT11	E1	3	3	1	1	10	7	2	2	14	7
ENT12	E1	3	3	1	1	18	7	3	2	9	7
ENT14	E1	3	4	1	1	10	5	2	2	9	7
ENT13	E2	3	2	9	1	4	7	3	2	4	7
ENT15	E2	3	2	9	1	4	7	2	1	4	7
ENT17	E2	2	2	9	1	4	7	2	1	4	7
ENT16	ENT16	4	1	8	8	4	5	2	1	4	9
Ts14	S1	3	3	1	1	10	7	2	1	9	8
Tss	S1	3	3	1	1	18	5	2	1	4	8
Ms8	S1	3	2	9	1	15	7	3	2	5	8
Ms3	S1	4	3	6	1	18	5	2	1	4	8
A2	S2	2	3	7	2	1	7	2	2	4	7
A3	S2	2	3	7	2	4	7	3	2	4	9
A5	S2	3	1	8	2	1	7	2	1	4	9
A6	S2	2	1	8	2	1	7	2	2	4	8
A9	S2	3	1	8	2	4	7	3	2	4	9
Ts2	S2	3	1	8	1	4	7	2	1	4	8
Ts3	S2	2	1	8	2	4	7	2	1	4	8
Ts5	S2	2	3	7	2	1	7	2	1	4	8
Ts6	S2	2	3	7	2	1	7	2	1	5	9
Ts7	S2	3	1	7	2	1	7	2	1	5	9
Ts9	S2	2	1	8	2	1	7	2	1	4	8
Ts10	S2	2	1	7	2	4	7	3	2	5	8
Ts11	S2	2	3	7	2	4	7	2	1	4	9
Ts13	S2	3	1	8	2	1	5	2	1	4	9
Ms1	S2	3	1	8	2	18	7	3	2	4	9
Ms2	S2	2	2	7	1	4	7	3	2	4	8
Ms6	S2	3	1	7	2	4	7	2	1	4	9
Ms7	S2	3	1	8	2	5	7	3	3	4	9
Ms9	S2	3	1	7	2	1	5	2	1	4	9
Ms10	S2	3	1	7	2	4	7	3	2	4	9
A1	S3	3	2	7	1	15	6	3	3	4	8
Ts1	S3	4	2	7	1	15	6	3	2	1	8
Ts4	Ts4	3	1	8	1	5	7	3	3	5	9
CET5	TV	3	4	1	1	10	7	3	2	9	7
SAM	MAB	2	3	6	1	18	7	2	2	4	9
CE7	SC	3	1	7	1	1	4	1	1	1	9
rav6	LIC	1	3	1	1	1	4	2	2	4	8

Table 4. Codified petrographic data of indigenous tableware ceramics from Entella, Adranone, and Saraceno, according to the variables defined in Table 3.

More than half of the samples of paste group A1 from Adranone clustered very close to the Marnoso-Arenacea Formation clay (Figure 9) and separate fairly well from the previous clusters of A2 and E1 samples. These sherds are characterized by low-medium packing and very fine aplastic inclusions, which mainly consist of quartz and variable amounts of foraminifera, with sporadic grog. The rest of the samples of paste group A1 cluster together with the samples assigned to paste group S2 from Saraceno, which form a very tight group (Figure 9). Interestingly, the clay sample from the Licata Formation, which crops out not far away from the settlement of Saraceno is plotted more or less within the same cluster. The samples assigned to paste groups E2, S1, and S3 exhibit a broader vertical pattern in the middle part of the MCA plot (Figure 9), with some of the individuals from groups E2 and S1 incorporated into the cluster formed by groups A1 and S2. The loner from Entella (sample Ts4) is centrally located within this broad pattern. Furthermore, it should be noted that all the ceramic artefacts assigned to paste group E2 exhibit relatively thicker walls (up to 10 mm) than the overall average.

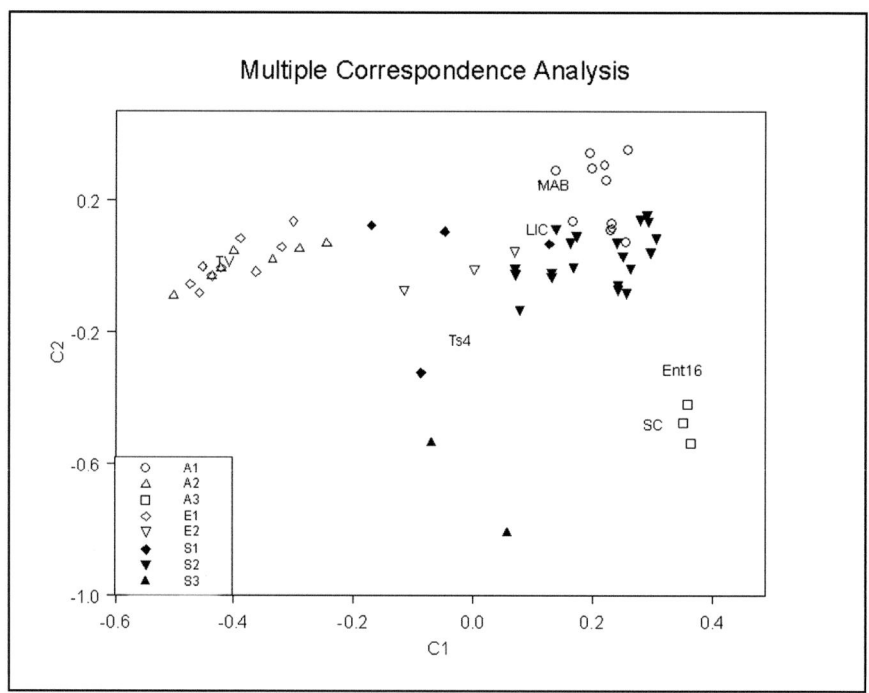

Figure 9. Two-dimensional component plot based on multiple correspondence analysis (MCA) of petrographic data collected from indigenous tableware and clay samples, with outliers identified during the first step of the statistical treatment removed. Entella paste groups = E1-E3, Adranone paste groups = A1-A3, Saraceno past groups = S1-S3, Loners = ENT16, Ts4, Terravecchia Formation clay – TV, San Cipirrello Formation clay = SC, Marnoso-Arenacea Formation clay – MAB, Licata Formation clay = LC.

The single loner from Entella (ENT16) and the clay-rich raw material sample from San Cipirrelo plot close to the samples assigned to paste group A3 of Adranone, forming a fairly distinct cluster that is well separated from the other main clusters. All the ceramic artefacts assigned to paste group A3 are characterized by an abundant calcareous component, represented mainly by foraminifera. This corresponds well with the Middle-Upper Miocene deposits of San Cipirrello, which also contain abundant microfossils.

Despite their similarities, the paste groups detected within the ceramics from Entella (E1 and E2) are well separated by the statistical classification. These two related paste groups could have been produced from a single ceramic clay source but with the addition of fine to medium sand temper in the case of the group E2 samples. Alternatively, the ceramics could have been made from different clay sources or beds within the same geological formation. The petrography of the loner sample ENT16 indicates that a similar paste could have been produced by employing clays procured from the San Cipirrello Formation, which crop out locally.

Discussion and Conclusions

The analysis Archaic Period indigenous tableware ceramics from the archaeological sites of Entella, Adranone, and Saraceno, western Sicily in this study has permitted their classification into several related paste groups that originated from different local production centres. Detailed petrographic analysis of these macroscopically and compositionally very similar ceramics using fine-scale mineralogical and textural criteria enabled them to be related to nearby sources of Cenozoic marine clay. The provenance interpretation of indigenous tablewares from the three sites provides important data with which to examine the exchange of this type of pottery in Archaic Sicily.

Pottery of paste groups E1 and E2 made at Entella from Terravecchia has been recognized at the nearby settlement of Adranone (group A2), which is located 15 km from Entella. Likewise, similar pottery also occurs at Saraceno in the form of paste group S1, at a distance of 90 km from its source. Paste group A1 of Adranone is considered to have been locally produced using Marnoso-Arenacea Formation clays, whereas paste group A3 from the same site is related petrographically to the San Cipirrello clays, which crop out near the settlement of Entella. Paste group S2 can be considered a local product of Saraceno having been made from Licata Formation clays. The partial convergence of the mineralogical and textural features of the aplastic constituents encountered through the statistical evaluation of the petrographic data could be explained in terms of the depositional history of both the Terravecchia Formation (Upper Tortonian–Lower Messinian) and the Licata Formation (Upper Langhian–Lower Messinian). These geologic formations were deposited almost contemporaneously, the latter being semi-pelagic sediments from the basin margin.

The petrographic analysis of indigenous tableware ceramics in this study demonstrate for the first time that this peculiar class of pottery was produced in more than one centre and widely distributed throughout western and central Sicily. The identification of this ceramic class in several Greek colonial settlements, which were already producing their own pottery, requires further attention and will be addressed in future research.

Acknowledgements

This article has greatly benefited by the detailed reviews of Dr. James M. Heidke and the Editor, Dr. Patrick S. Quinn. We would also like to thank Dr. Beth Miksa her comments on the text.

References

Alaimo, R., Giarrusso, R., Iliopoulos, I. and Montana, G. 2003. Risultati delle indagini mineralogico-petrografiche su reperti ceramici. In: Spatafora, F. (Ed.) *Monte Maranfusa un insediamento nella media valle del Belice, l'abitato indigeno*, Soprintendenza ai Beni Culturali e Ambientali, Servizio Beni Archeologici, Palermo: 425-441.

Campisi, L. 2003. La ceramica indigena a decorazione geometrica dipinta. In: Spatafora, F. (Ed.) *Monte Maranfusa, un insediamento nella media valle del Belice, l'abitato indigeno*, Soprintendenza ai Beni Culturali e Ambientali, Servizio Beni Archeologici, Palermo: 157-228.

Cau, M.A., Day, P.M., and Montana, G. 2002. Secondary calcite in archaeological ceramics: evaluation of alteration and contamination processes by thin section study. In Kilikoglou, V., Hein, A., and Maniatis, Y. (Eds.), *5th EMAC: Modern Trends in Scientific Studies on Ancient Ceramics*. International Series 1011. Oxford: British Archaeological Reports: 9-18.

Cau, M.A., Day, P.M., Baxter, M.J., Papageorgiou, I., Iliopoulos, I., and Montana, G. 2004. Exploring automatic grouping procedures in ceramic petrology. *Journal of Archaeological Science*, 31: 1325-1338.

Gargini, M. 1995. La ceramica indigena a decorazione geometrica dipinta. In: Nenci, G. (Ed.) *Entella I*, Pisa: 111-161.

Guglielmino, R. 2000. Entella: un'area artigianale extra-urbana di età tardo-arcaica. In: Corretti, A. (Ed.) *Atti delle Terze Giornate Internazionali di Studi sull'Area Elima*, Pisa 2000: 701-713.

Hilgen, F.J., Krijgsman, W., Langereis, C.G., Lourens, L.J., Santarelli, A., and Zachariasse, W.J. 1995. Extending the astronomical (polarity) time scale into the Miocene. *Earth and Planetary Science Letters*, 136: 495-510.

Krijgsman, W., Hilgen, F.J., Langereis, C.G., Santarelli A., and Zachariasse W.J. 1995. Late Miocene magnetostratigraphy, biostratigraphy and cyclostratigraphy in the Mediterranean. *Earth and Planetary Science Letters*, 136: 475-494.

Montana, G., Polito, A.T., Lavore, A.T., Caruso, A. and Trombi, C. 2007. Indagini archeometriche funzionali all'individuazione dei centri di produzione ceramica attivi in età arcaica nella Sicilia centro-occidentale: In: Four Italian Natioal Congress on Archaeometry, Pisa, 1-3 February 2006, Patron Editore, Bologna: 447-457.

Roda, C. 1965. I sedimenti plio-pleistocenici della Sicilia centro-meridionale. *Atti Accademia Gioenia di Scienze Naturali*, 17: 37-62.

Ruggeri, G. and Torre, G. 1974. Geologia delle zone investite dal terremoto del Belice. *Rivista Mineraria Siciliana*, 139/141: 27-48.

Spatafora, F. 1996. La ceramica indigena a decorazione impressa e incisa nella Sicilia centro-occidentale: diffusione e pertinenza etnica. *Sicilia Archeologica*, 29: 91-110.

Sprovieri, R. 1993. Pliocene-early Pleistocene astronomically forced and chronology of Mediterranean calcareous plankton bio-events. *Rivista Italiana di Paleontologia e Stratigrafia*, 99: 371-414.

Sprovieri, R., Di Stefano, E., and Sprovieri, M. 1996. High resolution chronology for Late Miocene Mediterranean stratigraphic events. *Rivista Italiana di Paleontologia e Stratigrafia*, 102: 77-104.

Sprovieri, R., Bonomo, S., Caruso, A., Di Stefano, A., Di Stefano, E., Foresi, L., Iaccarino, S.M., Lirer, F., Mazzei, R., and Salvatorini, G. 2002. An integrated calcareous plankton biostratigraphic scheme and biochronology for the Mediterranean Middle Miocene. *Rivista Italiana di Paleontologia e Stratigrafia*, 108: 337-353.

Trombi, C. 1999. La ceramica indigena dipinta della Sicilia. In: Barra Bagnasco, M., De Miro, E., and Pinzone, A. (Eds.), *Origine e incontri di culture nell'antichità Magna Grecia e Sicilia, Stato degli Studi e prospettive di ricerca*, Atti dell'Incontro di Studi, Messina 2–4 dicembre 1996, Catanzaro 1999: 275-293.

Trombi, C. 2000. *La ceramica indigena della Sicilia occidentale dalla metà del IX sec. a.C. al V sec. a.C. Tesi di Dottorato in Archeologia e Storia dell'Arte greco – romana*, XIII ciclo, Università degli Studi di Messina, Vols. I–III.

Whitbread, I.K. 1995. *Greek transport amphorae. A petrological and archeological study*. Fitch Laboratory Occasional Paper 4, The British School at Athens.

PETROGRAPHIC & MICROSTRATIGRAPHIC ANALYSIS OF MORTAR-BASED BUILDING MATERIALS FROM THE TEMPLE OF VENUS, POMPEII

Rebecca Piovesan

Department of Geosciences, University of Padova, Italy
(rebecca.piovesan@unipd.it)

Emmanuele Curti

Archaeological School of Matera, University of Basilicata, Matera, Italy

Celestino Grifa

Department of Geological and Environmental Studies,
University of Sannio, Benevento, Italy

Lara Maritan

Department of Geosciences, University of Padova, Italy

Claudio Mazzoli

Department of Geosciences, University of Padova, Italy

Introduction

Recipes for mortar-based building materials may change over time and differ in various construction and restoration phases. They normally reflect craftsmen's knowledge, availability of raw materials, and also the importance of the building in which they are found. The present research focuses on mortar-based materials from several construction and renovation phases of the Temple of Venus, at Pompeii, Italy, in order to identify any changes over time in production recipes.

The Temple of Venus, who was both the main and polyad divinity of Pompeii, is located on the southwestern side of the town (Figure 1), and underwent numerous reconstructions and renovations until the eruption of Vesuvius buried it under a thick layer of pumice in 79 AD. The site of the temple had probably been a holy place since Archaic times, connected with the Etruscan worship of Venus. The area was certainly occupied again in the late 4th-3rd centuries BC, when the Sannites entered it. It was completely redesigned in about 130 BC, during definitive Romanisation. The sanctuary of Roman Republic times was then renovated during the Julian and Claudian ages, but was almost completely destroyed during the earthquake of 62 AD. At the time of the eruption of Vesuvius in 79 AD, rebuilding was still under way, as attested by findings of building elements. A recent new hypothesis on the debated location of the harbour of Pompeii suggests that the Temple of Venus had not only religious but also trade connotations (Curti, 2007, 2008). This new interpretation places the harbour on the southwestern side of Pompeii, outside the town walls, near the market and right in front of the Temple. Therefore, the Temple of Venus becomes an extremely interesting case study with which to follow continual religious and political changes through architectural renovations in Pompeii.

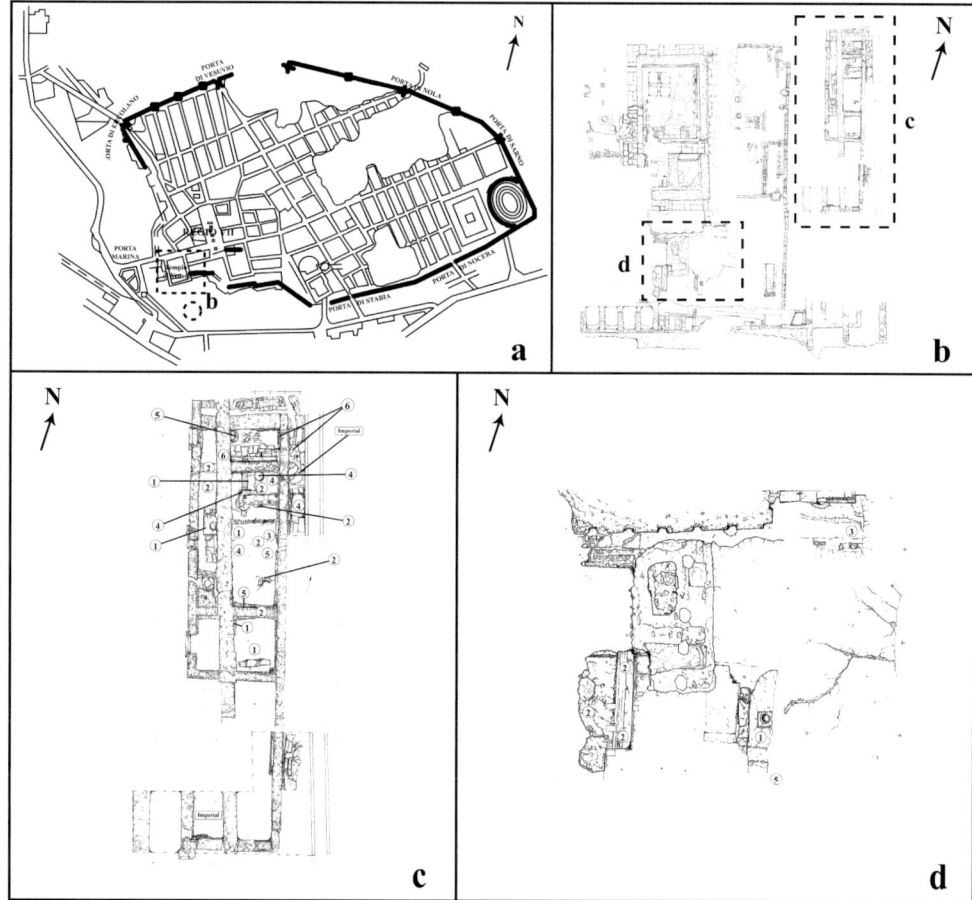

Figure 1. Map of Pompeii and the Temple of Venus. a) Plan of archaeological site, b) Temple of Venus, c) Details of northern excavated area, d) Details of southern excavated area.

Sample Selection

During excavations at the Temple of Venus by the Postgraduate School of Archaeology, University of Basilicata, between 2004 and 2008, thousands of fragments of mortar-based building materials were unearthed. In this study, we analysed a selection of 127 of these samples, mostly from the southern and northern areas of the site (Figure 1B). The samples were dated archaeologically from the end of the 4th century BC to the 1st century AD and divided into six age groups (Table 1).

Stratigraphic units were dated and related chronologically on the basis of ceramic type. The samples were also subdivided according to their architectural provenance into three groups: walls, floors, and hydraulic structures (conduits, wells and cisterns) (Table 1), and then analysed petrographically and microstratigraphically. Identified aggregate particles were also compared with samples of sand collected from 14 localities along the Neapolitan coastline, from Cuma to Castellammare di Stabia in order to determine the provenance of the raw materials (Figure 2).

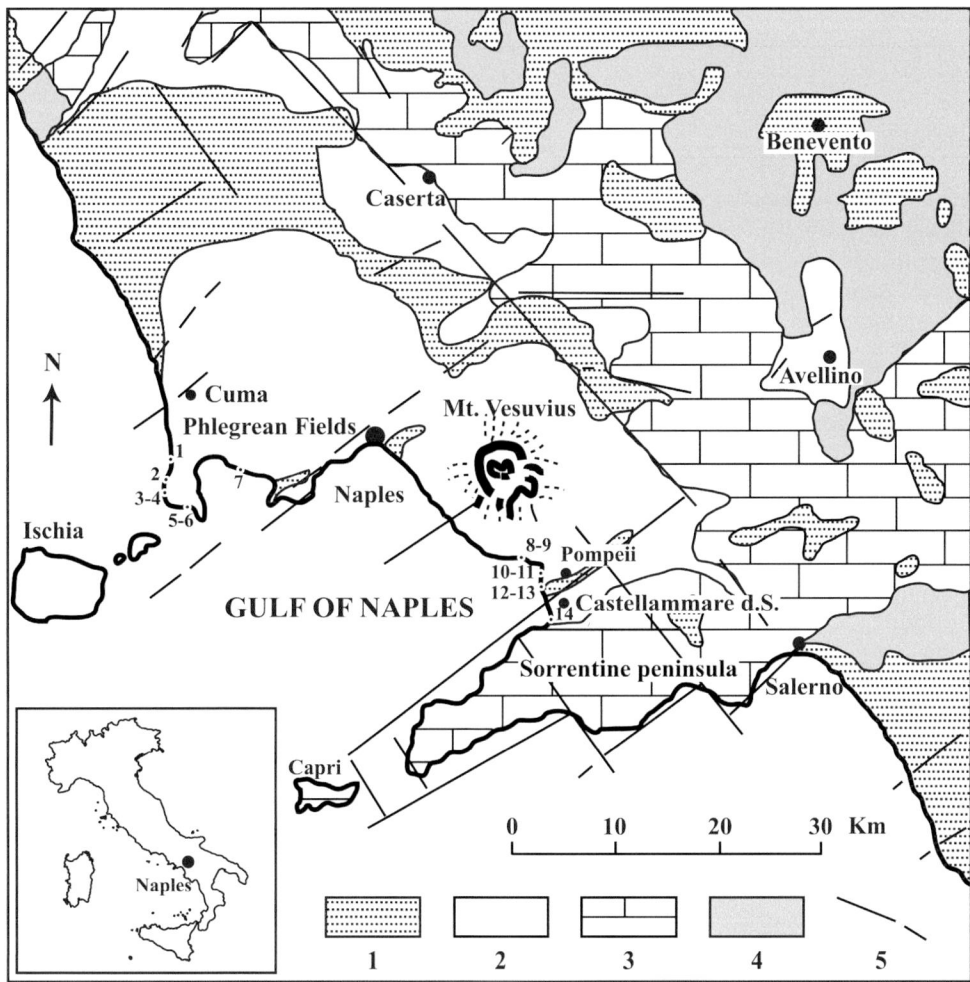

Figure 2. Geological sketch of Mount Vesuvius and surrounding areas. Modified after Revellino et al. (2004). 1 = Alluvial, lacustrine and coastal sediments, 2 = Potassic to ultrapotassic lavas and volcaniclastic deposits, 3 = Limestone and dolostone, 4 = Silico-clastic and carbonate deposits, evaporates, 5 = Faults.

Age		Walls					Floors		Hydraulic S.	
		Arriccio		Intonaco						
		VSRA	CRA	I	C	M	VSRF	CRF	VSRH	CRH
Republican times	End 4th-3rd BC	9	1	1	3	1	4	1	2	-
	Second half 2nd BC	25	3	3	11	8	12	7	5	4
	1st BC	8	1	2	2	7	-	-	-	-
	Augustan age	3	-	-	1	2	1	-	1	1
Imperial times	Julio-Claudian age	13	-	1	5	4	4	2	1	1
	Flavian age	2	-	-	-	-	-	-	-	-
Unknown age		18	1	-	1	-	4	4	-	-
	Samples									
Walls	89	93%	7%	15%	31%	54%	-	-	-	-
Floors	28	-	-	-	-	-	69%	31%	-	-
Hydraulic structures	10	-	-	-	-	-	-	-	58%	42%

Table 1. Time distribution of differing types of mortars and relative abundances in various architectural features of provenance structures. VSRA = Volcanic scoria-rich *arriccio*, CRA = Clinopyroxene-rich *arriccio*, I = *Intonachino*, C = *Cocciopesto*, M = *Marmorino*, VSRF = Volcanic scoria-rich floors, CRF = Ceramic-rich floors, = VSRH = Volcanic scoria-rich hydraulic structures, CRH = Ceramic-rich hydraulic structures.

Analytical Methods

All samples were analysed by optical microscopy, following macroscopic and microstratigraphic analytical procedures for study of mortar-based building materials described in UNI Norm 11176:2006 'Cultural heritage - Petrographic description of a mortar' proposed by the Italian Organization for Standardization, a member of the International Organization for Standardization (ISO). These procedures are applicable to mortars and plasters, where 'mortar' is a material composed of an inorganic binder plus an aggregate with dimensions of <5 mm (Prentice, 1990), and 'plaster' is a type of fine-grained, often multi-layered, mortar which provides a smooth coat to a wall or other surface.

The definitions, originally introduced by Vitruvius (1999), and used again by Mora *et al.* (1984) were adopted to define differing portions of the multi-layered plaster, normally composed of the following microstratigraphic sequence: scratch coat, *arriccio*, and *intonaco*. As defined by Mora *et al.* (1984) and Vitruvius (1999), the scratch coat is a very rough rendering applied to smooth the surface of a wall, *arriccio* is a sequence of "not less than three coats of sand and mortar, besides the rendering coat" (Vitruvius, 1999, p. 89) and *intonaco* is a series of finishing layers made of limewash and very fine sand, which may also be painted (Figure 3). Floors and hydraulic structures also display a multi-layer structure, but usually with a simpler microstratigraphy, composed of a preparation layer and one or two finishing layers (Figure 3). The filler, including various types of rocks and minerals such as carbonate rocks, ground ceramic materials and volcanic sand, was likely to have been chosen according to the required aesthetic and physical properties of the mortar-based materials, such as colour and brightness, and hydraulicity and weathering durability, respectively.

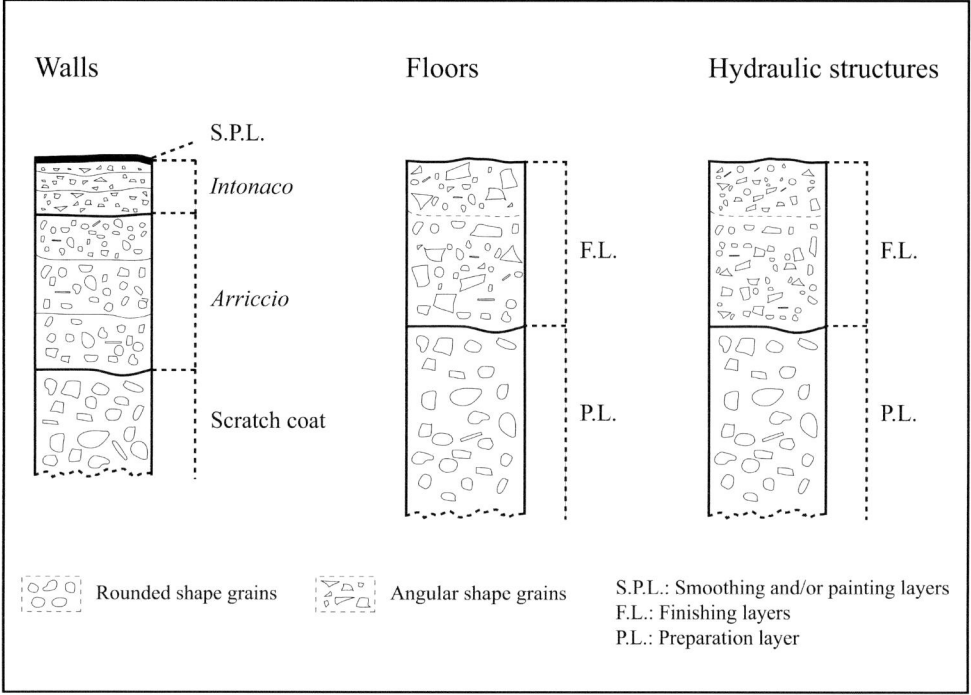

Figure 3. Microstratigraphy of walls, floors and hydraulic structures. Thicknesses of layers are not to scale.

Selected samples of the main petrographic groups were also studied by scanning electron microscopy (SEM) for microtextural and microchemical characterisation. Concentrations of the major elements Al, Fe, Si, Ca and Mg were also determined by analysis of selected areas of the binder with an energy-dispersive X-ray system (EDS). The hydraulicity of mortar, measured by the hydraulicity index (HI), is defined by Boyton (1966) as:

$$HI = \frac{Al_2O_3 + Fe_2O_3 + SiO_2}{CaO + MgO}.$$

According to this equation, the higher the index, the greater the property of the mortar to harden in a wet or water-saturated environment. The hydraulicity index should be lower than 1.2, which corresponds to the upper limit for quick-setting limes. Higher HI values in the analysis indicate the presence in the selected area of small silicate rock or mineral fragments, which increases Al_2O_3, Fe_2O_3 and SiO_2, without reflecting the real amount of hydraulic reaction products. Hydraulicity was also determined for lumps of lime, which were interpreted as a measure of the purity of the limestone used for its preparation (Charola and Henriques, 2000).

Petrography and Microstratigraphy

Petrographic and microstratigraphic analyses indicated that all samples were composed of a sequence of layers, mostly representing preparation and finishing layers of floors and hydraulic structures, and *arriccio* and *intonaco* on walls (Figure 3). The layers have various textural characteristics, aggregate compositions and matrix properties. The three classes of mortar-based building materials analysed here were found to have differing petrographic and microstructural features, and are described below.

Walls

Most of the wall samples are composed of several layers of preparatory plaster or *arriccio* covered by one or more poorly adherent *intonaco* layers. In some cases, smoothed and/or painted layers were also found, strongly adhering to the *intonaco* (Figure 3). The *arriccio* consists of strongly adherent grey plaster layers characterised by medium to fine sand-sized aggregate, medium to high porosity and low cohesion. The *intonaco* layers are composed of fine sand and limewash, and have low porosity and high cohesion.

On the basis of grain size, sphericity, roundness, and composition of aggregate plus the aggregate:binder ratio, two types of *arriccio* and three types of *intonaco* were identified. Their petrographic and microstructural features are listed in Table 2.

Arriccio

Two types of *arriccio* were distinguished, according to composition and relative abundance of various types of aggregate particles.

Volcanic scoria-rich *arriccio*

Most *arriccio* layers in wall samples belong to this type. It has a homogeneous matrix, consisting mostly of crypto- to microcrystalline calcite, with a high HI of 0.08 to 3.36, especially when the aggregate is medium silt-sized. In some cases, sub-millimeter lumps of lime, probably due to incomplete carbonation, were also identified. The aggregate:binder ratio is always about 1:1. The filler shows a wide grain-size distribution, ranging from granules to very fine sand, the coarse to medium sand fraction being most abundant. This fraction is mainly composed of rounded to well-rounded fragments of leucite-bearing volcanic rock, of leucititic or trachytic composition, and spherical scoria particles associated with abundant angular and sub-angular crystals of green and colourless diopside (Figure 4A), of medium sphericity. A few crystals of sanidine and plagioclase feldspar, black and yellow fragments of altered volcanic glass, rare flakes of biotite, and very rare crystals of Ti-rich andradite (melanite) also occur.

Clinopyroxene-rich *arriccio*

This plaster is very similar to the former type as regards the composition and grain size of its aggregate. It differs in terms of its higher aggregate:binder ratio, which ranges from 1.5:1-1:1, as well as the higher relative abundance of clinopyroxene crystals compared with fragments of volcanic rock and scoriae (Figure 4B).

Arriccio			
		Volcanic scoriae-rich *arriccio*	Clinopyroxene-rich *arriccio*
Aggregate	Aggregate to Binder	1:1	about 1:1
	Grain size	Granules to very fine sand	Very coarse to very fine sand
	Main fractions	Coarse to medium sand	Coarse to medium sand
	Sphericity	High to medium	Medium
	Roundness	VS and VRF: well-rounded to rounded grains. Cpx and Bt: sub-	Mainly sub-angular to sub-rounded
	Distribution	Homogeneous	Homogeneous
Matrix	Matrix	Micrite-like to spotted	Micrite-like to spotted
	Hydraulicity Index (HI)	From 0.08 to 3.36	From 0.14 to 4.10
Intonaco			
		Intonachino	*Cocciopesto*	*Marmorino*
Aggregate	Aggregate to Binder	about 1:1	1:1	about 1:1
	Grain size	From granules to very fine sand	From granules to coarse silt	From granules to coarse silt
	Main fractions	Coarse to medium sand	Coarse to fine sand	Medium to very fine sand
	Sphericity	High to medium	High to low	From high (L) to very low (SC)
	Roundness	VS and VRF: well-rounded to rounded	Very angular to rounded	From strongly angular (SC) to sub-rounded (L)
	Distribution	Homogeneous	Homogeneous	Homogeneous
Matrix	Matrix	Micrite-like	Spotted	Micrite-like
	Hydraulicity Index (HI)	From 0.14 to 0.19	From 0.19 to 0.38	From 0.04 to 0.06

Table 2. Classification of *arriccio* and *intonaco*. Grain size after Wentworth (1922). Cpx = Clinopyroxene, Bt = Biotite, VS = Volcanic Scoriae, VRF = Volcanic Rock Fragment, L = Limestone, SC = Sparry calcite.

Intonaco

Three types of *intonaco* were detected in wall plaster samples, and were classified, according to the composition of their aggregate particles, as *intonachino* (siliceous minerals and rock), *cocciopesto* (crushed pottery or bricks) and *marmorino* (limestone and calcite).

Intonachino

This type of *intonaco*, which is not common in wall samples, has an almost pure crypto- to microcrystalline calcite matrix, with sporadic lumps of lime, a low HI (0.14-0.19) and an aggregate:binder ratio of about 1:1. The filler shows homogeneous distribution, and grain size ranges from granules to very fine sand, coarse and medium sand classes being the most frequent. The filler is composed of fragments of volcanic scoriae and volcanic rock, frequently associated with sub-angular to angular crystals of diopside and, less often, sanidine, plagioclase, biotite and melanite (Figure 4C).

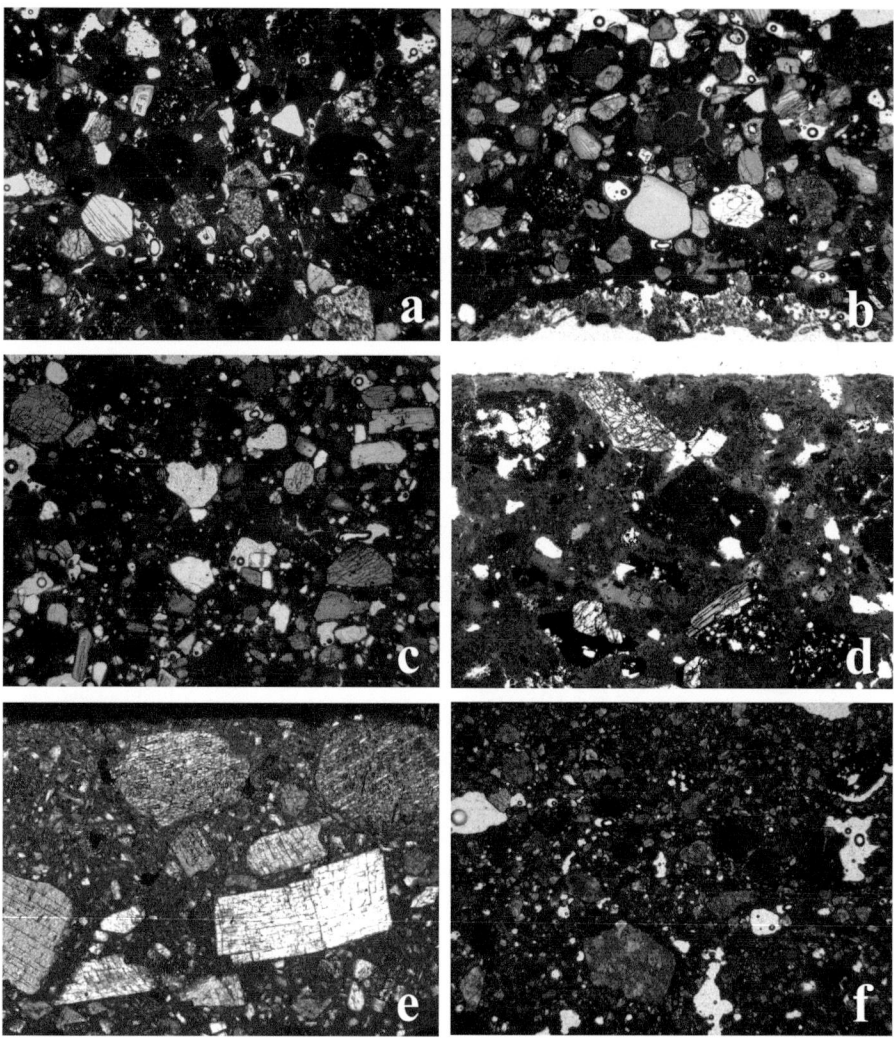

Figure 4. Polarising light micrographs of mortar samples. a) Volcanic scoria-rich *arriccio*, b) Clinopyroxene-rich *arriccio*, c) *Intonachino*, d) *Cocciopesto*, e) *Marmorino* with sparry calcite, f) *Marmorino* with limestone. All images taken in plane polarised light. Image width = 8.2 mm, except d and e = 3.9 mm.

Cocciopesto

Intonaco layers of *cocciopesto* are quite common in the wall samples. They have a spotted matrix (Figure 4D), composed of cryptocrystalline calcite and hydrated calcium silico-aluminates, with a relatively high HI (0.19-0.38; Table 2) and an aggregate:binder ratio of about 1:1. The aggregate grain size ranges from granules to coarse silt, with modal values of coarse to fine sand. Angular fragments of ground ceramic materials or 'grog' are most common, with a few well-rounded fragments of rock, scoriae, and altered glass, all of volcanic origin (Figure 4D). Rare crystals of diopside, sanidine, plagioclase and garnet and flakes of biotite also occur. Two types of grog were distinguished within the aggregate of the *cocciopesto intonaco*. The first contains rounded sand-sized inclusions of volcanic rock and volcanic scoriae, and relatively few angular crystals of diopside, plagioclase and opaque minerals, and the second contains quartz, feldspars and rare opaque minerals.

Marmorino

The third type of *intonaco, marmorino,* has a micrite-like matrix (Figure 4E,F) composed of crypto- and microcrystalline calcite, with a very low HI (0.04-0.06). The filler exhibits homogenous distribution within samples, and wide grain-size from granules to coarse silt, with maximum frequency in the medium to very fine sand classes. The aggregate consists of euhedral crystals of calcite, associated with occasional fragments of volcanic scoriae and rare crystals of diopside and feldspar (Figure 4E). In some cases, the carbonate fraction of the aggregate is composed of well-rounded fragments of micritic limestone rather than crystals of spathic calcite (Figure 4F).

Floors

Petrographic and microstratigraphic analysis of floor samples revealed that they consisted of preparatory and finishing layers with similar compositional and textural characteristics (Figure 3). According to the petrographic composition of the filler, two types of plasters were identified (Table 3).

Floors				
		Volcanic scoria-rich f.	Ceramic-rich floors	Carbonatic layers
Aggregate	Aggregate to Binder	About 1:1	1:1	About 1:1
	Grain size	From granules to coarse silt	Pebbles to very fine sand	From granules to coarse silt
	Main fractions	Coarse to fine sand	Medium to fine sand	Medium to very fine sand
	Sphericity	High to medium	Medium to low	High to medium
	Roundness	VS and VRF: well rounded to sub-angular grains Cpx and Bt: sub-angular to angular	Angular to sub-angular	Various
Matrix	Distribution	Homogeneous	Homogeneous	Homogeneous
	Matrix	Micrite-like to spotted	Micrite-like	Micrite-like
	Hydraulicity Index (HI)	From 0.16 to 1.60	From 0.10 to 0.17	From 0.01 to 0.07

Table 3. Classification of floor mortar. Abbreviations as in Table 2.

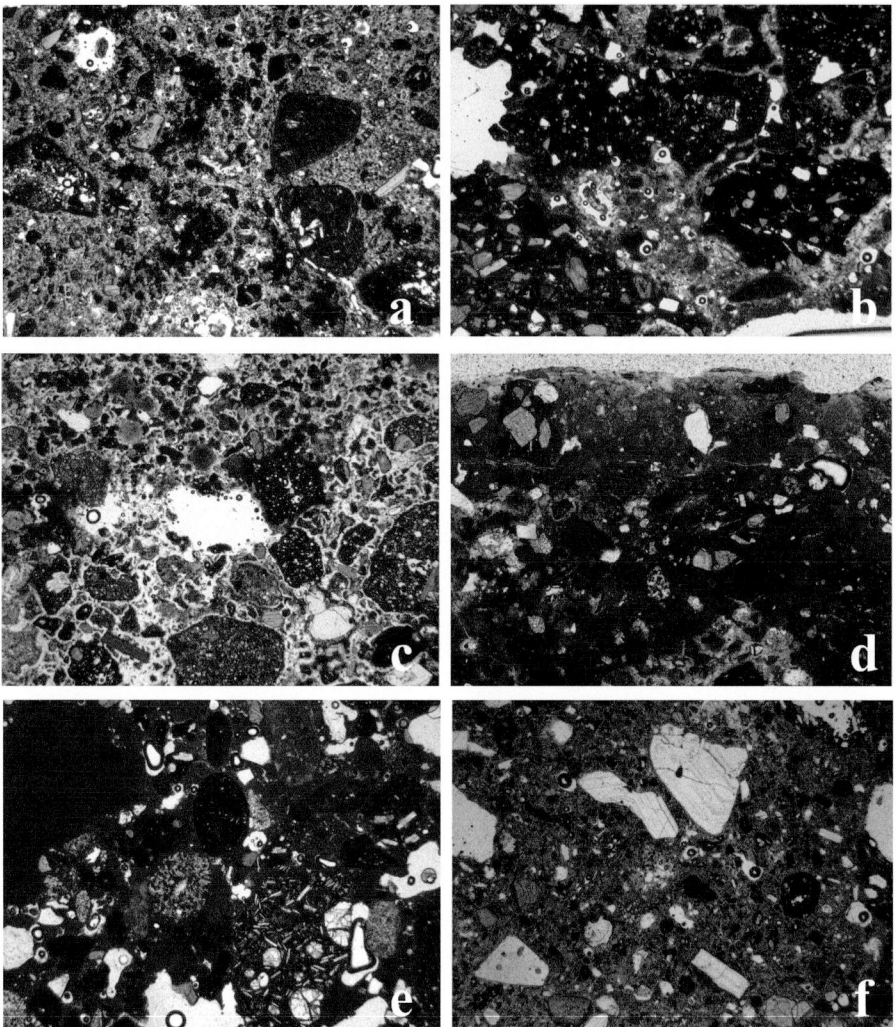

Figure 5. Polarising light micrographs of mortar samples. a) Volcanic scoria-rich floors, b) Ceramic-rich floors, c) Volcanic scoria-rich hydraulic structures, d) Ceramic-rich hydraulic structures, e) Micrite-like matrix with lime lumps, f) Spotted matrix. All images taken in plane polarised light. Image width = 8.2 mm, except d = 3.9 mm.

Volcanic scoria-rich floors

This type has a crypto- to microcrystalline calcite matrix, with an average HI lower than that of the preparatory layers of the walls, and an aggregate:binder ratio of 1.5:1 to 1:1. The aggregate grain size ranges from granules to coarse silt, with a maximum frequency in the fine sand class. It is predominantly composed of well-rounded, highly spherical, sand-sized grains of volcanic rock, scoriae and glass (Figure 5A). Angular crystals of diopside are also common, and are associated with rare crystals of plagioclase, sanidine, biotite and garnet.

Ceramic-rich floors

This type of plaster is characterised by an aggregate composed of angular, medium to fine sand-sized fragments of grog (Figure 5B). Two types of grog were identified, closely matching those observed in the *cocciopesto*. The fragments are embedded in a spotted matrix composed of cryptocrystalline calcite and hydrated calcium silico-aluminates, with a HI of between 0.10-0.17.

A finishing layer was also observed in two of the floor samples, characterised by a micrite-like calcite matrix with a low HI (0.01-0.07), an aggregate:binder ratio of 1.5:1-1:1, and aggregate composed of predominant spathic crystals of calcite and subordinate rounded fragments of micritic limestone.

Hydraulic Structure

Within the conduit, well and cistern samples, two types of plaster were identified, one rich in volcanic scoriae and one characterised by grog (Table 4, Figure 5C,D). These mortars show strong textural and compositional similarities to those used in the construction of the floors, but they have a higher HI. Both types of plaster were used for both preparatory and finishing layers in the hydraulic structures. However, volcanic scoria-rich plaster was used more frequently for preparatory layers, and ceramic-rich plaster mostly, but not exclusively, for the finishing layers.

Hydraulic structure		Volcanic scoria-rich h.s.	Ceramic-rich h.s.
Aggregate	Aggregate to Binder	About 1.5:1	About 1:1
	Grain size	From pebbles to coarse silt	From pebbles to coarse silt
	Main fractions	Coarse to medium sand	Medium to fine sand
	Sphericity	High to medium	Medium to low
	Roudness	VS and VRF: well rounded to rounded grains. Cpx and Bt: sub-angular to angular	Angular to sub-angular
	Distribution	Homogeneous	Homogeneous
Matrix	Matrix	Spotted	Spotted
	Hydraulicity Index (HI)	From 0.59 to 2.68	From 0.22 to 2.31

Table 4. Table 4. Plaster features in hydraulic structures. Abbreviations as in Table 1.

Raw Materials and Technology

Binders

Petrographic analysis of the various mortar-based building materials from the Temple of Venus revealed that they have a lime-based matrix. Hydraulicity index values and the micrite-like (Figure 5E) vs. spotted (Figure 5F) microscopic aspect of the matrix suggest differing contents of hydrated calcium silico-aluminates, implying either the use of a lime prepared from impure limestone, or pure lime which underwent hydraulic reactions with a pozzolanic aggregate. The use of impure limestone in the preparation of the mortars is not supported by chemical analysis of the lime lumps, which generally have a high degree of purity and low chemical variability, with very low HI values (c. 0.02), even when they are found in mortars with a spotted matrix and high HI, suggesting that pure limestone was selected and ignited to produce lime. Pozzolanic aggregates such as those containing volcanic scoriae, volcanic glass and ground fragments of ceramic materials (Elsen, 2006) must therefore have been involved in hydraulic reactions. Lime lumps in hydraulic structures have higher HI, with values between 0.08 and 0.27, indicating that true hydraulic lime was probably used only to construct these architectural features. The use of pure lime in most of the applications is also confirmed by the observation that mortars with a carbonate aggregate such as *marmorino* and the carbonate layers of floors always have a micrite-like matrix with low HI (< 0.07), whereas samples from hydraulic structures always have a spotted matrix and high HI values (> 0.22).

The systematic differences observed in the matrix of *cocciopesto intonaco* and ceramic-rich floors (i.e., spotted matrix and high HI vs. micrite-like matrix and relatively low HI, respectively), may be related to the grain size of the aggregate, the former including the fine-grained fraction which was hydraulically reacted with lime, and the latter the sifted coarse-grained fraction, thus giving rise to the different chromatic effects.

All the other types of mortars show very variable HI, suggesting that varying amounts of the fine-grained pozzolanic fraction were originally present in the raw material used as aggregate, rather than being added intentionally to modulate hydraulicity.

The occurrence of lime lumps in a large number of samples indicates that some of the lime often did not react completely with water during slaking or with atmospheric CO_2 after application (Hughes *et al.*, 2001). This provides strong evidence that lime, water and aggregate were mixed without due attention, perhaps because of workers' lack of technological skills or acceptance by buyers of such wares.

Aggregates

The composition of aggregates in the mortar samples shows that three types of filler were commonly used: volcanic rock, grog, and carbonate rock. The mineralogy and petrography of the volcanic aggregate is compatible with the products of the Somma-Vesuvius volcano (Santacroce, 1987) and matches the composition of beach sands collected from the Vesuvian area, suggesting that local materials was used for the aggregate. Differences in the composition of the volcanic aggregate in volcanic scoria-rich *arriccio* and clinopyroxene-rich *arriccio* may be due to differing sources of local sand.

Petrographic analysis of the grog inclusions in the crushed ceramic plaster (*cocciopesto*) suggests that the two types of ceramic were used indiscriminately. The presence of volcanic inclusions compatible with the Somma-Vesuvius complex in one type of grog, and the quartz and feldspar inclusions in the other, suggests that both locally produced and imported ceramics were used as fillers.

The carbonate-bearing plasters contain filler composed of euhedral spathic crystals of calcite. This material may have been ground from crystalline calcite veins occurring in limestone. As such, its origin is not easy to determine.

Conclusions

Archaeometric study of mortar-based building materials from the Temple of Venus at Pompeii has permitted several distinct mortar recipes to be identified, characterised by their microstratigraphy, petrographic features of the aggregate, and their matrix. These various types were deliberately prepared for specific applications, due to their different hydraulicity, or for aesthetic purposes. *Cocciopesto* plaster containing crushed ceramic was used in hydraulic structures, perhaps because of its superior hydraulic performance with respect to other types of plaster. It may have been used as *intonaco* on walls, due to its warm hues and resistance to damp.

The recipes used to construct the various mortar-based features at the Temple of Venus remained constant from the 4th century BC to the 1st century AD, suggesting the persistence of technological tradition (Table 1).

The ubiquitous presence of grains of volcanic origin, consistent with the volcaniclastic deposits of Somma-Vesuvius in many different types of plaster, clearly indicates that the raw materials were local in origin, probably alluvial or beach deposits in the Vesuvian area. The small grain size, high sphericity and roundness of the volcanic aggregate in many samples indicate great standardisation in the selection of the raw materials, which were probably quarried from identified sources as early as the 4th century BC.

Mortars used in hydraulic structures, in which specific performance was required, or on surfaces with specific aesthetic features such as *intonachino*, *cocciopesto*, *marmorino* and ceramic-rich floors, were produced by careful mixing of good-quality raw materials, and generally display relatively homogeneous textural features and hydraulicity. Other mortars, such as those used in *arriccio* and volcanic scoria-rich floors, generally covered by a finer finishing layer and a floor decoration (i.e., *opus signinum*), display greater variability in aggregate grain-size distribution, microscopic aspect of the matrix (i.e., micrite-like vs. spotted) and HI, suggesting that less attention was paid to their preparation. The finding of lime lumps in all types of mortars suggests that production was not sufficiently checked or that skilled workers were not readily available.

As regards the *marmorino* filler, the occurrence of carbonate sequences of pure limestone outcropping near Pompeii (Figure 2) indicates that these raw materials were locally available, although specific provenance markers are missing.

Acknowledgements

The authors are grateful to the staff of the Institute of Geosciences and Georesources, CNR, Padova, for their analytical support, and to G. Walton, who revised the English text.

References

Boyton, R.S. 1966. *Chemistry and technology of lime and limestone.* 2nd edition. John Wiley & Sons Inc., New York.

Charola, A.E. and Henriques, F.M.A. 2000. Hydraulicity in lime mortars revisited. In: Bartos, P., Groot, C. and Hughes, J.J. (Eds.) *Proceedings of the International RILEM-workshop "Historic Mortars: Characteristics and Tests", Paisley, 2000,* RILEM Publications: 5-105.

Curti, E. 2008. Il tempio di Venere Fisica e il porto di Pompei? In: Guzzo, P.G. and Guidobaldi, M.P. (Eds.) *Nuove ricerche archeologiche nell'area vesuviana (scavi 2003-2006)", Roma*: 47-60.

Curti, E. 2007. La Venere Fisica trionfante: un nuovo ciclo di iscrizioni dal santuario di Venere a Pompei. In: *Il filo e le perle. Studi per i 70 anni di Mario Torelli*, Venosa: 57-71.

Elsen, J. 2006. Microscopy of historic mortars – a review. *Cement and Concrete Research*, 36:1416-1424.

Hughes, J.J., Leslie, A. and Callebaut, K. 2001. The petrography of lime inclusions in historic lime based mortars. In: *Proceedings of the 8th Euroseminar on Microscopy Applied to Building Materials. Athens, Greece, September 2001*: 359-364.

Mora, P., Mora, L. and Philippot, P. 1984. *Conservation of wall paintings*, Butterworth & Co., London.

Prentice, J. E. 1990. *Geology of construction materials*, Chapman and Hall, London.

Revellino, R., Hungr, O., Guadagno, F.M., Evans, S.G. 2004. Velocity and runout simulation of destructive debris flows and debris avalanches in pyroclastic deposits, Campania region, Italy. *Environmental Geology*, 45:295-311.

Santacroce, R. (Ed.) 1987. *Somma-Vesuvius, Quaderni de "La Ricerca Scientifica".* Progetto finanziato "Geodinamica". Monografie finali, 8. CNR Roma.

UNI EN 11176. 2006. *Cultural Heritage - Petrographic description of a mortar.* Milano.

Vitruvius, P. 1999. *The ten books on architecture*. I.D. Rowland trans., Cambridge University Press.

Wentworth, C.K. 1922. A scale of grade and class terms for clastic sediments. *Journal of Geology*, 30: 377-392.

PROVENANCE & PRODUCTION TECHNOLOGY OF EARLY BRONZE AGE POTTERY FROM A LAKE-DWELLING SETTLEMENT AT ARQUÀ PETRARCA, PADOVA, ITALY

Lara Maritan

Department of Geosciences, University of Padova, Italy
(lara.maritan@unipd.it)

Claudio Mazzoli

Department of Geosciences, University of Padova, Italy

Marta Tenconi

Department of Geosciences, University of Padova, Italy

Giovanni Leonardi

Department of Archaeology, University of Padova, Italy

Stefano Boaro

Department of Archaeology, University of Padova, Italy

Introduction

Various types of settlements developed during the Bronze Age, according to geography, environment, and the need to adapt to the climate changes which occurred during the second millennium BC. Lake dwellings, settlements on drainage, *Terramarae*, hill settlements and, in a few cases, caves have been documented in the intra-moraine basins of Alpine lakes and in peat-bogs from several damp areas of northern Italy. The so-called 'Polada Culture' or 'Lake-dweller Culture' represents an archaeological facies that is typical of the Trentino, Lombardy and western Veneto areas, which developed between 2050-1600 BC and is characterised by lake-dwelling settlements.

The archaeological site of Arquà Petrarca (Veneto region, NE Italy) represents the maximum eastern expansion of the Polada Culture. This lake-dwelling settlement is located on the southern bank of the small Lago della Costa on the flank of the southeastern Euganean Hills (Figure 1A). The site, characterised by a peculiar type of bank drainage created by stratified accumulations (Balista and Leonardi, 1996), was active for a long time, from the end of the Eoneolithic-Early Bronze Age, throughout the Middle Bronze Age. A comprehensive study of the site has been difficult as it was investigated between the late 19th and early 20th centuries, and a reference stratigraphy is missing for the pottery collections, now preserved in the City Museum of Padova and the Anthropological Museum of the University of Padova. Added to this, Poladian pottery is still scarcely known and little studied and long-lasting shapes are relatively abundant. Correlations between ceramic type and chronology can only be achieved by analysing topographic data based on the locations of the excavated areas during the various surveys and archaeological campaigns carried out in the past, which are well documented. This chronology can be correlated with archaeometric data for validation. Changes in technological features are also clear-cut and cases of possible importation verified. Stylistic analogies in ceramic objects suggest possible exchanges among different settlements, although this hypothesis has not been supported so far by archaeometric data. Moreover, in the site at Arquà Petrarca, some shapes and decorative elements are still unique to the Poladian production, and are comparable to those of central-eastern Europe. There are also some large bowls and polypod bowls, the chronological and cultural attribution of which is still debated.

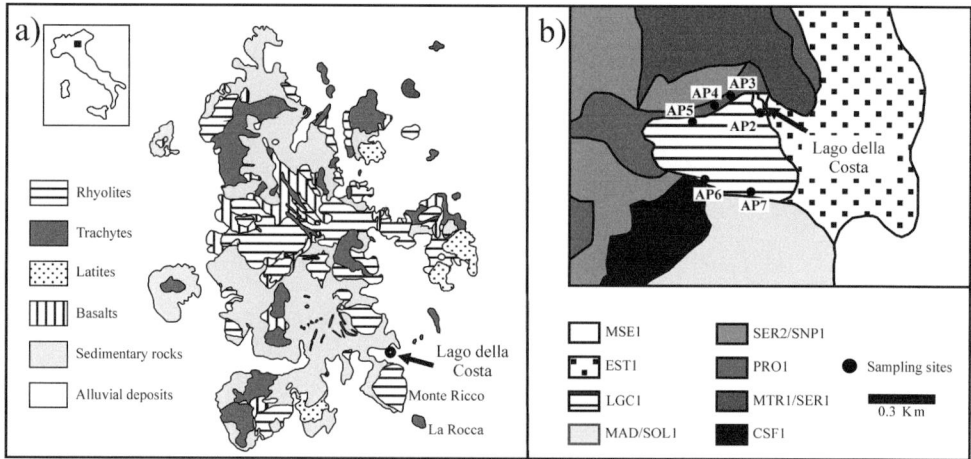

Figure 1. Location of the site of Arquà Petrarca and geology of the Euganean Hills. a) Geological sketch of Euganean Hills with location of Lago della Costa (modified after Piccoli *et al.*, 1981), b) Location of clay samples collected near Lago della Costa, with indication of cartographic soil units (modified from Veneto Agricoltura, 2001). MSE1 = soils on Adige River deposits in elevated areas, EST1 = soils on Adige River deposits in depressed areas, MAD/SOL1 = soils on acidic vulcanites, SER2/SNP1 = soils on limestone, PRO1 = soils on slope deposits, MRT1/SER1 = soils on limestone, CSF1 = soils on marls, LGC1 = soils on alluvial deposits of Adige near lake.

The present paper describes an archaeometric study of ceramics from the site of Arquà Petrarca, with the main aims of defining the provenance of the raw materials, identifying possible cases of imported pottery, and characterising the production technology of this pottery, from the preparation of paste to the firing process.

Materials and Methods

An archaeometric study was conducted on a sample of 50 potsherds, representing the main classes distinguished within the ceramic repertoire of the Lago della Costa site at

Arquà Petrarca, including jugs, jars, pans, vessels, and polypod bowls (Figure 2). The samples come from two areas, excavated in 1886 (Cordenons, 1887) and 1901 (Moschetti and Cordenons, 1901) and corresponding to two distinct and probably subsequent settlements, and from a field survey.

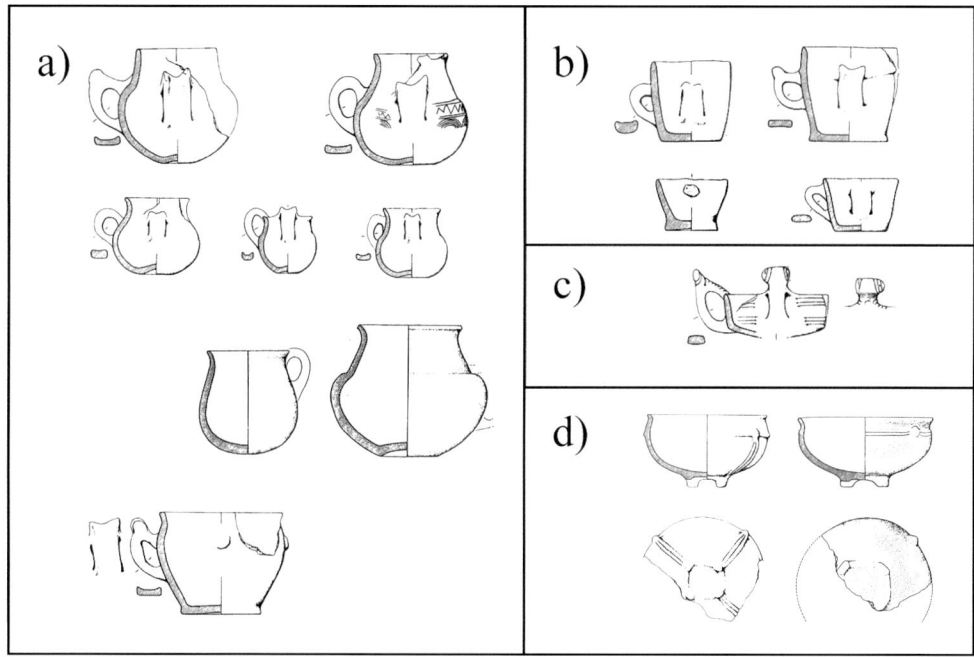

Figure 2. Some ceramic types found at the lake-dwelling settlement of Arquà Petrarca (modified from Salzani, 1982). a) Jugs, b) Cups, c) Small jugs, d) Hemispherical bowls.

All samples were studied petrographically, mineralogically and analysed geochemically. Petrographic analysis was carried out following the procedure and terminology of Whitbread (1986, 1989, 1995). Quantitative estimates of pores and inclusions were defined by optical comparison with the charts of Bacelle and Bosellini (1965). The mineralogical composition of the ceramic paste was determined by X-ray powder diffraction (XRD) on a Philips X-Pert PRO diffractometer. The presence or absence of specific mineral phases provided constraints on firing temperatures.

Quantitative chemical analyses of major, minor (SiO_2, TiO_2, Al_2O_3, Fe_2O_3, MnO, MgO, CaO, Na_2O, K_2O, P_2O_5) and trace elements (Sc, V, Cr, Co, Ni, Cu, Zn, Ga, Rb, Sr, Y, Zr, Nb, Ba, La, Ce, Nd, Pb, Th, U) were carried out by X-ray fluorescence (XRF) on a Philips PW2400 spectrometer, in order to define the chemical variability of the samples and their compatibility with local base clay. Samples were ground in an agate mortar, after removal of the surface layer with a micro-drill to avoid contamination, and then prepared as beads from calcined powder and $Li_2B_4O_7$, dilution ratio 1:10. A set of geological standards that have been analytically tested by the international scientific community (Govindaraju, 1994) was used for calibration. FeO content was determined by titration with potassium permanganate. Fe_2O_3 concentration was obtained by the difference from Fe_2O_{3tot} determined by XRF. Analytical data were processed with standard statistical tools such as Principal Component Analysis (PCA) and cluster analysis (CA) after standardisation of data according to procedures designed by Vitali and Franklin (1986) and Baxter (1999).

For further information on production technology and the source area of the raw materials, six clay samples were collected from various local soil units (Figure 1B), with reference to the pedological map of the Euganean Hills (Veneto Agricoltura, 2001). Sample AP2 was dredged from the lake, samples AP3 and AP4 were collected from clay pods within the calcareous formations outcropping west of the lake, samples AP5 and AP6 came from soils on alluvial deposits of the river Adige, which flowed immediately south of the Euganean Hills during the Bronze Age (Marcolongo and Zaffanella, 1987), and sample AP7 was collected from the alluvial deposits of the Adige, immediately north of the relief of Monte Ricco, composed entirely of rhyolite, and located south of the lake. These samples were studied in thin section and their chemical composition was determined by XRF.

Results

Petrography and mineralogy

The 50 potsherds from Arquà Petrarca exhibit considerable petrographic variability, with the inclusions differing in mineral and lithological composition, grain size, roundness, and absolute and relative frequencies (Tables 1 and 2). Three main petrographic groups were distinguished. Group 1 consists of 41 vulcanite-rich potsherds. This group includes most of the samples analysed, and is characterised by an abundance of volcanic rock fragments. Most of the samples of this group have a homogeneous groundmass, oriented parallel to the vessel walls and often optically active, with a thin striated birefringent fabric (b-fabric). Voids are medium to large in size (meso- and macro-voids), mainly with irregular or elongated shapes. Inclusions are variable in terms of coarse to fine (c:f) ratios, show a single-space porphyric-related distance and bimodal grain-size distribution. Fine sand-sized inclusions are angular and sub-angular in shape and coarser lithic fragments are mostly sub-angular to sub-rounded. They are mainly composed of coarse sand-sized fragments of volcanic rocks, the microstructure and mineral composition of which are referred to four rock types. Type 1 corresponds to a trachyte, with a microcrystalline intersertal groundmass characterised by elongated microlites, containing a few phenocrysts of plagioclase, sanidine and biotite (Figure 3A). Type 2 is a trachyte, with a hypocrystalline, almost glassy, texture, and a microlite-bearing and locally fluidal groundmass (Figure 3B). Type 3 is a trachyte with a very fine-grained intersertal texture, tabular microlites, and phenocrysts of biotite and plagioclase (Figure 3C). Type 4 corresponds to a rhyolite with a hypocrystalline xenomorphic texture, with phenocrysts of plagioclase and biotite and a quartz-rich groundmass, also containing sanidine microlites (Figure 3D). Fragments of strongly weathered vulcanites are occasionally found in samples of Group 1.

Provenance and Technology of Bronze Age Pottery from Arquà Petrarca, Padova, Italy

Sample	Ceramic type	Homo-genity	Groundmass Optical state (b-fabric)	Preferred orientation	%	Max size (μm)	Ch	PV	Ve	Vu	c:f ratio	c:f related distribution	Grain-size distribution	Max size (μm)	A	SA	SR	R	Sample
P460	jug	HOM	A(ST)	OR	10	1300	*			*	50:50	SS	B	1400		*	*	*	P460
P464	hemispherical bowl	HOM	A(ST)	OR	10	570	*		*	*	30:70	SS	B	1500			*		P464
P478	spherical jug	HOM	A(ST)	OR	10	350	*			*	40:60	SS	B	1400			*	*	P478
P481	bowl	HOM	A(ST)	IS	10	1200				*	40:60	SS	B	1100		*	*		P481
P485	spherical jug	HOM	A(ST)	OR	10	200	*				40:60	SS	B	1100		*	*		P485
P493	hemispherical bowl	HOM	IN	OR	5	2100	*			*	30:70	SS	B	1400		*	*		P493
P501	jar	HOM	A(ST)	IS	10	450			*	*	30:70	SS	B	1400		*	*	*	P501
P518	polypod bowl	HOM	A(ST)	OR	10	500	*				30:70	DS	B	1300		*	*		P518
P529	bowl	HOM	A(ST)	OR	10	700	*	*		*	20:80	DS	B	1000				*	P529
P534	vessel	HOM	A(ST)	OR	20	1800	*		*	*	30:70	SS	B	1150		*	*	*	P534
P558	pan	HOM	A(ST)	OR	10	510	*			*	40:60	SS	B	1260	*	*	*		P558
P565	bowl	HET	A(ST)	OR	20	750	*				40:60	SS	B	1100		*	*		P565
P583	hemispherical bowl	HOM	A(ST)	OR	20	1400	*				50:50	SS	B	1750		*		*	P583
P584	jug	HOM	A(ST)	OR	10	600	*				20:80	DS	B	1700		*	*	*	P584
P588	swell	HET	IN	OR	15	1300	*			*	50:50	SS	B	2700			*	*	P588
P594	jug	HOM	A(ST)	OR	5	350	*		*	*	30:70	SS	B	1050				*	P594
P599	polypod bowl	HOM	A(ST)	OR	5	400	*		*	*	20:80	DS	B	1300				*	P599
P610	cup	HOM	IN	OR	10	600			*	*	30:70	DS	B	1500		*	*	*	P610
P633	jar	HET	A(ST)	OR	10	650	*		*	*	40:60	SS	B	1500	*			*	P633
P636	polypod bowl	HOM	A(ST)	OR	10	300	*				40:60	SS	B	1700		*			P636
P647	bowl	HET	A(ST)	OR	10	300	*				30:70	SS	B	900		*	*	*	P647
P703	jar	HOM	A(ST)	OR	10	1000	*		*	*	30:70	SS	B	1100			*	*	P703
P515	jar	HOM	A(ST)	OR	10	900	*		*	*	30:70	SS	B	1150			*	*	P515
P559	swell	HOM	A(ST)	OR	10	1800	*			*	30:70	SS	B	1800	*	*	*	*	P559
P595	hemispherical bowl	HOM	IN	OR	10	530	*			*	30:70	SS	B	1150		*	*	*	P595
P632	jar	HOM	A(ST)	OR	10	1900	*	*		*	30:70	SS	B	1150		*			P632
S9-1	jar	HOM	A(ST)	OR	5	500	*				20:80	DS	B	900	*	*			S9-1
P511	jar	HOM	A(ST)	OR	10	250		*		*	40:60	SS	B	1000		*	*		P511
P513	jar	HOM	A(ST)	IS	5	1400	*		*	*	40:60	SS	B	1650	*	*	*	*	P513
P547	biconical vessel	HOM	A(ST)	OR	5	600	*				50:50	SS	B	1440		*	*		P547
P585	jar	HOM	A(ST)	IS	10	650			*	*	40:60	SS	B	1500				*	P585
P650	jar	HOM	A(ST)	IS	5	200	*		*		30:70	SS	B	1600				*	P650
P477	jar	HOM	A(ST)	OR	5	200	*		*		30:70	SS	B	1800		*	*	*	P477
S8-8	jar	HOM	A(ST)	OR	5	600	*				20:80	SS	B	900		*			S8-8
P461	jug	HOM	A(ST)	IS	5	300			*		30:70	SS	B	1450		*	*	*	P461
P556a	jar	HOM	A(ST)	OR	5	1500	*		*	x	30:70	SS	B	1550		*	*		P556a
P557	pan	HOM	A(ST)	OR	30	1000	*			x	40:60	SS	B	1600		*	*		P557
P560b	biconical vessel	HOM	A(ST)	OR	10	350	*		*	*	30:70	SS	B	1550	*	*	*	*	P560b
P620	jar	HOM	A(ST)	OR	10	660	*			*	40:60	SS	B	1300			*	*	P620
P621	jar	HOM	A(ST)	IS	5	2400	*			*	30:70	SS	B	1200			*	*	P621
P626	jar	HOM	A(ST)	OR	20	2300	*	*		*	30:70	SS	B	990	*	*	*		P626
P510	jar	HET	A(ST)	OR	15	760		*			20:80	DS	B	670	*				P510
P512	jar	HET	A(ST)	OR	15	740	*			*	20:80	DS	B	420		*		*	P512
P605	handled globular bowl	HET	A(ST)	OR	15	590		*		*	40:60	SS	B	550	*	*	*	*	P605
P642	spherical jug	HET	A(ST)	OR	10	500	*	*	*		40:60	SS	B	2000		*	*		P642
P521	jar	HET	IN	OR	40	2000	*	*		*	20:80	DS	B	550		*		*	P521
P554	hemispherical bowl	HOM	A(ST)	OR	20	950	*	*		*	30:70	SS	B	1750			*	*	P554
P574	carinated bowl	HOM	A(ST)	OR	15	1200	*				20:80	DS	B	1350		*		*	P574
P643	bowl	HOM	IN	OR	20	1400	*				20:80	DS	B	1300		*	*		P643
P654	bowl	HOM	A(CR)	OR	5	600			*	*	20:80	DS	U	600			*	*	P654

Table 1. Schematic petrographic description of Early Bronze Age pottery from Arquà Petrarca. HOM = homogeneous, HET = heterogeneous, A = active, IN = inactive, ST = striated, CR = crystallitic, Ch = channels, PV = planar voids, Ve = vesicles, Vu = vughs, SS = single-spaced, DS = double-spaced, B = bimodal, U = unimodal, A = angular, SA = sub-angular, SR = sub-rounded, R = rounded.

Sample	Qzt	Pl	Kfs	Bt	Ms	Op	Px	Am	Rt	Clay pellet	ARF	Grog	Chert	Q	VG	VR1	VR2	VR3	VR4	WVR	CM	CW	Shells
P460	xx	x	x	x	x	x				x				x		xxxxx	x						
P464	xx	x	x	x	x	x	x			x			x	x		xxxxx	x						
P478	xxx	xx	x	x	x	x	x			x						xxxxx							
P481	xx	x	x	x	x	x	x			x				xx		xxxxx				x	x		
P485	xx	x	x	x	x	x				x						xxxx				x			
P493	xx	xx	x	x		x						x				xxxx							
P501	xxx	xx	xx	x	x	xx	x	x		x				x	x	xxxxx				x			
P518	xx	xx	x	x	x	x				x				x		xxxxx							
P529	xx	x	x	x	x							xx	x		xx	xxxx	x			x			
P534	xxx	x	x	x	x	x				x						xxxx	x						
P558	xxx	xx	x	x	x	x				x				x	x	xxxx							
P565	xx	xx	xx	x	x	x										xxxx							
P583	xx	x	x	xx	x	x										xxxx							
P584	xx	x	x	x	x	x	x			xx				x	x	xxxxx							
P588	xx	x	x	x	x	x										xxxx							
P594	xx	x	xx		x	x	x			x		x		x		xxxx				x			
P599	xx	xx	x	x	x	x								xx		xxxx			xxxx				
P610	xx	xx	x	x	x	x										xxxx							
P633	xx	xx	x	xx		x				x						x		xxx					
P636	xxx	xx	x	x	x	x				x						xxxx				xx			
P647	xx	x	x	x	x	x	x			x						xxx		xxxx	x				
P703	xx	xx	x	x	x	x	x			x				x		xxxx							
P515	xx	x	x	x	x	x				x				xx				xxx		x	x		
P559	xxxx	xx	x	x	xx	xx					xxx			xx	x			xxx		x			
P595	xx	x	x	x	x					x				x	x			xxx		xx			
P632	xx	x	x	x	x	x				x				x	x	xxxx				x			
S9-1	xx	x	x	x	x	x		x	x				x	x		xxx							
P511	xx	xx	x	xx	x	xx											xxxx	x					
P513	xx	xxx	x	x	x	xx	x										xxxx		x				
P547	xx	x	x	x	x	x				x							xxxx						
P585	xx	x	x	x		x											xxxx						
P650	xx	x	x	x	x	x	x											xxx		xxx			
P477	xx	xx	xx	x	x	x				x	x								xxxx	x			
S8-8	xx	x	x	x	x	x	x			xx									xxxx				
P461	xx	x	x	x	xx												xxx	x		xxxx	xx		
P556a	xx	xx	xx	x	x	xx	x			xxx							xxx	xxxx					
P557	xx	x	x	x	xx							x				xx	xxxx						
P560b	xxx	xx	xx	x	x	x									xx	xxxx	xxxx						
P620	xx	xx	xx	x	x	x							x			xxxx	xxxx						
P621	xx	xx	x	x	x								x	x		xxxx	xxx						
P626	xx	xx	xx	x	x					x				x		xxx	xx						
P510	xxx	x	xx	x	x		x				xxx		x				xx						
P512	xx	x	x	x	xx	x					xxx			x	x	x	xxx						
P605	xxxx	x	x	x	xx	x				xx			xxxx	x		x					x		
P642	xxx	xx	xxx	x	x			x			xxxx						x	x			x		
P521	xx	x	x		x						xxxx			xxx				xxx			x		
P554	xx	xx	x	x	x	x					xx			x		xxx	xxx						
P574	x		x		x	x				x		xx					xxxx						
P643	xx	x	x		x				xx								xxx						
P654	x	x	x	x	x	xx				x	x			xx		xx						xxxx	x

Table 2. Schematic petrographic description of Early Bronze Age pottery from Arquà Petrarca continued. xxxxx = dominant (50-70%), xxxx = frequent (30-50%), xxx = common (15-30%), xx = few (5-15%), x = rare (<5%), Qtz = quartz, Pl = plagioclase, Kfs = K-feldspar, Bt = biotite, Ms = muscovite, Op = opaque minerals, Px = pyroxene, Am = amphibole, Rt = rutile, ARF = argillaceous rock fragments, Q = quartzite, VG = volcanic glass, VR1 = volcanic rock type 1, VR2 = volcanic rock type 2, VR3 = volcanic rock type 3, VR4 = volcanic rock type 4, WVR = weathered volcanic rock, CM = carbonate mudstone, CW = carbonate wackestone.

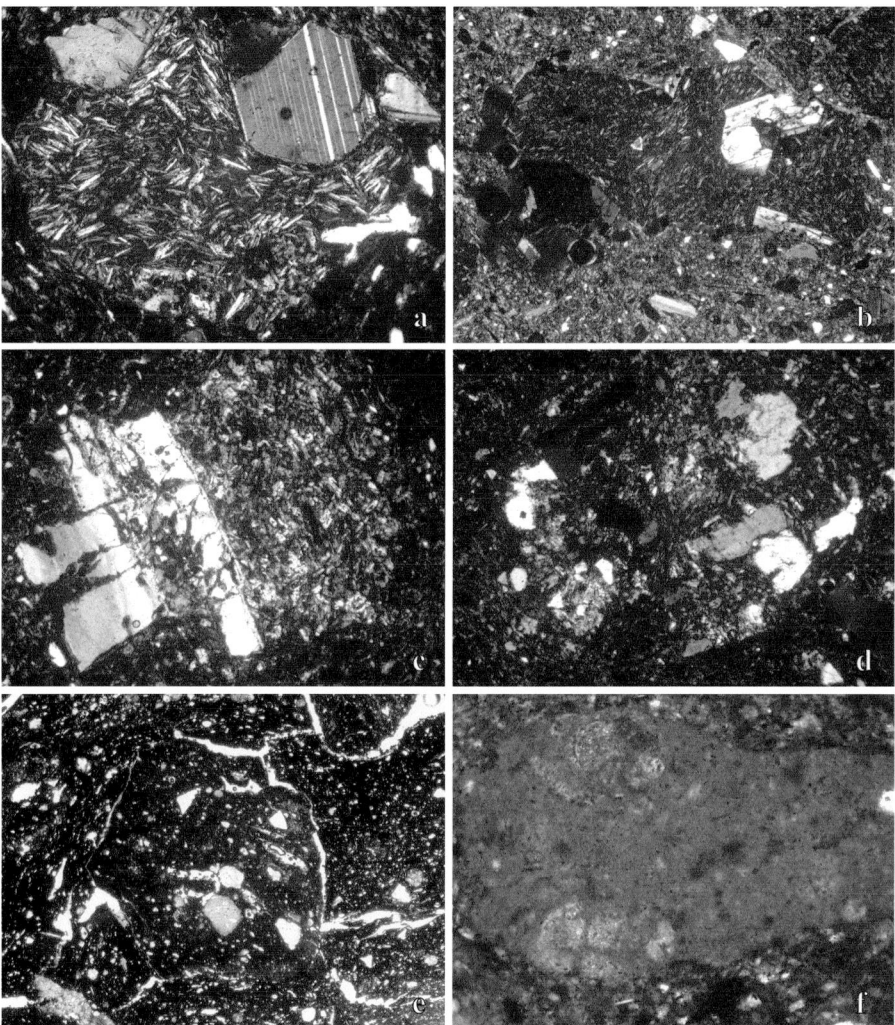

Figure 3. Photomicrographs of Early Bronze Age ceramics from Arquà Petrarca analysed in this study. a) Inclusion of volcanic rock type 1 in sample P518, b) Inclusion of volcanic rock type 2 in sample P515, c) Inclusion of volcanic rock type 3 in sample P547, d) Inclusion of volcanic rock type 4 in sample S8-8, e) Fragments of grog in sample P510, f) Limestone fragment in sample P654. Images a-d taken in crossed polars and e-f in plane polarised light. Image width = 1.5 mm, except a,f = 1 mm.

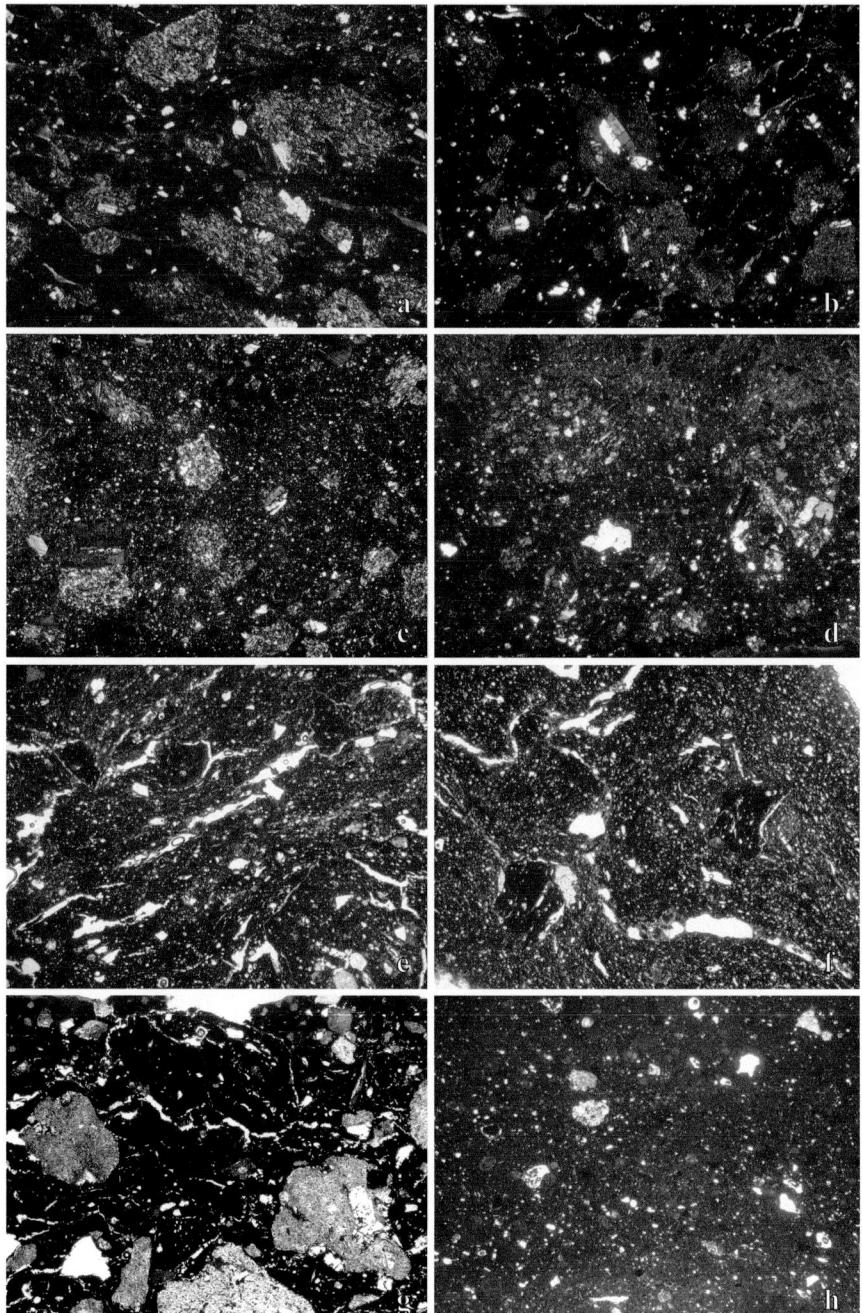

Figure 4. Photomicrographs of Early Bronze Age ceramics from Arquà Petrarca analysed in this study, demonstrating the main petrographic groups. a) Subgroup 1.1, sample P460, b) Subgroup 1.2, sample P632, c) Subgroup 1.3, sample P513, d) Subgroup 1.4, sample S8-8, e) Subgroup 2.1, sample P510, f) Subgroup 2.2, sample P605, g) Subgroup 2.3, sample P462, h) Group 3, sample P460. Images a-d taken in crossed polars and e-h in plane polarised light. Image width = 1.5 mm.

The volcanic rock fragments in Group 1 are associated with minor amounts of fine sand-sized angular to sub-angular crystals of quartz, plagioclase, sanidine, biotite, white mica and opaque minerals. They may also contain very rare fine sand-sized angular crystals of pyroxene, anorthoclase, green amphibole, rutile, and occasionally medium sand-sized sub-angular to sub-rounded fragments of argillaceous rocks (ARF), radiolarian chert, strongly weathered volcanic rocks, and well-rounded clay pellets. Five subgroups were defined according to the relative abundances of the various inclusions (Tables 1 and 2). Subgroups 1.1, 1.2, 1.3 and 1.4 are characterised by the prevalence of predominant volcanic rock types 1, 2, 3 and 4, respectively, and samples belonging to Subgroup 1.5 contemporaneously contain abundant fragments of the various volcanic rock types (Figure 4A-D).

Group 2 consists of eight grog-rich potsherds. These sherds are characterised by abundant large fragments of grog and volcanic rocks. Most have a heterogeneous groundmass, due to impregnative amorphous concentration features (ACF). The micromass often appears optically active, with a thin striated b-fabric, oriented parallel to the vessel walls. Meso- and micro-voids predominantly occur as vughs and channels. Inclusions are abundant, angular and sub-angular in shape, and show single-space porphyric-related distance and bimodal grain-size distribution. They are mainly composed of submillimetric to millimetric grog fragments, the groundmass of which is very similar to the surrounding paste, especially as regards the texture and minero-petrographic composition of the fine sand-sized inclusions (Figure 3E). Coarse sand-sized fragments of volcanic rocks, fine sand-sized angular to sub-angular crystals of quartz and, in small amounts, plagioclase, sanidine, biotite, white mica and opaque minerals, and fragments of radiolarian chert also occur. Samples also occasionally contain rare fine sand-sized angular crystals of pyroxene, green amphibole, rutile, clay pellets, ARFs, and rounded fragments of carbonate mudstone and bioclast wackestone. In view of the differences in the absolute and relative abundances of inclusions, three subgroups were distinguished. The samples of Subgroup 2.1 contain fewer inclusions, among which grog fragments prevail (Figure 4E), those of Subgroup 2.2 have more inclusions, with similar relative abundances of grog and volcanic rock fragments (Figure 4F), and those of Subgroup 2.3 are richer in volcanic inclusions (Figure 4G).

Group 3 is composed of a single carbonate-rich potsherd (sample P654), characterised by a homogeneous, isotropic and highly calcareous groundmass with a crystallitic b-fabric (Figure 4H). Voids are scarce, mostly meso-voids, although micro-voids also occur, and predominantly vughs. Inclusions have a coarse:fine (c:f) ratio of 25:75, a continuous grain-size distribution, single-space porphyric-related distance, and rounded to sub-rounded shape. They are mainly composed of rounded sand-sized fragments of bioclast wackestone, consisting of a matrix-supported micrite, sometimes bearing foraminifer tests (Figure 3F). Single isolated foraminifers are also found in the groundmass. Fragments of mollusc shells, volcanic rock type 1 inclusions, radiolarian chert, fine sand-sized crystals of quartz, plagioclase, sanidine, opaque minerals, and clay pellets are also occasionally found.

The samples exhibit different mineralogical associations based on the XRD (Table 3). Most of have quartz, sanidine, plagioclase and illite, the peak relative intensities of which distinguish three different compositions. The first is characterised by a prevalence of quartz, abundance of sanidine and plagioclase, and a scarcity of illite, the second by the prevalence of sanidine and plagioclase, abundance of quartz, and scarcity of illite, and the third by comparable quantities of quartz, sanidine and plagioclase, and a scarcity of illite. Other samples display diffractometric patterns similar to the first, but differ in the absence of illite and the presence of small quantities of calcite. Sample P654 is quite different from all the others, as its XRD pattern indicates prevalent calcite, abundant quartz and scarce sanidine, plagioclase and illite.

Minerals					Samples
Qtz	Pl	Sa	Ill	Cc	
xxx	xx	xx	x		P464, P478, P481, P485, P510, P511, P512, P521, P529, P534, P554, P556a, P558, P559, P565, P574, P595, P599, P605, P610, P621, P626, P632, P636, P643, P647, S8-8
xx	xxx	xxx	x		P585, P588, P594, P620
xx	xx	xx	x		P460, P461, P477, P478, P515, P518, P547, P557, P560b, P583, P584, P650, P703
xxx	xx	xx			P493, P501, P513, P633
xxx	xx	xx		x	P642
xx	x	x	x	xxx	P654

Table 3. Mineral assemblage identified in Early Bronze Age pottery from Arquà Petrarca by XRD. Qtz = quartz, Pl = plagioclase, Sa = sanidine, Ill = illite, Cc = calcite, xxx = very abundant, xx = abundant, x = scarce.

	Potsherds							Clay materials					
	A	SD	MIN	IQ	MED	IIIQ	MAX	AP2	AP3	AP4	AP5	AP6	AP7
SiO_2	63.55	3.38	44.53	62.69	63.65	65.47	68.15	47.26	20.76	37.61	55.53	58.58	62.98
TiO_2	1.01	0.27	0.61	0.81	0.96	1.10	1.82	0.54	0.30	0.36	0.76	0.68	0.80
Al_2O_3	18.38	1.66	11.43	17.62	18.27	19.22	22.95	10.07	5.09	7.60	20.77	14.46	16.57
Fe_2O_{3tot}	6.13	1.05	3.88	5.45	6.04	6.74	9.25	3.98	2.21	3.49	9.04	4.81	5.39
MnO	0.11	0.03	0.04	0.10	0.11	0.13	0.19	0.21	0.10	0.08	0.14	0.08	0.08
MgO	1.30	0.25	0.65	1.13	1.24	1.45	1.90	6.71	0.95	1.45	1.66	1.90	1.48
CaO	2.81	4.49	1.09	1.89	2.10	2.43	33.69	23.91	68.14	45.95	7.60	12.50	5.52
Na_2O	2.10	0.79	0.63	1.56	2.17	2.77	3.43	1.20	0.19	0.12	0.34	1.54	1.64
K_2O	3.38	0.51	1.80	3.10	3.46	3.80	4.19	2.34	0.89	1.66	2.22	3.21	3.40
P_2O_5	0.32	0.27	0.11	0.20	0.26	0.36	2.00	0.20	0.17	0.15	0.11	0.40	0.27

Table 4. Geochemical analysis of Early Bronze Age pottery from Arquà Petrarca by XRF. Average (A), standard deviation (SD), minimum (MIN), first quartile (IQ), median (MED), third quartile (IIIQ) and maximum (MAX) values of major and minor elements in studied potsherds, and chemical composition of clay samples from vicinity of Lago della Costa. Full analytical data available from authors on request.

Provenance and Technology of Bronze Age Pottery from Arquà Petrarca, Padova, Italy

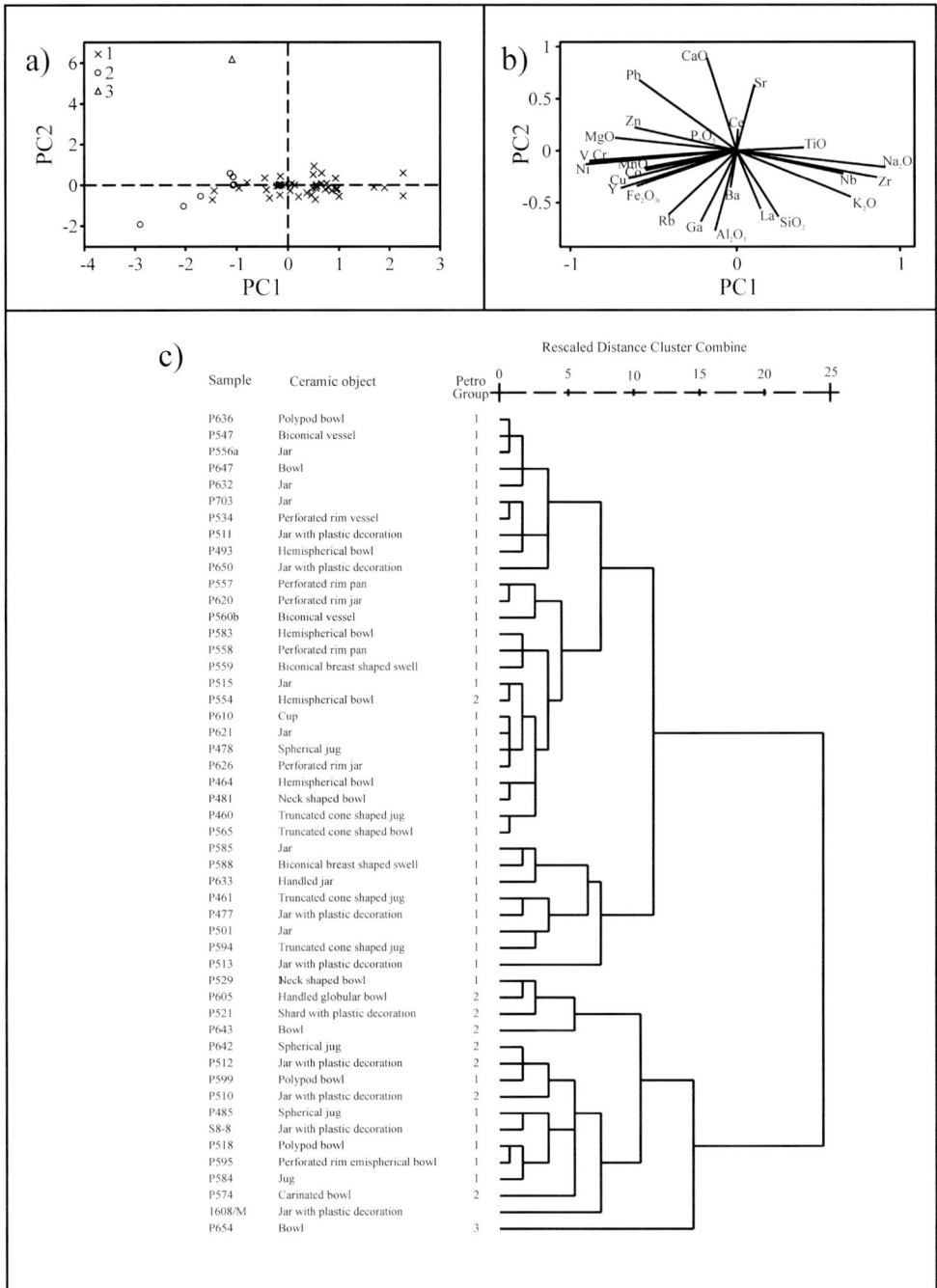

Figure 5. Statistical classification of geochemical data collected from Early Bronze Age ceramics from Arquà Petrarca. a) Score plot of principal components PC1 and PC2, explaining 32% and 18% of total variance, respectively, relative to PCA for potsherds classified according to petrographic groups, b) Loading plot of PC1 and PC2, c) Dendrogram of cluster analysis. Types of pottery and petrographic groups are also shown.

Geochemistry

Statistical analysis of the geochemical data shows that the samples are compositionally heterogeneous, and their concentration values display high standard deviations and wide distribution intervals for most of the chemical elements (Table 4). Chemical heterogeneities are well correlated to differences in the abundance and petrographic nature of the inclusions (e.g. vulcanites vs. grog vs. carbonate fragments), and in micromass composition. In more detail, the PCA score plot (Figure 5A) shows that petrographic Group 3 is clearly discriminated by its higher CaO content (Figure 5B), whereas groups 1 and 2 partially overlap, having very similar PC2 values and differing from each other especially in the higher concentration in alkali in the former and higher Fe_2O_3 in the latter. In addition, the CA dendrogram (Figure 5C) shows that the samples of Groups 1 and 2 sometimes plot in the same cluster, indicating a low dissimilarity level. This behaviour is due to the presence in both groups of volcanic rock fragments with similar composition, and to the likeness of grog composition with that of the surrounding groundmass, as inferred from microscopic observations in terms of textural features and minero-petrographic composition of inclusions. The presence of grog fragments compositionally similar to the base clay provides information on production technology, but does not modify the bulk chemical composition with respect to that of the clay material. This means that the geochemical differences between petrographic Groups 1 and 2 are only related to the absolute abundance of volcanic rock fragments in the various samples.

Interpretation

Provenance

Most of the pottery contains fragments of volcanic rocks such as trachyte and rhyolite. In northern Italy, these outcrop exclusively in the Euganean Hills (Figure 1A), where the most representative rocks are Late Eocene to Oligocene trachyte, rhyolite, latite, and basalt (Piccoli, 1966; De Vecchi *et al.*, 1976; Borsi *et al.*, 1969; Zantedeschi, 1994). Petrographic comparison of the volcanic rocks identified in the pottery with a reference collection of thin sections (Schiavinato, 1944) and data from Capedri *et al.* (2000) indicates that these inclusions are very similar, in terms of texture and mineral assemblage, to some types of trachyte and rhyolite outcropping in the southeastern Euganean Hills. This evidence strongly suggests that the pottery of Groups 1 and 2 was produced locally. The carbonate inclusions in sample P654 (Group 3) are similar to the bioclast carbonate-wackestones of the Scaglia Rossa Formation, an Upper Cretaceous-Lower Eocene limestone, which also commonly outcrops in the Euganean Hills.

Local clay materials are petrographically composed of 5-20% of rounded to sub-rounded silt- to fine sand-sized crystals of quartz, sanidine, plagioclase, illite, chlorite, biotite and opaque minerals, in a clay-rich matrix. The birefringence of the matrix also suggests the presence of carbonates in samples AP3 and AP4. Sample AP2 collected from the lake, turned out to match the petrographic features of potsherds belonging to Groups 1 and 2, in terms of both mineral assemblage and texture, including the relative abundances of minerals in the silt- to fine sand-sized inclusions. The lack of coarse sand-sized fragments of volcanic rocks in these clay materials suggests that they were deliberately tempered. The chemical composition of sample AP2 was compared with that of potsherds of Groups 1 and 2 (Table 4), and turned out to have far higher contents of CaO (23.58 wt%) and MgO (6.71 wt%). These geochemical differences may be due to the presence of a temper made of volcanic inclusions, or to the occurrence of secondary post-depositional calcite in the micromass of the clay. Secondary calcite may indeed represent the contribution of limestone from outcrops north and west of the lake, which probably modified the original mineralogical composition of the alluvial deposits of the Adige. An estimate of the calcite content was made by comparing the CaO wt% with loss on ignition (LOI), which was well correlated (R^2 = 0.97) (Figure 6). Assuming that calcite and plagioclase are the main CaO-bearing phases and that Na_2O is mainly located in plagioclase, we can use the CaO/Na_2O ratio of calcite-free sample AP7 to estimate CaO in silicates, the excess CaO being located in calcite. The excess CaO was combined with an adequate amount of CO_2 to calculate calcite, which turned out to be 27 wt% for sample AP2, 72 wt% for sample AP3, 55 wt% for sample AP4, 9 wt% for sample AP5, and 10 wt% for sample AP6. Subtracting the CO_2 from LOI, we obtained an estimate of the water released by clay minerals (assuming negligible organic matter) which was about 8 wt% in calcite-rich samples AP3 and AP4, 12 wt% in AP2, and between 16 and 18 wt% in AP5, AP6 and AP7. Although these estimates are only semi-quantitative and error may be relatively high, the resulting values are reasonable, and support the idea that excess CaO is indeed related to calcite, so that CaO should not be considered in PCA.

Principal component analysis of chemical data showed that sample AP2 does not overlap with potsherds of Groups 1 and 2 in the score plot (Figure 7A,B) but samples AP5, AP6 and AP7 do. This result may be misleading when attempting to identifying base clay materials, since petrographic analysis had shown that clay AP2 is very similar to the micromass of potsherds of Groups 1 and 2 in terms of the nature and abundance of inclusions. Clay sample AP7 also shows some analogies only with potsherds of Subgroup 1.4 and Group 2, due to the presence of fragments of magmatic rock type 4, although with a lower *c:f* ratio, whereas AP5 and AP6 are very different. The chemical misfit of AP2 may be due to tempering with magmatic rock fragments, from the colluvium deposits of the southeastern Euganean Hills. This is also supported by the PCA score plot (Figure 7A), in which potsherds of Groups 1 and 2 are located between clay AP2 and the composition of a rhyolite from Monte Ricco and a trachyte from the nearby relief of La Rocca (Zantedeschi and Zanco, 1993).

The chemical composition of clay AP4 shows a certain similarity to potsherd P654 of Group 3 (Figure 7A), although petrographic analysis revealed fewer limestone inclusions and no trachyte fragments. Therefore, the raw material used to produce this piece is probably not the clay sample, but may be material collected from alluvial deposits enriched in limestone fragments after mountainside run-off.

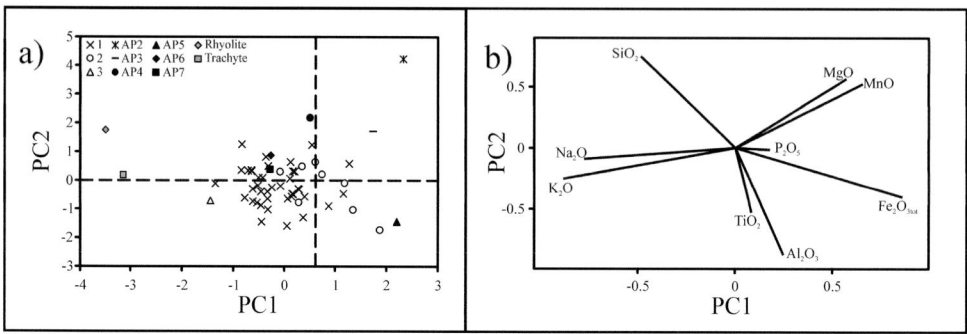

Figure 7. a) Score plot of principal components PC1 and PC2, explaining 36% and 26% of total variance, respectively, relative to PCA for potsherds classified according to petrographic groups, and clay materials, b) loading plot of PC1 and PC2. Petrographic groups are also shown.

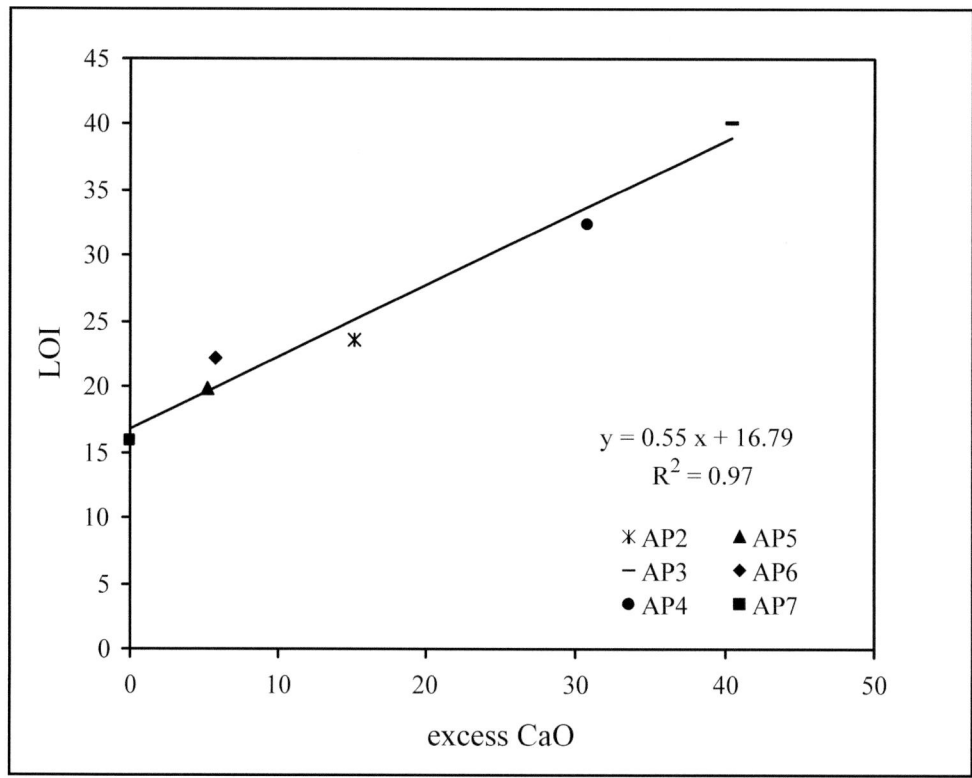

Figure 6. Excess CaO vs. LOI diagram. Symbols as in Figure 7. See text for calculation of excess CaO.

Provenance and Technology of Bronze Age Pottery from Arquà Petrarca, Padova, Italy

Production technology

The textural features of the pottery, the petrography of inclusions and the base clay provide indications of the technique used to prepare the Early Bronze Age clay pastes at Arquà Petrarca. The presence of coarse sand-sized inclusions, showing a dimensional gap with the silt- to fine sand-sized classes, causing bimodal grain-size distribution in the inclusions, suggests that the clay used for pottery production was tempered. On the basis of petrographic groups, three different recipes were used in terms of both types of inclusions and micromass. In particular, potsherds belonging to petrographic Group 1 were produced by tempering non-calcareous clay with fragments of various kinds of vulcanites, probably from the colluvium deposits at the foot of Monte Ricco or alluvial deposits from one of the temporary streams which flowed from the hills around the lake, as indicated by the sub-rounded shape of these inclusions. Potsherds belonging to petrographic Group 2 were prepared according to a similar recipe, but with the addition of grog, attesting to recycling of ceramic material. The single sample of Group 3 was produced by yet another recipe, probably using untempered calcareous clay and probably very similar, in terms of the texture and the petrography of inclusions, to clay AP5.

The distribution of potsherds in the various petrographic groups (Figure 8A), based on chemical composition in the PCA score plot (Figure 5B) and the CA dendrogram (Figure 5C), shows no correlation with pottery type. This means that several recipes were indiscriminately used to produce various types of pottery. This evidence suggests that the pottery production at Arquà Petrarca was not standardised and was probably family-based.

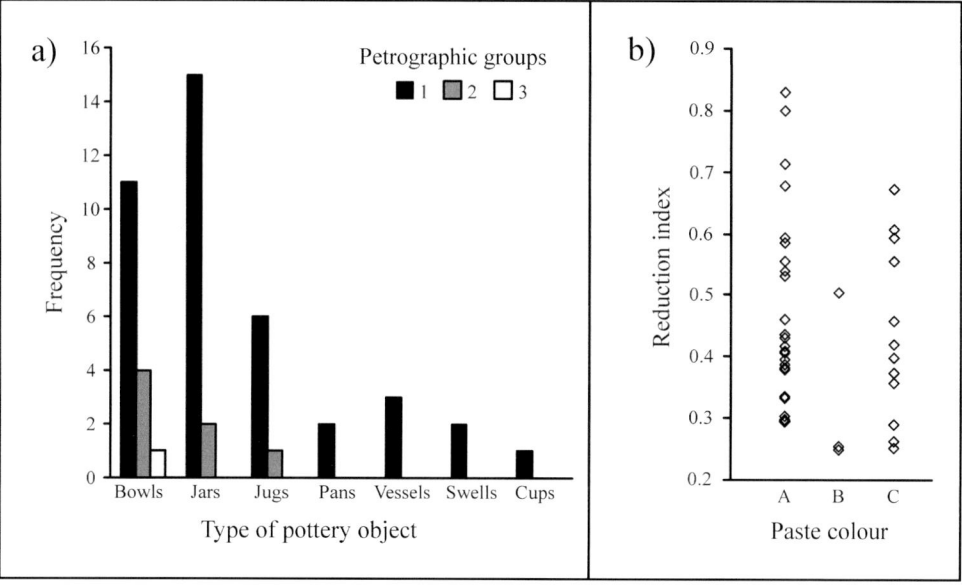

Figure 8. a) Frequency distribution diagram of petrographic groups among types of pottery objects, b) frequency distribution diagram of reduction index (RI) as a function of potsherd paste colour. A = greyish-black, B = pale brown-brown, C = grey, with pale grey core.

Firing temperatures were defined on the basis of mineral assemblages recorded by XRD. As the material used for pottery production was chlorite- and illite-rich clay, the absence of any chlorite peak in the diffraction patterns of all samples indicates firing temperatures above 650°C. The presence of illite in most of the samples and calcite in two of them also indicates that the firing temperature did not exceed the decomposition temperature of these mineral phases, which is 850-900°C in open-pit firing (Maritan *et al.*, 2006; Nodari *et al.*, 2007). Therefore, all samples were probably fired between 700-850°C.

As regards redox conditions during firing, the potsherds have a wide range of colours and sometimes chromatic differences within the same sample (Figure 8B). In addition, the reduction index, defined as FeO/FeO_{tot} by Maggetti and Galetti (1981), displays a wide distribution interval, between 0.25-0.83 (Figure 8B). These values are always higher than 0.2 and, according to the above authors, suggest reducing conditions, although redox conditions were probably quite heterogeneous, fitting a pre-industrial technology such as open-pit firing (Rye, 1981; Rice, 1987).

Conclusions

The present archaeometric study of the Bronze Age pottery found at the lake-dwelling settlement of Arquà Petrarca shows that the complete ceramic repertoire was locally produced. The presence of key inclusions is consistent with the geology of the Euganean Hills, and compositional analogies with some clay materials from the surroundings of Lago della Costa indicate local production. The absence of imported ceramics, even of artefacts of true aesthetic value, indicates that the site was self-sufficient, at least with regard to pottery production. Analogies in terms of stylistic features with the contemporaneous pottery from central-eastern Europe, as in the case of sample S9-1, which also appears to be produced locally, may be interpreted as a duplicate of ceramic objects circulating in the Veneto region but coming from central-eastern Europe.

Although there are no archaeometric studies of Poladian pottery from other sites, and therefore general conclusions regarding the technology of pottery production cannot be drawn for the Polada culture, comparisons with other coeval sites indicate only a few technical similarities. For example, grog, frequently adopted elsewhere (Levi *et al.*, 2006), is here found in only one-fifth of the artefacts. In addition, the absence of any parallels between petrographic groups and pottery types indicates that the raw materials were used independently of ceramic class. All these pieces of evidence points to pre-industrial, non-specialised, local ceramic technology.

From an archaeological viewpoint, two main observations arise from the archaeometric data collected here from the pottery of Arquà Petrarca. On one hand, the absence of a clear relation between ceramic class, evolution and chronology and, on the other, the technological features of production provides further support to present ideas and knowledge of methods and forms of ceramic production in this historic period, which was probably family-based, without specialisation or pressure to innovate. This is also confirmed by hybridism, gradual changes in type often observed among ceramic classes, and the constant repetition of similar morphological and decorative features, with repeated use of different combinations of the same basic elements.

From a chronologic and a cultural point of view, the proven absence of imported objects among the samples studied here leaves many questions unanswered, but does contribute to guiding discussion to a clearly defined level. Although accidental convergences, which may be excluded by findings of imported pottery, are possible, the occurrence in the eastern Poladian area of shapes and decorative styles typical of the Danubian and more generally of the central-eastern trans-Alpine area, should be interpreted as indicating circulation of ideas, tastes, and perhaps even habits, directly connected to groups of people or individuals sometimes moving from place to place, rather than to the existence of trade. This fact, if confirmed, may contribute with new arguments to the discussion on the forming elements of the 'Polada Culture' and its chronological and cultural relations with the archaeological facies typical of the end of the Eneolithic.

Acknowledgements

The authors are grateful to the Istituto di Geoscienze e Georisorse, CNR, Padova, for analytical support, and to G. Walton, who revised the English text.

References

Bacelle, L. and Bosellini, A. 1965. Diagrammi per la stima visiva della composizione percentuale nelle rocce sedimentarie. *Annali Università di Ferrara Sez. IX Scienze Geologiche e Paleontologiche* 1: 59–62.

Balista, C. and Leonardi, G. 1996. Gli abitati di ambiente umido nel Bronzo antico dell'Italia settentrionale. In Cocchi Genick D. (Ed.) L'antica età del Bronzo, Firenze: 199-228.

Baxter, M.J. 1999. Detecting multivariate outliers in artefact compositional data. *Archaeometry*, 41: 321-338.

Borsi, F., Ferrara, G. and Piccoli, G. 1969. Età delle eruzioni euganee: determinazione col metodo K/Ar. *Rendiconti della Società Italiana di Mineralogia e Petrologia*, 25: 27-34.

Capedri, S., Venturelli, G. and Grandi, R. 2000. Euganean trachytes: discrimination of quarried sites by petrographic and chemical parameters and by magnetic susceptibility and its bearing on the provenance of stone of ancient artefacts. *Journal of Cultural Heritage*, 1: 341-364.

Cordenons, F. 1887. Antichità preistoriche anariane della regione euganea. *Atti della Società Veneto-Trentina di Scienze Naturali*, 11: 67-101.

De Vecchi, G., Gregnanin, A. and Piccirillo, E.M. 1976. Tertiary volcanism in the Veneto: magmatology, petrogenesis and geodynamic implications. *Geologische Rundschau*, 62: 701-710.

Govindaraju, K. 1994. Compilation of working values and sample description for 383 geostandards. *Geostandards Newsletter* 18, Special Issue: 1-158.

Levi, S., Sonnino, M. and Jones, R.E. 2006. Eppur si muove... Problematiche e risultati delle indagini sulla circolazione della ceramica dell'Età del Bronzo in Italia. In Cocchi Genick D. (Ed.) *Atti della XXXIX Riunione Scientifica su Materie prime e scambi nella preistoria italiana*, Firenze, 2004. Istituto Italino di Preistoria e Protostoria, Firenze: 1093-1111.

Maggetti, M. and Galetti, G. 1981. Archäometrische Untersuchungen an spätlateinezeitlicher Keramik von Basel-Gasfabrik und Sissach-Brühl. *Archäologisches Korrespondenzblatt*, 11: 321-328.

Marcolongo, B. and Zaffanella, G.C. 1987. Evoluzione della pianura veneta atestina-padana. *Athesia*, 1: 131-145.

Maritan, L., Nodari, L., Mazzoli, C., Milano, A. and Russo, U. 2006. Influence of firing conditions on ceramic products: experimental study on clay rich in organic matter. *Applied Clay Science*, 31: 1-15.

Moschetti, A. and Cordenons, F. 1901. Relazione degli scavi archeologici eseguiti sulle sponde del lago di Arquà a cura e a spese del Museo Civico di Padova, dal giorno 18 aprile al giorno 8 maggio 1901. *Bollettino del Museo Civico di Padova*, 4: 5-6.

Nodari, L., Marcuz, E., Maritan, L., Mazzoli, C. and Russo, U. 2007. Hematite nucleation and growth in the firing of carbonate-rich clay for pottery production. *Journal of the European Ceramic Society*, 27: 4665-4673.

Piccoli, G. 1966. Studio geologico del vulcanismo paleogenico veneto. *Memorie dell'Istituto di Geologia e Mineralogia dell'Università degli Studi di Padova*, 26: 1-100.

Piccoli, G., Sedea, R., Bellati, R., Di Lallo, E., Medizza, F., Girardi, A., De Pieri, A., De Vecchi, G.P., Gregnanin, A., Piccirillo, E.M., Norinelli, A. and Dal Prà, A. 1981. Note illustrative della carta geologica dei Colli Euganei alla scala 1:25.000. *Memorie di Scienze Geologiche dell'Università degli Studi di Padova*, 34: 523-566.

Rice, P.M. 1987. *Pottery Analysis*. The University of Chicago Press, Chicago.

Rye, O.S. 1981. *Pottery Technology: Principles and Reconstruction*. Taraxacum, Washington DC.

Salzani, L. 1982. Arquà Petrarca (Padova). In: Aspes A. (Ed.) *Palafitte: mito e realtà*. Museo Civico di Storia Naturale, stampa, Verona: 222-224.

Schiavinato, G. 1944. Studio chimico-petrografico dei Colli Euganei. *Memorie dell'Istituto Geologico dell'Università di Padova*, 15: 1-59.

Veneto Agricoltura 2001. *I suoli dei Colli Euganei*. Imprimendo, Padova.

Vitali, V. and Franklin, U.M. 1986. New approaches to characterisation and classification of ceramics on the basis of their elemental composition. *Journal of Archaeological Science*, 13: 161-170.

Whitbread, I.K. 1986. The characterisation of argillaceous inclusions in ceramic thin sections. *Archaeometry*, 28: 79-88.

Whitbread, I.K. 1989. A proposal for the systematic description of thin sections towards the study of ancient ceramic technology. In Maniatis Y. (Ed.) *Proceedings of the 25th International Symposium of Archaeometry, Athens, 1986*. Elsevier, Amsterdam: 127-138.

Whitbread, I.K. 1995. *Geek transport amphorae: a petrological and archaeological study*. Fitch Laboratory Occasional Paper, 4, British School at Athens.

Zantedeschi, C. 1994. New Rb-Sr radiometric data from Colli Euganei (North-Eastern Italy). *Memorie di Scienze Geologiche dell'Università degli Studi di Padova*, 46: 17-22.

Zantedeschi, C. and Zanco, A. 1993. Distinctive characterisation of Euganean trachytes for their identification in ancient monuments. *Science and Technology for Cultural Heritage*, 2:1-10.

CERAMIC TECHNOLOGY & SOCIAL PROCESS IN LATE NEOLITHIC HUNGARY

Attila Kreiter

Field Service for Cultural Heritage, Budapest, Hungary
(attila.kreiter@kosz.gov.hu)

György Szakmány

Department of Petrology and Geochemistry, Eötvös Loránd University
Budapest, Hungary

Miklós Kázmér

Department of Palaeontology, Eötvös Loránd University, Budapest, Hungary

Introduction

By means of ceramic petrography this paper examines technological aspects of Late Neolithic Lengyel Culture ceramics from three settlements in southern Hungary (Figures 1 and 2) and considers their implications in terms of social complexity. The petrography of several vessel types is compared in order to assess possible similarities and differences in manufacturing technology at the Neolithic settlements. The results show that there are extensive ceramic technological similarities between the sites, particularly within some cups, mugs, bowls and jars. It seems that not only was it important that these vessels should look similar, but that they were made in a similar manner. These findings indicate that potters at the three different sites had similar understanding of the properties of raw materials and also had a high degree of common knowledge and a similar approach to the fabrication of these vessels. The technological similarity between some of the cups, mugs, jars and bowls suggests that they were specialised products and their production may have been standardised. Other pottery styles such as pedestalled bowls and storage or cooking vessels, exhibit differences in technological practice between similar vessel types, suggesting the existence of intra-site technological traditions.

Ceramic Technology and Social Theory

In this paper ceramic technological tradition is examined in order to assess social relationships between communities. Technological practices are viewed not only as a means of delivering an end product but are also an essential part of complex, dynamic social strategies that may reflect social boundaries (Dobres, 2000). The processes that lead to the production of ceramic objects are a result of cultural choices as technology is part of a cultural system. This notion suggests that potting technology may well convey equally significant information about the society as the object itself

(Lemonnier, 1992). If consistent patterns in technological choices can be recognised at different archaeological sites, then the nature of this patterning provides an important means of investigating relationships between social groups.

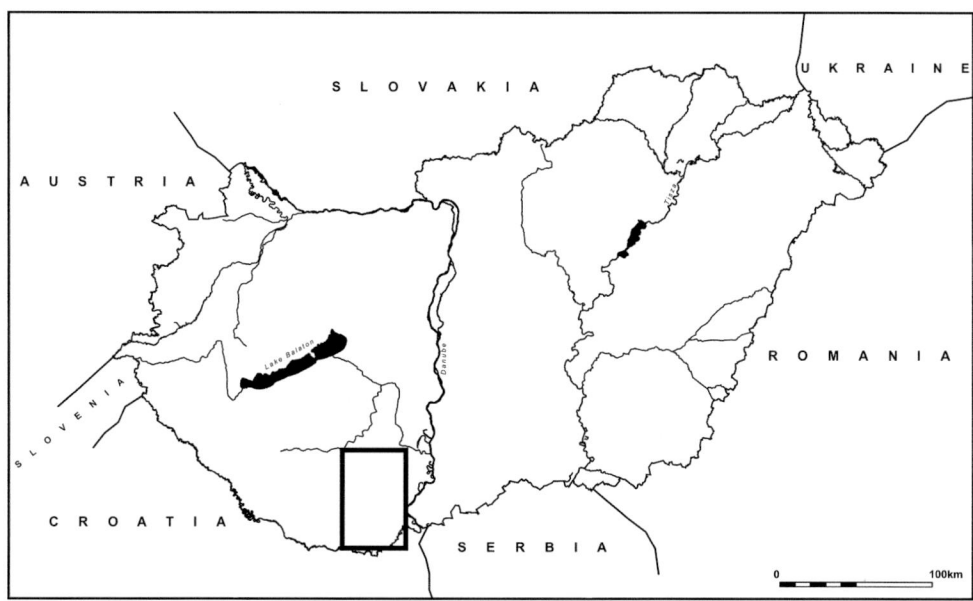

Figure 1. Map of Hungary with the location of the study area of Baranya County.

This project aims to highlight the important role of ceramic petrography in studying ceramic technology to reveal information about social organisation. The production sequences of Hungarian Late Neolithic vessels are deconstructed, and each stage that can be examined macroscopically and through ceramic petrography are examined and compared in the framework of technology as a meaningful and dynamic concept.

The Lengyel Culture emerged during the 5th millennium BC in an area south of Lake Balaton in southwestern Hungary and spread towards Austria, western Slovakia, southern Moravia and even as far north as southern Poland, gradually occupying a large geographical area during the Late Neolithic to the Middle Copper Age (Figure 1). Lengyel communities produced a large variety of pottery, particularly in terms of shape and decoration (Kozłowski and Raczky, 2007). Understanding the process of pottery production and possible technological similarities between the different sites may help to shed light on the social organisation of material production during this period.

Materials and Methods

In this study, the pottery of three Lengyel phase II sites, Szemely, Zengővárkony and Belvárdgyula, from Baranya County in South Hungary have been examined (Figure 2). An initial study of the ceramic assemblages from these sites was carried out in hand specimen, based on the distinguishing criteria of vessel type, shape, building techniques, surface treatments, color, decoration, firing conditions and fabric. Groups were formed according to hand specimen fabric in order to investigate the relationship between the different aspects of the pottery assemblages, for example, whether a specific fabric appeared in a number of wares. Detailed description of these fabrics are presented elsewhere (Kreiter and Szakmány 2008a,b). From the macroscopic examination all of the available sherds, 46 were selected for subsequent petrographic analysis, including 15 samples from Szemely, 15 from Zengővárkony and 16 from Belvárdgyula. These samples included many common Lengyel vessel types such as handmade cups, mugs, jars, bowls, pedestalled bowls and handmade storage and cooking vessels (Table 1).

Thin section petrographic analysis was used to characterize similarities and differences between the raw materials, fabric preparations and tempering practices of the Lengyel Culture ceramics. To distinguish between grog and clay fragments in thin section the criteria described by Whitbread (1986) and Cuomo di Caprio and Vaughan (1993) were used. Hand specimen examination and petrographic analysis together helped to reconstruct the technology of the examined vessels

Geological Background

The geological setting of Baranya County is varied. Zengővárkony is situated close to the southeastern foreland of the Mecsek Mountains, which consist of Lower Jurassic argillaceous limestone, marl and sandstone formations, Palaeozoic crystalline formations of the Mórágy Block and Pannonian and Pleistocene-Holocene clastic sediments (Figure 2). The sites of Szemely and Belvárdgyula are situated south of the Mecsek Mountains on Tertiary-Quaternary sediments that contain constituents of the Mórágy Block (Figure 2) (Gyalog, 2005).

The Mórágy block of the southeastern Mecsek Mountains is mainly comprised of granitoid and parametamorphic rocks (Fülöp, 1994; Király and Koroknai, 2004). The granitoid exhibits mainly acidic-neutral plagioclase, microcline, quartz, biotite and rarely amphibole, with accessory minerals of titanite, zircon, allanite, epidote and muscovite. The parametamorphic rocks include metamorphosed sandstone, siltstone and shale (Jantsky, 1979; Király and Koroknai, 2004) composed predominately of quartz, plagioclase, biotite, K-feldspar and muscovite and accessory minerals such as tourmaline, opaque minerals, epidote-clinozoisite, zircon and apatite. Lower Jurassic-Lower Cretaceous sedimentary formations appear on the surface in the southeastern part of the Eastern Mecsek Mountains.

Extensive Miocene and Pannonian fluvial-lacustrine sand sheets surround the Mecsek Mountains. It is highly probable that the older Miocene, fossiliferous marine sediments with abundant bryozoans, oyster and echinoid fragments were re-deposited in the Pannonian sediments by erosion, forming a natural mixture of siliclastic sand and fossil debris (Császár, 2005; Gyalog, 2005).

Sample Number	Site	Vessel type	Fabric	Period
1. (Ő 2006.27.542) Fig. 3C	Szemely	cup	1	Lengyel II.
2. (Ő 2006.27.543)	Szemely	mug	1	Lengyel II.
3. (Ő 2006.27.384) Fig. 3D	Szemely	jar	1	Lengyel II.
4. (Ő 2006.27.467) Fig. 3A	Szemely	jar	1	Lengyel II.
5. (Ő 2006.27.476) Fig. 3B	Szemely	jar	1	Lengyel II.
6. (Ő 2006.27.559)	Szemely	jar	1	Lengyel II.
7. (Ő 2006.27.282)	Szemely	bowl	3	Lengyel II.
8. (Ő 2006.27.502)	Szemely	bowl	4	Lengyel II.
9. (Ő 2006.27.375) Fig. 7E	Szemely	pedestalled bowl	3	Lengyel II.
10. (Ő 2006.27.155)	Szemely	storage vessel	3	Lengyel II.
11. (Ő 2006.27.177)	Szemely	storage vessel	4	Lengyel II.
12. (Ő 2006.27.203)	Szemely	storage vessel	5	Lengyel II.
13. (Ő 2006.27.268)	Szemely	storage or cooking vessel	6	Lengyel II.
14. (Ő 2006.27.458)	Szemely	storage or cooking vessel	3	Lengyel II.
15. (Ő 2006.27.513)	Szemely	storage or cooking vessel	3	Lengyel II.
16. (N 1/26-1949) Fig. 4B	Zengővárkony	mug	1	Lengyel II.
17. (N 1/421A-1947)	Zengővárkony	mug	1	Lengyel II.
18. (N 1/421B-1947)	Zengővárkony	mug	1	Lengyel II.
19. (N 1/421C-1947) Fig. 4A	Zengővárkony	mug	1	Lengyel II.
20. (N 1/593-1947) Fig. 4D	Zengővárkony	mug	1	Lengyel II.
21. (N 1/421D-1947)	Zengővárkony	jar	1	Lengyel II.
22. (N 1/421E-1947) Fig. 7B	Zengővárkony	storage vessel	8	Lengyel II.
23. (N 1/156-1947) Fig. 4C	Zengővárkony	bowl	1	Lengyel II.
24. (N 1/365-1947) Fig. 6C	Zengővárkony	bowl	9	Lengyel II.
25. (N 1/354-1947) Fig. 6A	Zengővárkony	bowl	9	Lengyel II.
26. (N 1/364-1947) Fig. 6B	Zengővárkony	pedestalled bowl	9	Lengyel II.
27. (N 1/465-1947)	Zengővárkony	pedestalled bowl	9	Lengyel II.
28. (N 1/25-1949) Fig. 7A	Zengővárkony	storage vessel	8	Lengyel II.
29. (N 1/376-1947)	Zengővárkony	storage vessel	9	Lengyel II.
30. (N 1/421F-1947) Fig. 6D	Zengővárkony	storage vessel	9	Lengyel II.
31. (225/2777)	Belvárdgyula	jar	2	Lengyel II.
32. (207/2418)	Belvárdgyula	pedestalled bowl	2	Lengyel II.
33. (207/2418)	Belvárdgyula	storage vessel	2	Lengyel II.
34. (322/1606) Fig. 5A	Belvárdgyula	jar	1	Lengyel II.
35. (162/1053) Fig. 5C	Belvárdgyula	cup	1	Lengyel II.
36. (477/4724)	Belvárdgyula	storage or cooking vessel	7	Lengyel II.
37. (225/2777) Fig. 7C	Belvárdgyula	storage vessel	8	Lengyel II.
38. (65/2045) Fig. 7D	Belvárdgyula	bowl	2	Lengyel II.
39. (174/869) Fig. 5D	Belvárdgyula	mug	1	Lengyel II.
40. (68/226)	Belvárdgyula	bowl	2	Lengyel II.
41. (27/961)	Belvárdgyula	bowl	2	Lengyel II.
42. (27/961) Fig. 5B	Belvárdgyula	cup	1	Lengyel II.
43. (126/213)	Belvárdgyula	cup	1	Lengyel II.
44. (126/213)	Belvárdgyula	bowl	2	Lengyel II.
45. (207/2418)	Belvárdgyula	cup	1	Lengyel II.
46. (267/1223)	Belvárdgyula	jar	2	Lengyel II.

Table 1. Details of the Late Neolithic Lengyel Culture ceramics analysed in this study, including their petrographic fabric classification.

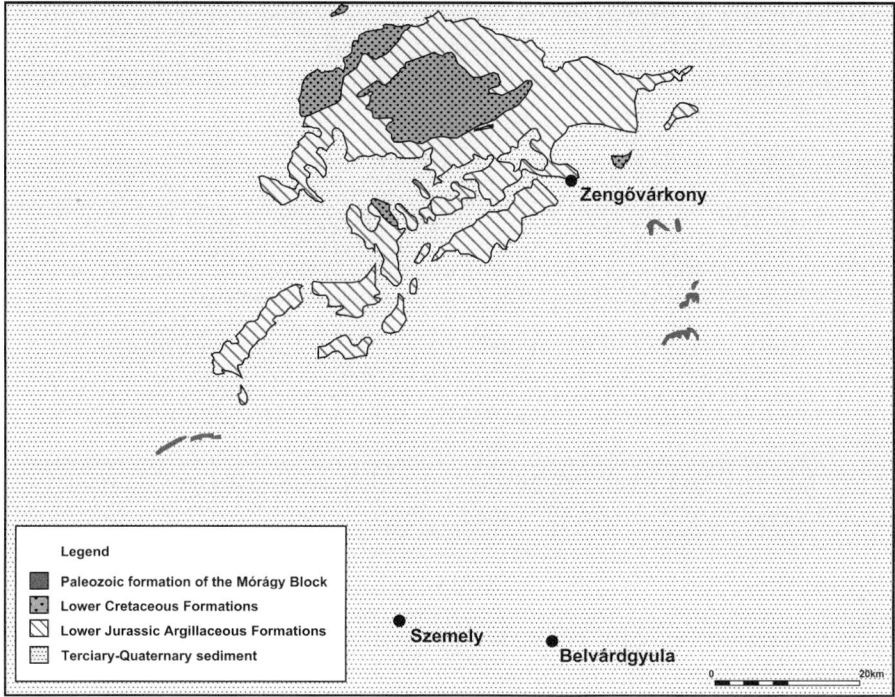

Figure 2. Geological map of the study area with the location of the Late Neolithic Lengyel Culture sites analysed in this study.

Results

A total of nine different petrographic fabric groups could be distinguished within the 46 Lengyel Culture ceramics, according to their most characteristic non-plastic inclusions (Table 1; Figures 3-7). A description of each fabric group can be found in Table 2.

The reconstruction of the steps involved in ceramic manufacture indicates that at each site the majority of vessels were made in a very similar manner (Table 1; Figures 3-5). Petrographic analysis reveals that a considerable number of vessels (19 out 34) including cups, mugs, bowls and jars from the three different sites (Figures 3-5) belong to Fabric 1. These vessels possess a uniform very fine grained, dense and oriented fabric. The very fine fabric makes the intentional introduction of temper unlikely. The raw material used to make these vessels was therefore, either naturally occurring very fine-grained clay or clay that was levigated by the potters prior to its use. The process of levigation seems very likely since the fabric of some vessels is dense and the very fine inclusions are well-sorted. The dominant inclusion type within the above-mentioned vessels is monocrystalline quartz, although constituents of granitoid containing plagioclase and K-feldspar also appear. K-feldspar and plagioclase appear less frequently in the samples from Zengővárkony and Belvárdgyula than those from Szemely. The non-plastic inclusions are very well sorted and angular to sub-angular in most samples, although their abundance varies between and within the sites studied. In these vessels muscovite mica is not common, although in one cup from Belvárdgyula it occurs more frequently. Some samples from each of the three sites contain rare rounded limestone inclusions.

Fabric Group	Description
Fabric 1	Very fine (< 0.1 mm), dense fabric, the non-plastic inclusions show parallel orientation to the vessel wall. The main inclusions are monocrystalline quartz. Rare (1-2%) amounts of biotite, K-feldspar, plagioclase and limestone also appear together with mainly rare (1-2%) or sparse (3-10%) amounts of muscovite. The main characteristic of this group is that the fabric is composed mainly of very fine monocrystalline quartz, showing a dense and serial fabric, and no tempering material could be identified.
Fabric 2	Very fine (< 0.1 mm) fabric although fine (0.1-0.25 mm) and medium (0.25-1 mm) grains also appear in rare (1-2%) amounts. The main inclusion is monocrystalline quartz. Rare (1-2%) amounts of muscovite also appear. The main characteristic of this group is that the raw material was tempered with rare (1-2%) to common (20-30%) amounts of medium (0.25-1 mm) to coarse (1-3 mm) hard clay fragments.
Fabric 3	Very fine (< 0.1 mm) fabric, although fine (0.1-0.25 mm) and medium (0.25-1 mm) grains also appear. The main inclusion is very fine (< 0.1 mm) monocrystalline quartz although rare (1-2%) amounts of muscovite, biotite, K-feldspar and plagioclase also appear in similar size ranges to the quartz. Rare (1-2%) or sparse (3-10%) amounts of fine to coarse (0.1-3 mm) limestone also appear. The main characteristic of this group is that the raw material was tempered with moderate (10-20%) to common (20-30%) amounts of medium to coarse (0.25-3 mm) grog and hard clay fragments.
Fabric 4	Very fine (< 0.1) fabric, the main inclusion is monocrystalline quartz. Rare (1-2%) amounts of K-feldspar and plagioclase also appear in similar size range to the quartz. Rare (1-2%) amounts of medium to coarse (0.25-3 mm) fossils and limestone inclusions also appear. The main characteristic of this group is that the raw material was tempered with common (20-30%) amounts of coarse (1-3 mm) hard clay fragments and organic (chaff) material.
Fabric 5	Very fine to medium (< 0.1-1 mm) fabric. The main inclusion is monocrystalline quartz although granitoid fragments also appear in similar size ranges to the quartz. Rare (1-2%) amounts of fine to medium (0.1-1 mm) limestone and fossils are also present. The size distribution of the non-plastic inclusions is hiatal. The main characteristic of this group is that the raw material was tempered with medium sand and with rare (1-2%) amounts of coarse (1-3 mm) grog and clay fragments.
Fabric 6	Mainly fine to medium (0.25-1 mm) fabric although coarse (1-3 mm) grains are also present. The characteristic non-plastic inclusions are mono- and polycrystalline quartz, graphic granite with simplectitic structure and rhyolithic lithoclast with quartz phenocryst. There are also biotite, plagioclase and K-feldspar. The main characteristic of this group is that the raw material was tempered with coarse sand.
Fabric 7	Very fine to medium (< 0.1-1 mm) fabric. The main inclusions are monocrystalline quartz and muscovite. Rare (1-2%) amounts of coarse (1-3 mm) polycrystalline quartz, K-feldspar and limestone also appear. The fabric is hiatal. The main characteristic of this group is that the raw material was tempered with medium to coarse (0.25-3 mm) sand and with moderate (10-20%) amounts of coarse (1-3 mm) hard clay fragments.
Fabric 8	Coarse (1-3 mm) fabric. The fabric mainly shows sparse amounts of very fine (< 0.1 mm) monocrystalline quartz and common (20-30%) amounts of fine to coarse (< 0.1-3 mm) fossils (bryozoans, oyster and echinoid). Rare (1-2%) amounts of very fine (< 0.1 mm) muscovite and biotite also appear. Sparse (3-10%) to moderate (10-20%) amounts of medium to coarse (0.25-3 mm) mono- and polycrystalline quartz, and granitoid fragments are also present. The fabric is hiatal. The main characteristic of this group is that the raw material was tempered with coarse sand exhibiting mainly fossils, quartz and granitoid fragments.
Fabric 9	Medium to very coarse (0.25->3 mm) fabric. There are sparse amounts (3-10%) of very fine (< 0.1 mm) monocrystalline quartz. The main composition of the fabric is moderate to very common (10-40%) amounts of medium to very coarse (0.25->3 mm) mono- and polycrystalline quartz and granite. Rare (1-2%) amounts of fine to coarse (0.1-3 mm) biotite, plagioclase and microclinealso appear together with rare to sparse (1-10%) amounts of fine to medium (0.1-1 mm) limestone. The fabric is hiatal. The main characteristic of this group is that the raw material was tempered with mainly coarse and very coarse granite.

Table 2. Description of petrographic fabric groups of Late Neolithic Lengyel Culture ceramics analysed in this study.

In the above-mentioned samples there are differences in the naturally occurring non-plastic inclusions between the sites studied. For example, zoisite-clinozoisite and biotite occur in ceramics from Szemely appear, biotite, zircon and epidote are present in samples from Zengővárkony and at Belvárdgyula ceramics contain biotite, epidote, zoisite-clinozoisite, rutile, tourmaline, zircon and chert. The non-plastic inclusions that occur in the vessels examined in this paper are generally indicative of local pottery production and the minor variations between the fabrics revealed by the petrographic analysis can be attributed to the local geology around the sites.

Hand specimen macroscopic examination of the fine ware ceramics indicates that they were formed using a slab building technique. After forming, the vessels were then smoothed or burnished both inside and out. Most vessels exhibit a fully reduced black surface indicating that the firing conditions in the majority of cases were well controlled. Since no kiln structures have been reported from the Hungarian Neolithic, it might be assumed that the vessels were fired in a pit or in an open fire.

The general form of the cups, mugs, jars, and bowls appears to be similar within and between the sites studied, although their size and proportions vary. The vessels can be painted or unpainted. Red painting is present on either or both surfaces and the motifs do not show a correlation with a particular vessel type. Horizontal and vertical bands or geometric motifs may cover the neck or the complete lower part of the vessel can be painted red.

It seems that whilst the shape, proportion and decoration of the vessels vary widely, their manufacturing techniques were strongly standardised. The raw material preparation and production of these vessels exhibits a level of skill and standardisation consistent with manufacture by specialists. Minor petrographic differences detected between the fine ware ceramics at the three sites are probably due to the use of local raw materials at each site.

In spite of the technological similarities between the Lengyel Culture ceramics from the three sites, some important differences were also been observed. For example at Belvárdgyula very fine clay was in some cases tempered with medium to coarse hard clay fragments, giving it a bimodal grain-size distribution (Fabric 2) (Figure 7D). In some cases the clay fragments were unevenly distributed in the fabric, indicating that they were intentionally added, but not thoroughly mixed. Such a practice appears among bowls, jars, pedestalled bowls and storage vessels.

At Szemely both hard clay fragments and grog were added as temper to some vessels (Fabric 3) (Figure 7E). Grog could be identified both macroscopically and in thin section. Some grog inclusions exhibit fire clouding and relic burnished vessel surfaces. The aligned internal microstructure that has been proposed by Cuomo di Caprio and Vaughan (1993, p. 29) as an important feature for identifying grog was not observed in the Szemely ceramics, indicating that the grog derived from a less refined ware. It must be noted that at Szemely grog was not used exclusively as temper, but appears in vessels that were also tempered with sand (Fabric 5) and organic material (Fabric 4).

The remains of organic material are present in the form of phytoliths (Kreiter and Szakmány, 2008a, figure 3, p. 60) and were not observed in the ceramics analysed from either Zengővárkony or Belvárdgyula.

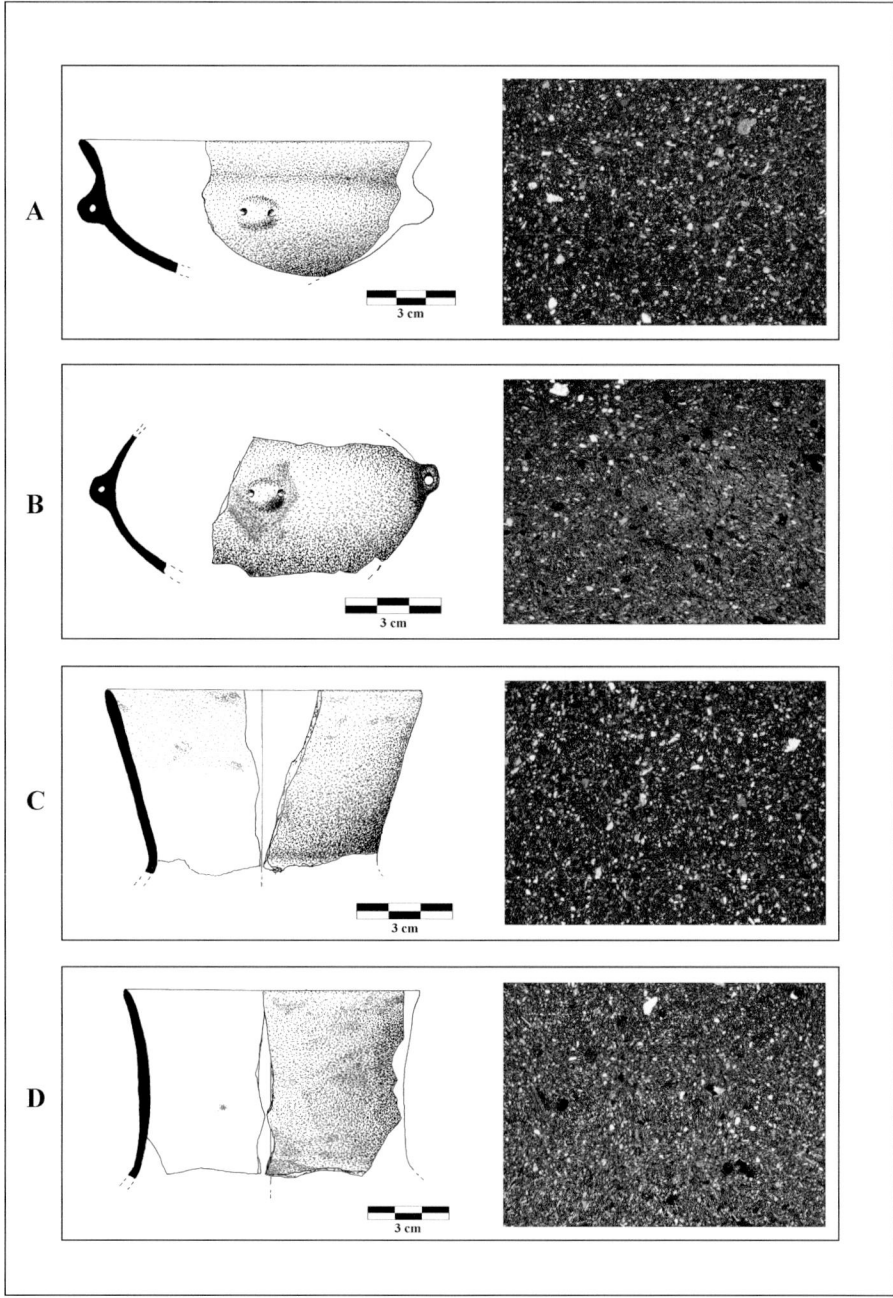

Figure 3. Illustrations and photomicrographs of the Late Neolithic Lengyel Culture ceramics from the site of Szemely-Hegyes with very fine fabric. a) Bowl, b) Cup, c,d) Jars. All photomicrographs taken in crossed polars. Image width = 3.1 mm.

Ceramic Technology and Social Process in Late Neolithic Hungary

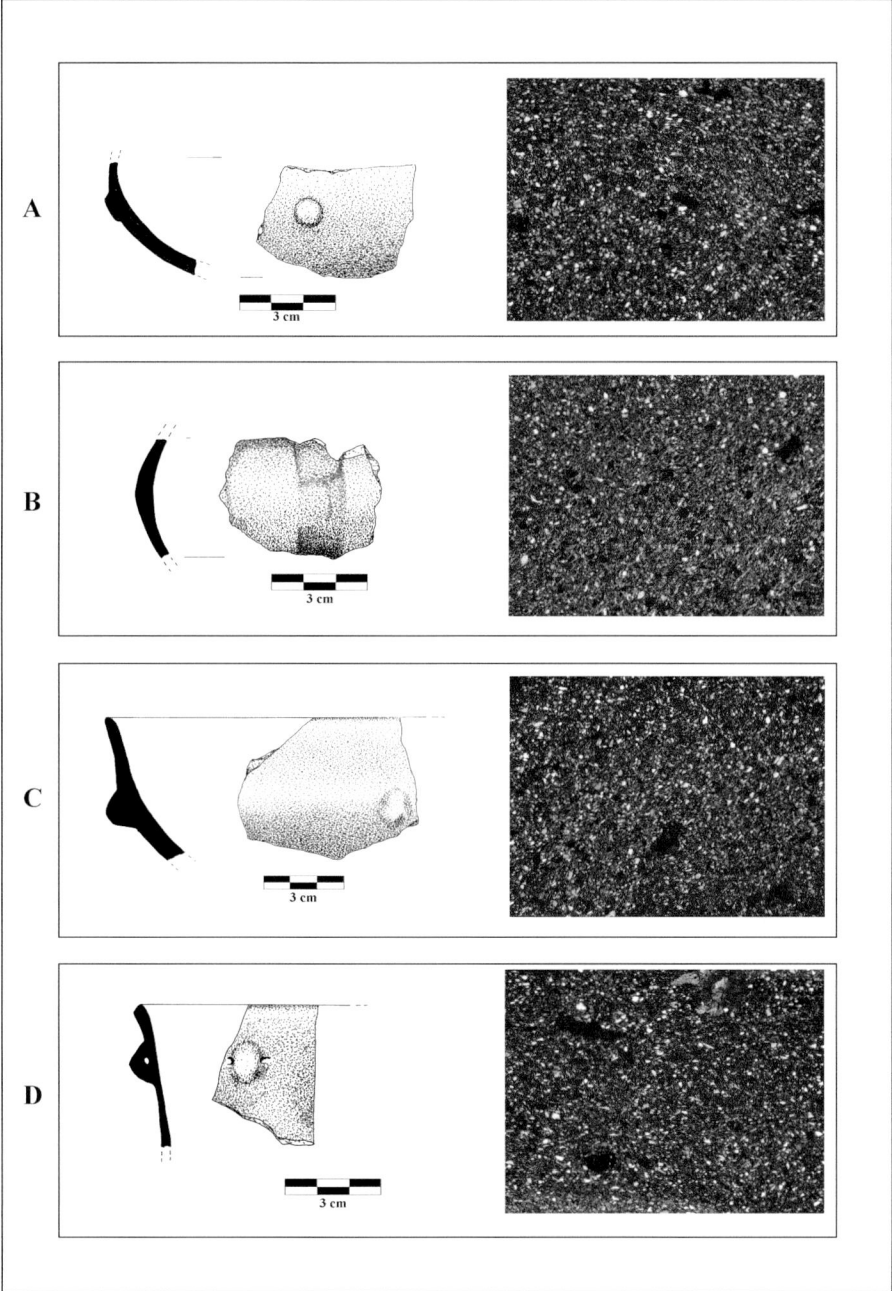

Figure 4. Illustrations and photomicrographs of the Late Neolithic Lengyel Culture ceramics from the site of Zengővárkony with very fine fabric. a,b) Cups, c) Bowl, d) Jar. All photomicrographs taken in crossed polars. Image width = 3.1 mm.

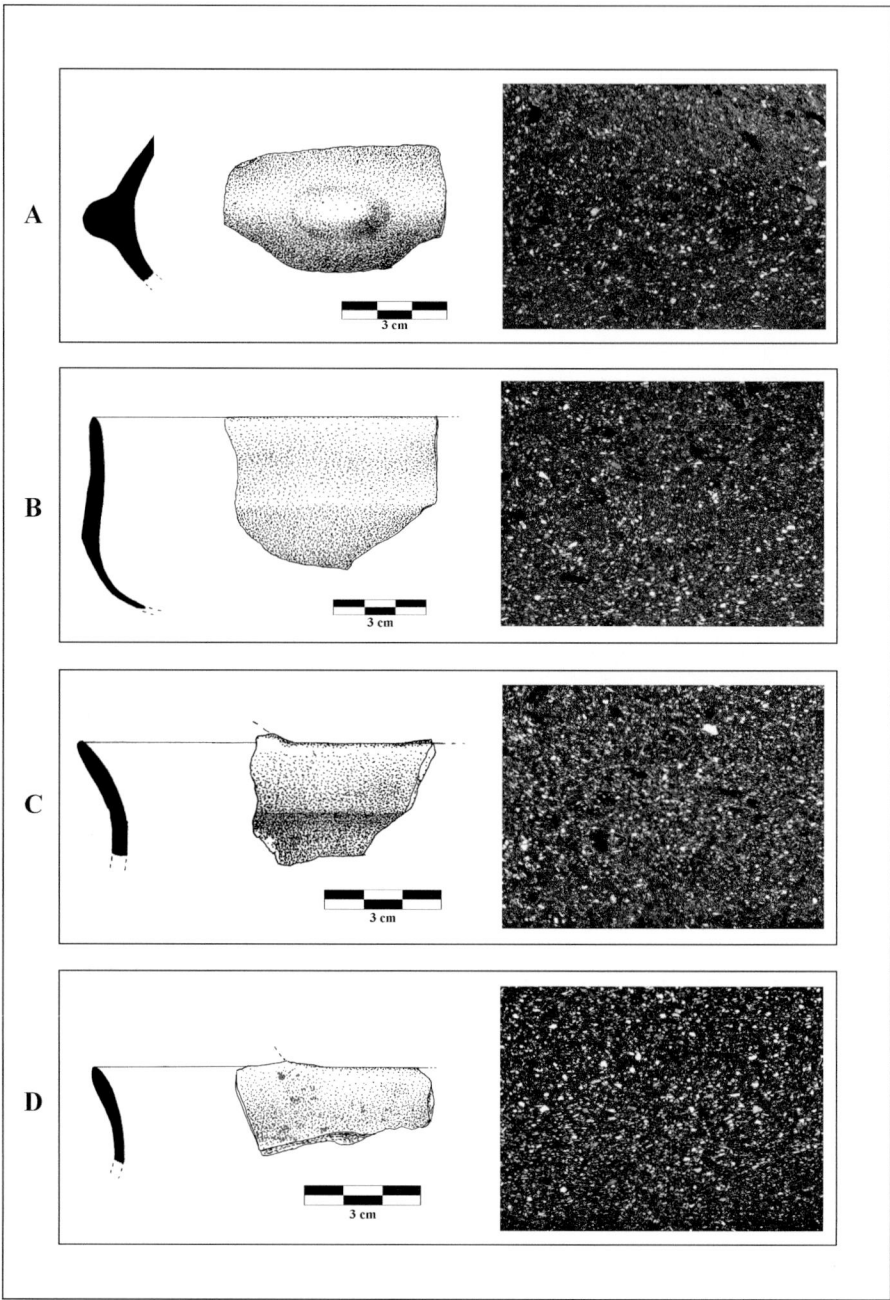

Figure 5. Illustrations and photomicrographs of the Late Neolithic Lengyel Culture ceramics from the site of Belvárdgyula-Szarkahegy with very fine fabric. a) Cup, b) Bowl, c,d) Mugs. All photomicrographs taken in crossed polars. Image width = 3.1 mm.

Ceramic Technology and Social Process in Late Neolithic Hungary

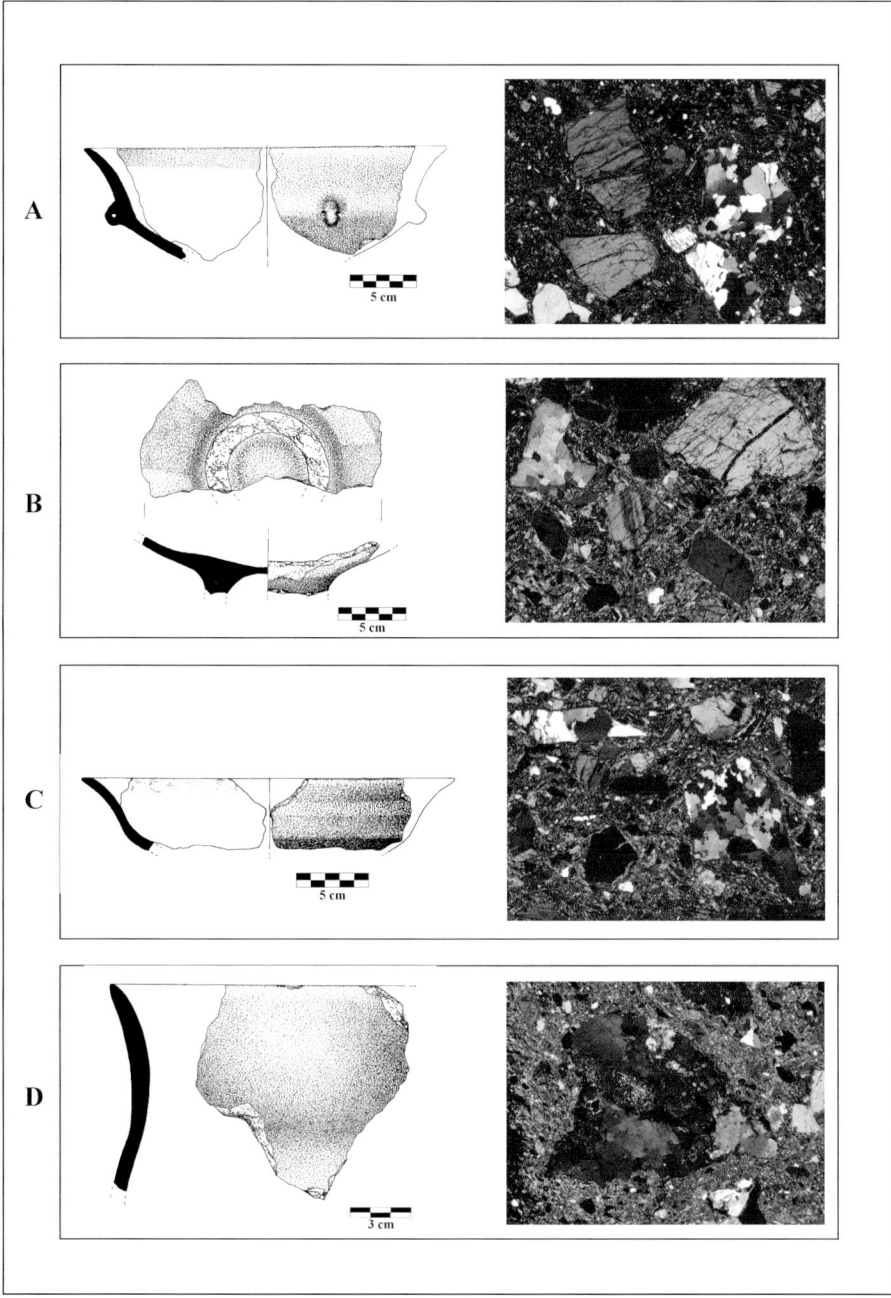

Figure 6. Illustrations and photomicrographs of the Late Neolithic Lengyel Culture ceramics from the site of Zengővárkony with coarse granite tempering. a) Bowl, b) Pedestalled bowl, c) Bowl, d) Storage vessel. All photomicrographs taken in crossed polars. Image width = 3.1 mm.

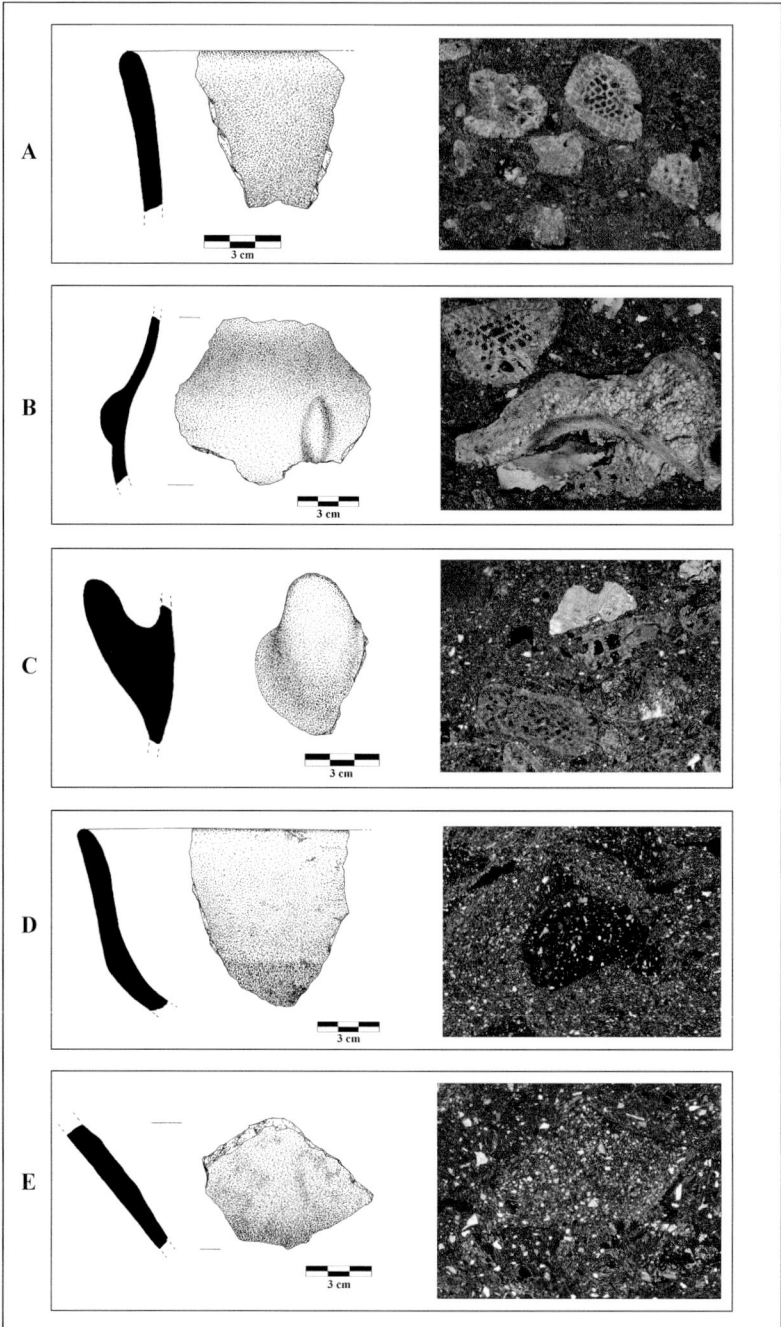

Figure 7. Illustrations and photomicrographs of the Late Neolithic Lengyel Culture ceramics from the sites of Zengővárkony and Belvárdgyula. a) Storage vessel from Zengővárkony, b) Jar from Zengővárkony, c) Storage vessel from Belvárdgyula, d) Bowl from Belvárdgyula with clay fragment temper, e) Pedestalled bowl from Szemely with grog tempering. All photomicrographs taken in crossed polars. Image width = 3.1 mm.

Another difference between the sites is that at Zengvárkony not only storage vessels, but bowls and pedestalled bowls were also tempered with coarse granitoid fragments (Fabric 9) (Figure 6), while among the cups, mugs, bowls and jars from Szemely this practice does not appear and in Belvárdgyula coarse tempering is also characteristic for storage or cooking vessels (Fabrics 7-8). At Zengővárkony and Belvárdgyula a few of the analysed samples contain shell fragments of bryozoans, oysters, echinoids, as well as granitoid inclusions (Figure 7A-C). The hiatal size distribution of these inclusions indicates that they were added as temper at the two sites.

The petrographic analysis of the Late Neolithic ceramics from the sites of Szemely, Zengővárkony and Belvárdgyula indicates that a large number of vessels of various shapes and sizes were manufactured in a single standardised way, whereas other less common vessels were produced using techniques that differ from one another and the dominant group. In the following section an attempt is made to interpret this data within a theoretical archaeological framework.

Discussion

Specialisation and standardisation has been an intriguing topic in archaeology for several decades. Many archaeological models of specialisation posit a direct link between craft specialisation and the emergence of social complexity, implying a strong correlation between product specialisation and political administration (Brumfiel and Earle, 1987). These subjects have been thoroughly examined from many viewpoints such as economy, cost effectiveness and improved efficiency (Feinman *et al.*, 1984; Abrams, 1987), risk avoiding or conservative tactics of potters insisting on using proven raw materials and technological practices (Balfet, 1966; Rice, 1991), socio-political factors and power (Rice 1981), function, ecological/environmental factors (Arnold, 1985), the role of ideology and religion in production (Carter, 2007; Hruby, 2007) and so on. In this paper the standardisation of Late Neolithic Lengyel Culture ceramic production is examined in terms of reconstructing social relationships between communities.

The similarity in fabric and manufacturing technology between many of the cups, mugs, bowls and jars of all three sites suggests some form of standardised production. A uniform, very fine raw material was used to produce most of the vessels, which may have been levigated. The vessels were constructed using the slab building technique and then burnished. The very fine fabric and the reduction-firing technology emphasises the production of fine wares with nearly identical textures in hand specimen.

The consistency observed in technique indicates that a high level of skill was employed in their production and is considered to be a form of standardisation, as it leads to a more homogenous product (Rice 1991, p. 268). This is a crucial point to the way in which vessels are perceived, as the similarities between each vessel type in terms of technology may have also promoted similarity between their producers and users.

The similarity in the very fine fabric of the vessels is intriguing as potters perception of resource suitability affects their selection or rejection of raw materials and may affect the variability of the resulting paste. Ceramic pastes are described by archaeologists using mineralogical and chemical categories, but potters are concerned with more relevant properties such as colour, the presence of crystalline substance, whether the clay feels too sticky or tastes salty, and its performance characteristic during drying and firing (Arnold, 1971, 1972, 1993; Rye, 1976). Potters' learn about these properties from experience and use them to select appropriate raw materials. With this in mind the similarities in the fabrics may suggest that potters at the different sites may have had similar ways of assessing the suitability of raw materials for the production of particular vessel types. The reasons for this selective use of raw materials and production techniques may be cultural since the clay used for making the examined vessels does not appear to be better suited to the production of small vessels, since even similar wares show very distinct fabrics (e.g. Fabric 9) (Figure 6).

On the other hand, the existence of different fabric groups at each settlement in this study, even within similar vessel types can be considered to reflect potters' choices. We might view these choices as being made by individuals according to a tradition into which they have been socialised. The different technological choices within each settlement might be considered to be a part of intra-site traditions. At this stage of the research the scale of pottery production or identification of areas of production could not be assessed since no pottery production sites have yet been found anywhere within the Lengyel Culture in Hungary. For this reason the scale of production is not directly addressed here but rather the process and practice of production and what we can infer from this partial data.

The level of technical complexity and consistency may indicate that pottery was made in household contexts, albeit by specialists (Hagstrum, 2001; Peregrine, 2001) since it has been argued that fine craftsmanship is not contrary to household specialisation (Cameron, 2001; Hagstrum, 2001, p. 51). Ethnoarchaeological studies have demonstrated that one of the factors influencing standardisation may be the demand of customers for a unified, recognisable product with specific dimensions, volume, colour or other technological characteristics (Arnold and Nieves, 1992; Longacre, 1999; Longacre et al., 2000). It must be noted that what appears to be standardised in the eyes of one group of people may not be so in the eyes of others. Thus standardisation is socially defined and contextually variable (Rice, 1991). Since the Neolithic potters of the three examined sites appear to have used standardised methods and reproduced material culture in a similar way, it is considered that the reproduction of tradition may have been organised through social networks. Similarities in material culture production create and concretise social similarities between groups and it reinforces social cohesion by defining social roles and creating a common connection between individuals.

In addition to the standardised products of the very fine fabric, there is a range of other vessels such as bowls, pedestalled bowls, storage and cooking vessels, that were made differently and in some cases were more elaborately decorated (Figure 6B,C). The variability observed within these vessels is not considered to be the result of functional differences, but instead a social process or significance is being sought. It has been argued that in weakly specialised economies there is a greater competition, which leads to increased energy expenditure in the form of product elaboration as specialists try to differentiate their wares from the products of their competitors and thereby attract consumers (Foster, 1966; Feinman *et al.*, 1984; Costin, 1993). It is also a possibility that in the case of the Late Neolithic ceramics analysed in this study, there may have been restrictions to certain resources, since the raw materials and tempers of some bowls were not best suited to make fine-walled products. For example, several bowls made from raw materials tempered with coarse inclusions, were also elaborately burnished, suggesting that the potter tried to make the wall as smooth as possible but the added coarse inclusions hindered the process (Figure 6A-C).

Increased labour investment may not directly reflect characteristics of economic systems, but instead energy may be expended because the objects carry social information that the maker and/or the user wished to broadcast (Longacre *et al.*, 2000). The occurrence of elaborately painted vessels made from inappropriate raw materials may suggest that producers were competing with each other, indicating that material culture production did play a role in social relations. It seems that a group of potters had a particular, standardised tradition allowing them to concretise on one particular technique. The appearance of bowls and pedestalled bowls, which do not have standardised techniques, but in some cases have more elaborated painted decorations may therefore indicate that social order faced challenges to the understood social norm. Faced with uncertainty, communities may respond by investing increased labour into material culture production or consolidating existing material culture repertoires (Hodder, 1982; Braithwaite, 1984; Budden, 2007, p. 286). Such a strategy may have played a role in signaling identity in order to resist changes in social order, or it may also be an attempt to challenge the concepts and ideas that accompany destabilising change (Sterner, 1989; Budden, 2007, p. 286).

Since standardised practices consistently appear at the examined sites they may have constituted social meaning and related to socially informed decisions. It is considered that these practices worked to protect the continuity of a prestige item and hence the continuity of their social performance. It is further considered that a particular range of vessels may relate to the articulation of social identity, or expressing status through the visual display of socially important items, thus playing a role in negotiating social relationships. This highlights the individual social role of a particular vessel group and the social status of their producer.

Conclusions

The analysis of Late Neolithic ceramics from the three Lengyel Culture settlements in this study has revealed that some of the consumption wares such as cups, mugs, jars and bowls were made in a similar way. They exhibit a homogeneous, very fine, dense, perhaps levigated raw material, were built using the slab technique and were fired under reducing conditions. The technological characteristics of these vessels clearly distinguish them from those of the other examined vessels. The observed similarities in technological practices in the three examined sites may indicate that potters had a similar conception about standardisation and how a particular group of vessels should be made in a standardised manner. The similarities in technological practices may indicate that these vessels were viewed similarly and were incorporated similarly into daily activities.

The particular technological order revealed by this research provides a contrast to other vessel groups and seems to offer confirmation of the social importance of some of the wares, indicating that highly prescribed technological strategies were present in the examined Lengyel Culture settlements. This strategy ensured the continuity of a standardised and culturally determined range of vessel types and reflects the highly structured nature of the Lengyel society. Similar technological strategies reproduced categories of objects that are deeply embedded in the mediation of cultural and social relationships.

Acknowledgements

The authors would like to thank Gábor Bertók, Csilla Gáti, Olga Vajda, Olivér Gábor and the Museums of Baranya County for providing the samples for the analysis and Márta Lakó for the illustrations. The research presented in this paper was supported by a grant (NK 68255) from the Hungarian Scientific Research Fund.

References

Abrams, E.M. 1987. Economic specialization and construction personnel in Classic Period Copan, Honduras. *American Antiquity*, 52: 485-499.

Arnold, D.E. 1971. Ethnomineralogy of Ticul, Yucatan potters: etics and emics. *American Antiquity*, 36: 20-40.

Arnold, D.E. 1972. Mineralogical analyses of ceramic materials from Quinua, department of Ayacucho. *Archaeometry*, 14: 93-101.

Arnold, D.E. 1985. *Ceramic theory and cultural process*. Cambridge University Press, Cambridge.

Arnold, D.E. 1993. *Ecology and ceramic production in an Andean community*. Cambridge University Press, Cambridge.

Arnold, D.E. and Nieves, A.L. 1992. Factors affecting ceramic standardization. In: Bey III, G.J. and Pool, C.A. (Eds.) *Ceramic production and distribution: an integrated approach*. Westview Press, Oxford: 93-113.

Balfet, H. 1966. Ethnographical observations in North Africa and archaeological interpretation: the pottery of the Mahgreb. In: Matson, F.R. (Ed.) *Ceramics and man*. Methuen & Co Ltd, London: 161-177.

Braithwaite, M. 1984. Ritual and prestige in the prehistory of Wessex c. 2200-1400 BC: a new dimension to the archaeological evidence. In: Miller, D. and Tilley, C. (Eds.) *Ideology, Power and Prehistory*. Cambridge University Press, Cambridge: 93-110.

Brumfiel, E.M. and Earle, T. 1987. Specialization, exchange and complex societies: an introduction. In: Brumfiel, E.M. and Earle, T. (Eds.) *Specialization, exchange and complex societies*. Cambridge University Press, Cambridge: 1-9.

Budden, S.A. 2007. *Renewal and reinvention: the role of learning strategies in the Early to Late Middle Bronze Age of the Carpathian Basin*. Unpublished doctoral dissertation, University of Southampton.

Cameron, C.M. 2001. Pink chert, projectile points, and the Chacoan regional system. *American Antiquity*, 66: 79-101.

Carter, T. 2007. The theatrics of technology: consuming obsidian in the Early Cycladic burial arena. *Archeological Papers of the American Anthropological Association*, 17: 88-107.

Costin, C.L. 1993. Craft specialization: issues in defining, documenting, and explaining the organization of production. *Journal of Archaeological Method and Theory*, 3: 1-56.

Császár, G. 2005. *Magyarország és környezetének regionális földtana*. I. Paleozoikum-pleogén. ELTE Eötvös Kiadó, Budapest.

Cuomo di Caprio, N. and Vaughan, S.J. 1993. An experimental study in distinguishing grog (chamotte) from argillaceous inclusions in ceramic thin sections. *Archeomaterials*, 7: 21-40

Dobres, M.A. 2000. *Technology and social agency*. Blackwell, Oxford.

Feinman, G., Kowalewski, S.A. and Blanton, R.E. 1984. Modelling ceramic production and organizational change in the pre-Hispanic valley of Oaxaca, Mexico. In: van der Leeuw, S.E. and Pritchard, A.C. (Eds.) *The many dimensions of pottery: ceramics in archaeology and anthropology*. Universiteit van Amsterdam, Amsterdam: 297-337.

Foster, G.M. 1966. The sociology of pottery: questions and hypotheses arising from contemporary Mexican work. In: Matson, F.R. (Ed.) *Ceramics and man*. Methuen & Co Ltd, London: 43-61.

Fülöp, J. 1994. *Magyarország geológiája. Paleozoikum II.* Akadémiai Kiadó, Budapest.

Gyalog, L. 2005. *Magyarázó Magyarország fedett földtani térképéhez (1: 100 000).* Magyar Állami Földtani Intézet, Budapest.

Hagstrum, M. 2001. Household production in Chaco Canyon society. *American Antiquity*, 66: 47-55.

Hodder, I. 1982. *Symbols in action. Ethnoarchaeological studies of material culture.* Cambridge University Press, Cambridge.

Hruby, Z.X. 2007. Ritualized chipped-stone production at Piedras Negras, Guatemala. *Archeological Papers of the American Anthropological Association*, 17: 68-87.

Jantsky, B. 1979. A mecseki gránitosodott kristályos alaphegység földtana (Géologie dusocle cristallin granitisé de la montagne Mecsek*). Annales of the Geological Institute of Hungary*, 60: 1-385.

Király, E. and Koroknai, B. 2004. The magmatic and metamorphic evolution of the north-eastern part of the Mórágy Block. *Annual Report of the Geological Institute of Hungary*: 299-310.

Kozłowski, J.K. and Raczky, P., (Eds.) 2007. *The Lengyel, Polgár and related cultures in the Middle/Late Neolithic in Central Europe.* Kraków.

Kreiter, A. and Szakmány, G. 2008a. Preliminary report on the petrographic analysis of Late Neolithic ceramics from Szemely-Hegyes és Zengővárkony. *Archeometriai Műhely*, 5: 55-68

Kreiter, A. and Szakmány, G. 2008b. Preliminary report on the petrographic analysis of Late Neolithic ceramics from Belvárdgyula-Szarkahegy. *Archeometriai* Műhely. 5: 65-74.

Lemonnier, P. 1992. *Elements for an anthropology of technology. Anthropological Papers No. 88.* Ann Arbor: Museum of Anthropology, University of Michigan.

Longacre, W.A. 1999. Standardization and specialization: what's the link? In: Skibo, J. M. and Feinman, G. (Eds.) *Pottery and people.* University of Utah Press, Salt Lake City: 44-58.

Longacre, W.A., Xia, J.F. and Yang, T. 2000. I want to buy a black pot (Philippine techniques). *Journal of Archaeological Method and Theory*, 7: 273-293.

Peregrine, P.N. 2001. Matrilocality, corporate strategy, and the organization of production in the Chacoan world. *American Antiquity*, 66: 36-46.

Rice, P.M. 1981. Evolution of specialized pottery production: a trial model. *Current Anthropology*, 22: 219-240.

Rice, P. 1991. Specialization, standardization, and diversity: a retrospective. In: Bishop, R.L. and Lange, F.W. (Eds.) *The Ceramic Legacy of Anna O. Shepard*. University of Colorado Press, Niwot: 257-279.

Rye, O.S. 1976. Keeping your temper under control: materials and the manufacture of Papuan pottery. *Archaeology and Physical Anthropology in Oceania*, 11: 106-137.

Sterner, J. 1989. Who is signalling whom? Ceramic style, ethnicity and taphonomy amongst the Sirak Bulahary. *Antiquity*, 63: 451-459.

Whitbread, I.K. 1986. The characterisation of argillaceous inclusions in ceramic thin sections. *Archaeometry*, 28: 79-88.

EARLY POTTERY TECHNOLOGY & THE FORMATION OF A TECHNOLOGICAL TRADITION: THE CASE OF THEOPETRA CAVE, THESSALY, GREECE

Areti Pentedeka

Laboratory of Prehistoric Archaeology, Department of Archaeology,
Aristotle University of Thessaloniki, Greece

Anastasia Dimoula

Laboratory of Prehistoric Archaeology, Department of Archaeology,
Aristotle University of Thessaloniki, Greece
(adimoula@hist.auth.gr)

Introduction

Theopetra Cave, situated at the NW end of the Thessalian plain, is one of the rare cases in the Aegean where a continuous sequence of deposits from the Middle Palaeolithic (c. 50,000 BP) to the end of the Neolithic period (c. 3000 BC) is attested (Kyparissi-Apostolika, 2000a; Facorellis and Maniatis, 2000; Facorellis et al., 2001). At this site the Mesolithic period and its uninterrupted transition into the Neolithic, is for the first time verified in Thessaly, where habitation during the Neolithic saw an exceptional flourish in comparison to the rest of mainland and insular Greece (Kyparissi-Apostolika, 2000a; Kyparissi-Apostolika, 2003). To date, Theopetra Cave is the only cave site known in Thessaly, an area where the prevailing settlement form is that of a tell, along with very few extended sites (see Andreou et al., 2001 for a recent review).

The continuous Neolithic habitation of the Theopetra Cave left behind a rich ceramic assemblage containing a variety of wares characteristic of all phases of the Thessalian Neolithic. This serves as an outstanding case study of the initiation of ceramic production at the beginning of the Neolithic period, a topic that has been of interest for many researchers, mainly with regard to the indigenous character of the Greek Neolithic. In this paper the subject of early pottery technology is examined through the detailed petrographic analysis of selected sherds of the EN-LN assemblage of the site. Particular attention is given to the nature of the ancient potters' choices in pottery manufacture throughout the Neolithic and of the formation of a local technological tradition at Theopetra Cave. The study also contributes to unravelling the production sequence and use of pottery in the only inhabited cave set within a cultural landscape in which open-air settlements predominate. Finally, the degree and character of intraregional contacts between the inhabitants of Theopetra and the rest of Thessaly is explored, mainly through the nature of pottery exchange networks.

The Site of Theopetra Cave

The cave of Theopetra is located in the prefecture of Trikala in western Thessaly on the north face of a limestone hill that rises on the route leading from Trikala to Kalambaka (Figure 1). As it lies between the edge of the Thessalian plain and the foothills of the east Pindus Mountains, it is characterised by attributes of both environments. The cave has a roughly quadrangular plan, covering approximately 500 m^2 and a large arched entrance (17x3m), allowing for natural light to enter the interior (Kyparissi-Apostolika, 2000a, p. 17).

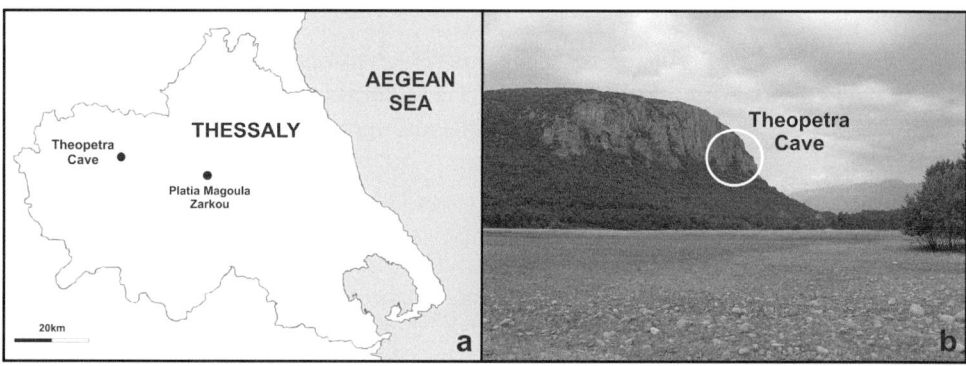

Figure 1. Theopetra Cave and Neolithic sites in Thessaly. a) Location of Theopetra Cave and Platia Magoula Zarkou, b) The Theopetra limestone outcrop, viewed from the north.

Between 1987 and 1998, an area of 144m^2 was excavated within the cave. The finds that were recovered which suggest an uninterrupted habitation from the Middle Palaeolithic to the end of the Neolithic period that is supported by ^{14}C dates (Facorellis and Maniatis, 2000; Facorellis *et al.*, 2001). Middle Palaeolithic finds consist of stone tools of the Mousterian type, but, more importantly, of layers of fire remains alternating with silty sediments, where human footprints are preserved (Manolis *et al.*, 2000, p. 81). Deposits of the following Upper Palaeolithic period yielded stone tools and bones, including a complete human skeleton, along with at least fifteen hearths, between which clay lumps were found (Kyparissi-Apostolika, 2000a, p. 22-23; Kyparissi-Apostolika, 1999). The end of the Palaeolithic period coincides with the Last Glacial which is overlain by the Mesolithic deposit, a distinct yellowish-brown humic sediment with sporadic fire remains. The Mesolithic finds include mainly lithics, animal and human bone, including the burial of a young woman in a shallow pit, as well as archaeobotanical and anthracological material, some unbaked clay lumps and a few monochrome sherds, representing the beginning of pottery technology in the cave (Kyparissi-Apostolika, 2000c, p. 134-137).

The beginning of the Neolithic period in Theopetra Cave is estimated as between 7000-6700 BC, according to radiocarbon dating of samples deriving from the limits between the Mesolithic and the Neolithic deposits (Facorellis and Maniatis, 2000, table 3.1). The Neolithic deposits cover the entire excavated area, though variability is observed in both the thickness of the deposits and the density of the finds contained within. The thicker deposits (c. 1.5 m) are located in the central, well-lit area of the cave, while on its western and eastern sides Neolithic strata are reduced to as little as 0.5 m (Kyparissi-Apostolika, 2000b, p. 181-182).

The stratigraphic sequence of the Neolithic deposits, especially in the centre of the cave, was highly disturbed due to large rocks brought in by the water during periods of high rainfall through the carstic aquifers of the limestone rock of the cave. Additionally, large rocks must have fallen from the roof during a period of increased seismic activity. Finally, towards the end of the Neolithic period, water stagnated on the whole surface of the cave, as evidenced by a horizontal line running along the walls of the cave at a height of 2 m above the contemporary surface of the Neolithic deposits. This indicates that the actual thickness of Neolithic strata may have reached 3.5 m (Kyparissi-Apostolika, 2000b, p. 182).

The Neolithic Ceramic Assemblage

The lack of undisturbed stratigraphy for most of the Neolithic deposits in the Theopetra Cave is counterbalanced by the wealth of portable finds, representing all aspects of the Neolithic material culture, including stone tools (ground, polished and chipped), bone tools, spindle whorls and other weaving implements, figurines and jewellery, and pottery (Kyparissi-Apostolika, 2000b, p. 182-183).

The ceramics of the cave contain many wares known from other Neolithic settlements in Thessaly, thus enabling their attribution to the various sub-phases of the Neolithic on the basis of morphotypological criteria (Figure 2). Pottery ascribed to the Early Neolithic (EN) is present in small quantities in the assemblage compared to the following periods. Undisturbed EN deposits are located in the west area of the cave, stratigraphically above the underlying Mesolithic deposits, from which, as mentioned above, some early monochrome sherds derived (Kyparissi-Apostolika, 2000b, p. 183-184). The EN pottery is primarily monochrome (light or dark brown, red, orange and grey colours), with coarse, polished, burnished or slipped surfaces. The vessel shapes include mainly bowls and rarely closed vessels, often bearing small handles and lugs. The decorated pottery, which appears later in the course of the EN, includes the so-called 'Early Painted' and Black-topped wares, common on open vessels of medium-large size (Kyparissi-Apostolika, 2000b, p. 184-185; Kaznesi, in press), along with incised and impressed wares, common on open vessels and rarely on closed ones (Kyparissi-Apostolika, 2000b, p. 192-194; Kaznesi, in press).

Middle Neolithic (MN) pottery includes monochrome and decorated wares, with Red Monochrome, 'Scraped' and Red-on-White wares being the most common finds. Open vessels predominate in all wares, with the most common shapes being the bowl, the cup and the basin, while in Red Monochrome (and less frequently in 'Scraped' ware) closed spherical vessels and jars are also recognised (Kyparissi-Apostolika, 2000b, p. 186-187; Kaznesi, in press).

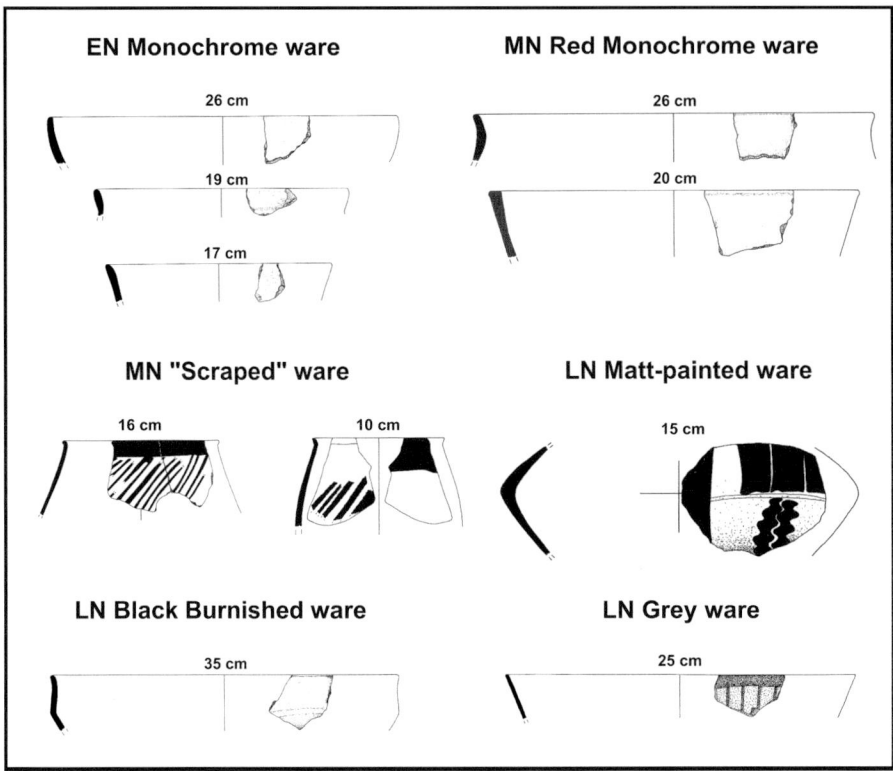

Figure 2. Illustration of typical Neolithic ceramic vessels from the Theopetra Cave (after Kyparissi-Apostolika, 2000b, figs. 14.2, 14.6, 14.7, 14.8, 14.9).

Pottery dated to Late Neolithic (LN) includes characteristic wares of all LN sub-phases until the end of the Neolithic. However, judging from the quantity and diversity of wares and shapes, it seems as though the earliest LN phase, 'Tsaggli-Larissa', is the major or better-attested habitation phase of Theopetra Cave (Kyparissi-Apostolika, 2000b, p. 187-198; Kaznesi, in press). The main ceramic styles of this sub-phase are plain monochrome and decorated Black Burnished ware, which occur most commonly as small-medium carinated bowls (Kyparissi-Apostolika, 2000b, p. 188; Katsarou, 2000, p. 244-245), Matt-painted ware, with dark coloured matt decoration on plain surface, including mainly jars, along with few bowls and/or basins of medium to large size (Kyparissi-Apostolika, 2000b, p. 189; Kaznesi, in press), plain and decorated Grey ware (widely known as Grey-on-Grey painted ware), with the cup being the most common shape (Kyparissi-Apostolika, 2000b, p. 187-188; Kaznesi, in press), and lastly Polychrome ware, with black and red decoration on white background or black and white decoration or red background, encountered mainly on large open vessels (Kyparissi-Apostolika, 2000b, p. 190-191; Kaznesi, in press).

Research Objectives and Methodology

Theopetra Cave is a rather exceptional case for the Greek Neolithic, as it presents continuous habitation from the Mesolithic to the end of the Neolithic period. Its pottery assemblage, therefore, is very important, as it preserves not only the later and better-known sub-phases of MN and LN, but also the transition from the Mesolithic to the Neolithic. The EN phase is characterised by the initiation of pottery manufacture and use, after which pottery production stands as a hallmark for the Neolithic way of life not only at Theopetra Cave or Thessaly, but throughout Greece. As mentioned above, the ceramic assemblage of Theopetra Cave, in terms of typology, does not differ in any substantial way from assemblages known from other Neolithic settlements in Thessaly (Kyparissi-Apostolika, 2000b, p. 183-197). In this context, ceramic analysis in this project was mainly aimed at studying the initiation of pottery technology at the very start of the Neolithic period, and the implications that this might have regarding the indigenous character of the Greek Neolithic. An additional goal was to examine the ancient potters' choices in pottery manufacture throughout the Neolithic period at Theopetra and the potential formation of a local technological tradition, as well as to discern similarities and differences between pottery production characterising cave habitation and that of open-air settlements, mainly tells, which dominate the cultural landscape of Neolithic Thessaly. Additionally, the analysis aimed to shed light on the role that Theopetra played in a regional perspective, mainly through its participation in pottery exchange networks, and the level and character of intraregional contacts between the inhabitants of Theopetra and the rest of Thessaly, through comparison with the results of similar analysis on assemblages of other Thessalian sites (Pentedeka, 2008).

In order to better investigate the beginning of pottery technology and the level of contact with the rest of Thessaly, this paper focuses on pottery dating mainly to the Mesolithic, EN I (c. 7000-6700 BC), the later part of MN (MN III, to the extent that such discrimination was feasible due to the disturbed stratigraphy, c. 5500-5300 BC) and the beginning of LN, namely the 'Tsaggli-Larissa' phase (c. 5300-5000 BC), which seems to be the major or better attested habitation phase of Theopetra (Kyparissi-Apostolika, in press). In total, 110 sherds were selected from the ceramic assemblage of Theopetra Cave, representing the main wares that were macroscopically identified (Table 1).

Since the present study focuses on both provenance and technological issues, petrographic analysis was considered to be the most suitable analytical technique applicable to the pottery assemblage from Theopetra, which is rather coarse in texture (Whitbread, 1995, p. 391-394; Whitbread, 2001, p. 451). All samples were thin-sectioned and examined under the polarising microscope, following the systematic description and explanatory scheme proposed by Whitbread (1986, 1989, 1995, p. 365-396). Along with petrographic analysis, the samples were also subjected to refiring tests at 900°C and 1050°C in controlled conditions in the laboratory. Refiring of the ceramic samples, in a fully oxidising atmosphere and at higher temperature than the original firing, has aimed to distinguish different clay and slip/paint compositions reflected in clay colour, by eliminating any variation caused by ancient firing conditions (Whitbread, 1995, p. 390-391; Kiriatzi, 2000, p. 90-92; Daszkiewicz and Schneider, 2001).

Number of samples	Date	Ware	Shape
TH1-TH35 (35/110)	EN	Monochrome of variable surface colour and surface treatment	Bowls prevail, along with a few closed vessel shapes of various sizes
TH36-TH51 (16/110)	MN	Red Monochrome	Bowls of various forms and sizes, closed spherical vessels and jars
TH52-TH65 (14/110)	MN	"Scraped"	Mainly bowls and few closed spherical vessels of small to medium size
TH66-TH81 (16/110)	LN	Black Burnished, monochrome and decorated	Mainly carinated bowls of small-medium size
TH82-TH95 (14/110)	LN	Matt-painted	Mainly jars and few bowls and/or basins of medium to large size
TH96-TH105 (10/110)	LN	Grey monochrome and Grey painted	Cups and bowls of small size
TH106-TH110 (5/110)	LN	Coarseware	Jars or pithoi of large size
Number of raw material samples		Description	
GS3, GS4, GS11-GS14, GS20, GS22 (8/23)		Rock fragments (ophiolitic, chert, sandstone, limestone)	
GS5, GS6, GS17 (3/23)		Sand	
GS1, GS2, GS7-GS10, GS15, GS16, GS18, GS19, GS21, GS23 (12/23)		Sediments of variable plasticity	

Table 1. Summary table presenting the main attributes of ceramic and geological samples analysed

The integrated analytical approach adopted in this project also included geological prospection, in order to obtain a clearer picture of the raw materials available in the area. Twenty-three geological samples (GS) of sediments and rocks were collected from the vicinity of Theopetra Cave (Figure 3). All sediments were diluted with distilled and deionised water, mixed within beakers and left to settle out. Those showing plasticity were further processed in order to produce four briquettes, three of which were fired, each at a different temperature (700°C, 900°C, 1050°C). All fired briquettes, sand samples and rock fragments were thin-sectioned and examined under the polarising microscope.

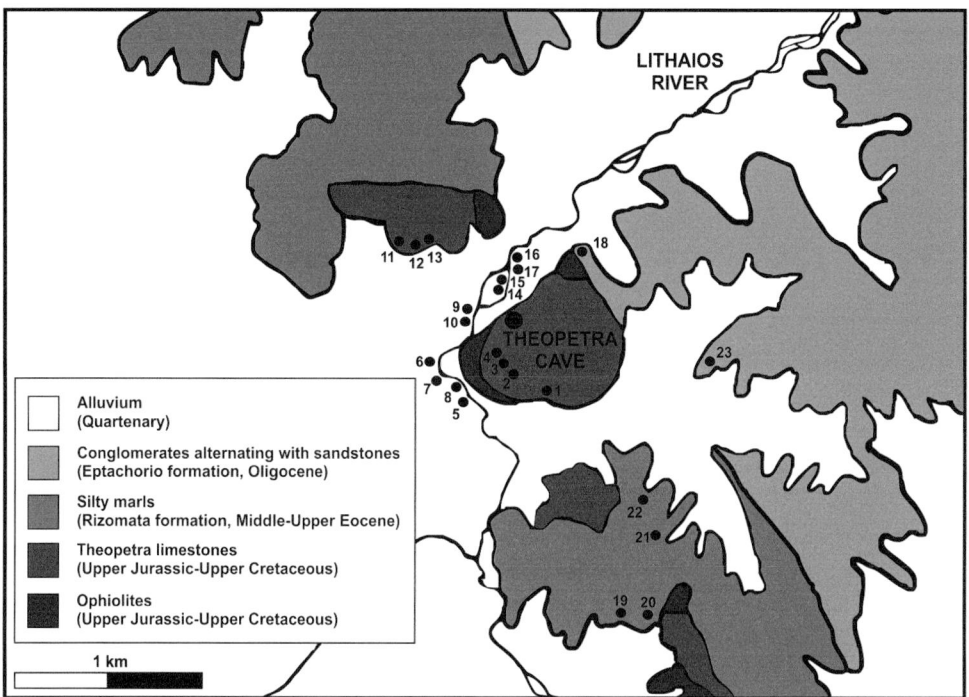

Figure 3. Geological map of the vicinity of Theopetra Cave with the geological sampling locations (adapted from Savoyat and Lalechos, 1972).

Petrographic Results

Based on the evidence provided by the petrographic analysis and refiring tests, seven major fabric groups (THFG - Theopetra Fabric Groups) have been distinguished. For detailed descriptions of all fabric groups as well as geological samples see Pentedeka and Dimoula (in press). Characterisation of inclusions frequency, sorting and roundness was based on comparative charts by Bullock *et al.* (1985, figure 24, p. 24-25; figure 27, p. 27; figure 31, p. 31).

THFG1 is a coarse to medium-coarse fabric group, characterised by poorly sorted sub-angular to sub-rounded inclusions, of medium to coarse sand grain size. The fabric group consists of gneiss/quartzite fragments and their constituents, in particular sericitised and/or cloudy orthoclase and microcline feldspar and quartz, along with very few other rock inclusions such as metamorphic rocks, chert, sandstone and very altered ophiolitic rock fragments (Figure 4A-C). Almost 70% of the samples examined in this study fall into THFG1, including all shapes and wares of all periods represented.

THFG2 is characterised by moderately to poorly-sorted sub-angular to sub-rounded inclusions, of medium to coarse sand grain size. These include quartzite and biotite schist fragments, along with polycrystalline quartz, fresh and commonly poikiloblastic alkali feldspar, and few biotite mica. Only a few samples, from 'Scraped' and Grey wares fall into THFG2 (dating to MN and LN respectively), all of which are open shapes of small to medium size (Figure 4E).

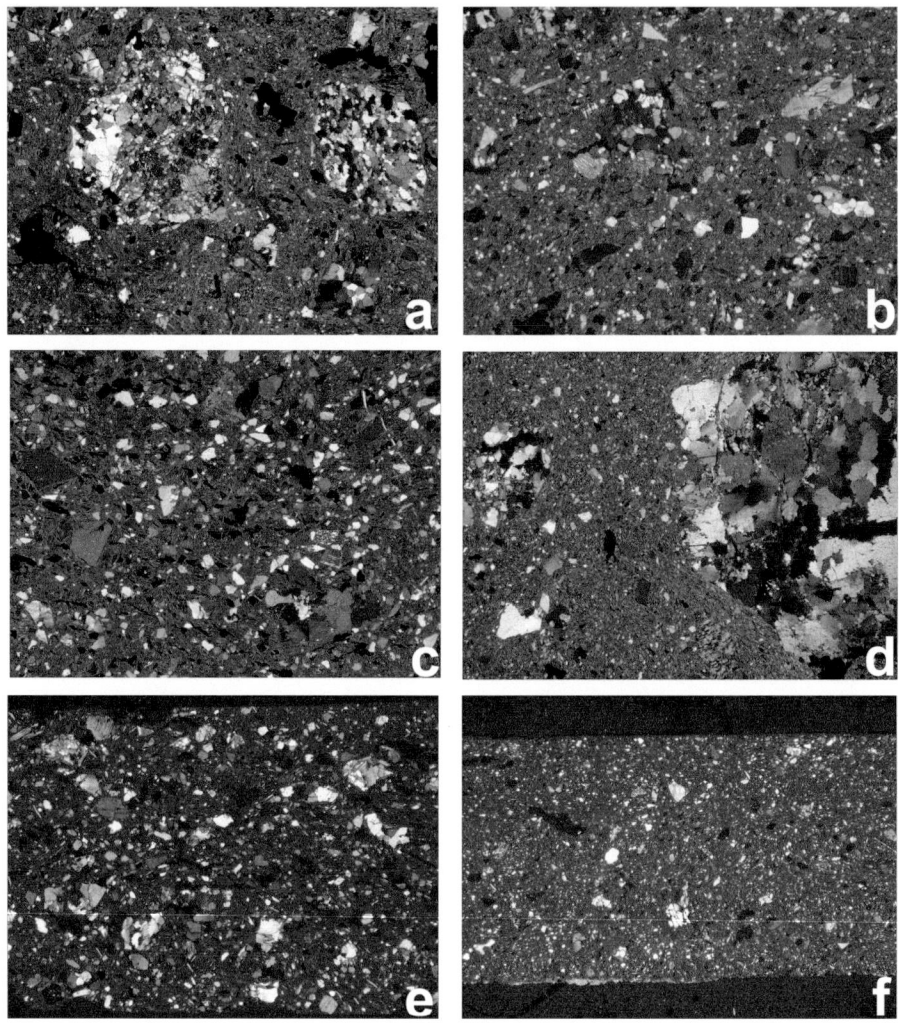

Figure 4. Thin section micrographs of the Theopetra Cave ceramics analysed in this study a) Sample TH21 (EN), coarse-grained example of THFG1, showing gneiss/quartzite fragments, b) Sample TH40 (MN), medium-coarse grained example of THFG1, showing gneiss fragments and mineral constituents, c) Sample TH72 (LN), medium-coarse grained example of THFG1, showing better inclusion sorting, d) Geological sample GS21 (unrefined, fired at 700°C), showing compositional resemblance to THFG1, e) Sample TH101 (LN), coarse-grained example of THFG2, showing quartzite and biotite schist fragments, f) Sample TH56 (MN), medium-grained example of THFG3, showing gneiss/quartzite fragments and their main constituents. All images taken in crossed polars. Image width = 5.9 mm.

THFG3 is a medium to medium-coarse fabric group, characterised by moderately to poorly sorted sub-angular to sub-rounded inclusions, of medium sand grain size. It contains gneiss/quartzite fragments and their main constituents such as sericitised alkali feldspars and quartz, along with micas. Only a few samples from 'Scraped' (MN) and Grey (LN) wares fall into THFG3, all being open shapes of small to medium size (Figure 4F).

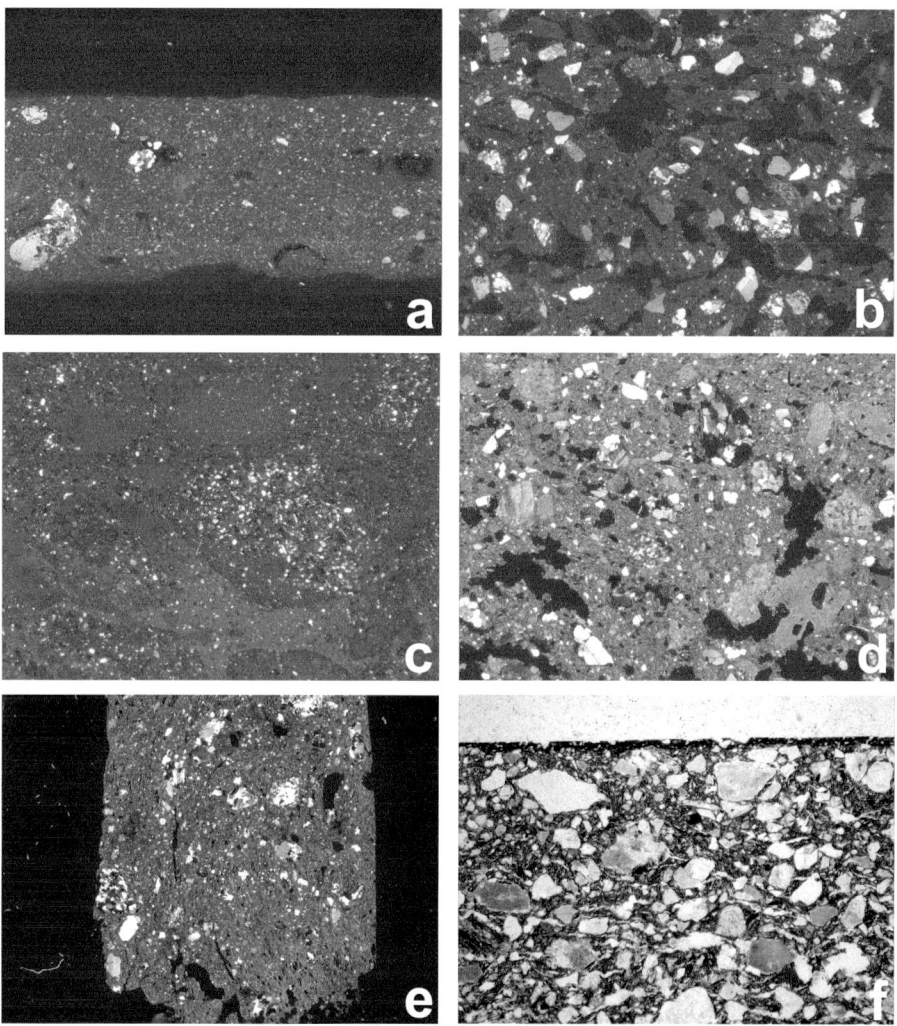

Figure 5. Thin section micrographs of the Theopetra Cave ceramics analysed in this study. a) Sample TH97 (LN), medium-fine grained example of THFG4, showing gneiss/quartzite fragments and carbonate marls, along with secondary calcite deposition, b) Sample TH94 (LN), medium-coarse grained example of THFG5, showing serpentinite, chert, mudstone and gneiss/quartzite fragments, c) Sample TH82 (LN), coarse-grained example of THFG6, showing mudstone/siltstone and sandstone fragments, d) Sample TH16 (EN), coarse-grained example of THFG7, showing limestone and gneiss/quartzite fragments, e) Sample TH21 (EN), example showing concentric arrangement of voids and inclusions, indicating the use of the coiling technique, f) Sample TH45 (MN), showing layer of fine-grained clay identified as slip. All images taken in crossed polars. Image width = 5.9 mm. except e = 18.5 mm.

THFG4 is a medium-fine fabric group, characterised by moderately well sorted sub-rounded to sub-angular inclusions, of fine sand grain size. The inclusions consist of marl, gneiss/quartzite fragments, sericitised and/or cloudy orthoclase and microcline feldspar, micas and monocrystalline quartz. A distinctive characteristic of this fabric group is the occurrence of dark regions that are probably due to the carbonization of organic matter during firing. Few samples fall into this fabric, the majority belong to Grey ware (LN) (Figure 5A).

THFG5 is a medium-coarse fabric group, characterised by moderately well to well sorted sub-angular to sub-rounded inclusions and bimodal grain size distribution with coarse fraction of medium-coarse sand grain size. The inclusions consist of serpentinite fragments (also few altered ophiolitic fragments), radiolarian chert, mudstone and gneiss/quartzite fragments, along with quartz and sericitised and/or cloudy alkali feldspars. All THTG5 samples belong to the Matt-painted ware (LN) (Figure 5B).

THFG6 is a medium to coarse fabric group, with moderately well sorted sub-angular to sub-rounded inclusions of medium to very coarse sand grain size and a weakly bimodal grain size distribution. The inclusions contain mudstone/siltstone and sandstone fragments, along with monocrystalline quartz, cloudy alkali feldspars and quartzite rock fragments. All samples belong to the Matt-painted ware (LN) (Figure 5C).

THFG7 is a coarse fabric group, characterised by poorly sorted rounded to sub-angular inclusions, of coarse sand grain size and unimodal to weak bimodal grain size distribution. It consists of micritic and sparitic limestone fragments, gneiss/quartzite fragments, quartz and alkali feldspars. All THFG7 samples belong to EN Monochrome ware (Figure 5D).

Provenance of the Neolithic Pottery

The majority of the fabric groups detected in the ceramics studied are considered to be local to Theopetra Cave. The hill on which the site is situated sites within the northeastern part of an alluvial plain formed by Pineios River and its tributaries. The geology of the area is characterised by mollassic formations of the Mesohellenic Trough (Figure 3) as well as gneiss, schist and flysch of the crystalline Sub-Pelagonian basement. The hill itself comprises a large limestone outcrop overlying chert and ophiolite formations. It is flanked by flyschoide facies of the Rizomata Formation, consisting of micaceous silty marls, marly siltstones and frequent thin sandy layers. The other major lithology in the area adjacent to the site is the alternating conglomerates and sandstones of the Eptachorio Formation, which containing coarse and fine sandstones and micaeous marls (Savoyat and Lalechos, 1972; Ferrière, 1982, p. 348-349; Caputo, 1990, p. 59, 104-110).

The largest fabric group THFG1, which makes up 77 out of the 110 samples analysed, is characterised by a mineralogical composition that is compatible with the local geology and strongly resembles many of the geological samples collected from the vicinity of the site, especially fluvioterrestrial clayey sediments GS15 and GS21

(Figure 4D). A local source could also be argued for the raw materials of THFG3, THFG5, and THFG7. Ceramics of THFG3 are very similar to THFG1 in thin section although more micaceous and THFG5 presents striking textural and compositional resemblance to the serpentinised ophiolitic rock fragment GS4. THFG7, which contains limestone fragments, is not particularly diagnostic in terms of provenance, but could be tentatively considered as local, indicating the exploitation of a different raw material source, possibly within the broader area of the cave.

A small number of sherds (13) are most probably imported to Theopetra Cave. For example, THFG2, which is incompatible with the local geology of the region and THFG4 both have petrographic compositions that are observed throughout Thessaly and characterise Grey ware, with Platia Magoula Zarkou being the most probable production centre (Pentedeka, 2008, p. 201-203; Schneider *et al.*, 1991, p. 22, 43). The composition of THFG6, is not entirely incompatible with the local geology of the cave, but its inclusions point to a rather different sedimentary environment than the sandstones in the geological samples collected. The use of a different raw material source could be argued in this case and given that this fabric is observed only as Matt-painted ceramics, a non-local origin might be tentatively proposed.

In summary, the majority of the samples (92 out of 110) are believed to have been locally produced, as they are compatible with the geology of the broader area of the cave, and bear strong similarities to the geological samples collected. The EN material seems to be entirely locally produced, indicating the exploitation of a very limited number of clay sources. In the MN and LN, more raw material sources seem to be in use, possibly deriving from the broader vicinity of Theopetra Cave. Moreover, during these periods imported fabrics make their appearance, indicating the participation of the cave inhabitants/users in exchange networks that are active in Thessaly. In general, though, local production of pottery remains the common practice throughout the Neolithic period.

Neolithic Pottery Technology

The integrated research methodology applied in this study has facilitated the reconstruction of the manufacturing process of the Theopetra Cave ceramics and shed light on the choices made by the potters at various stages in the production sequence. The majority of clays selected for pottery production were non-calcareous and rich in silicate inclusions. Nevertheless, exploitation of a range of different raw material sources from the vicinity of the cave is indicated by the variety of local fabric compositions that are in use throughout the Neolithic period. The EN material is characterised by the use of coarse clays, with high inclusion frequency and diversity in grain size, suggesting the utilisation of unrefined, naturally occurring clay sediments. The uneven distribution of inclusions and the weak bimodality of the coarse fraction might indicate the further addition of sand as temper, whilst the high percentage of voids and clay pellets tentatively suggests lack of thorough kneading during clay paste preparation. Similar clays seem to be used for both the MN and LN material, though these may have been better processed. The smaller range in grain size implies either the intentional removal of larger inclusions through sieving or levigation or the use of naturally levigated clays (e.g. washed by rainwater). Tempering was practised for the paste preparation of THFG5, and is also sporadically discerned in a few samples from THFG1 and THFG6.

The use of the coiling as a method of vessel forming is observed throughout the Neolithic period in Theopetra, as indicated by the commonly well developed preferred orientation of elongated voids and inclusions parallel to vessel walls in all fabric groups. In one EN sample voids and inclusions remained concentrically arranged, possibly due to minimal further processing of the coil (Figure 5E). Macroscopic evidence suggests that flattened coils may have been used for some ceramics in the manner reported at the site of Sesklo (Kotsakis, 1983, p. 121-124; Pentedeka, 2002, p. 103-104).

A range of surface treatments have been identified in the Theopetra Cave ceramics including smoothing, burnishing, slip and painted decoration. The EN Monochrome ware is mostly burnished and occasionally slipped, the MN Red Monochrome and 'Scraped' wares are almost always slipped and burnished, the LN Black Burnished ware is predominantly slipped, while the LN Matt-painted ware is moderately burnished or well smoothed and then decorated with a manganese-rich paint. The clays used for most of the slips were fine-grained, probably resulting from levigation of clay with similar, non-calcareous, composition to the one used for the vessel body. Exceptions to this are a few EN samples, which have calcareous slips and one LN Black Burnished ware sample where the slip contains different inclusions to the clay body.

A general estimate of the original firing conditions of the ceramics was based on the optical activity of the groundmass and the textural condition of various minerals (Whitbread, 1995, p. 394; Zisi, 2005, p. 55-62). The majority of the samples, especially the monochrome wares, were probably fired below 850°C. Early Neolithic samples belonging to THFG7 may have been fired below 800°C, as suggested by the unaltered calcite crystals and the optical activity of the groundmass. Some samples of 'Scraped' ware, most Grey ware samples and all Matt-painted samples were optically inactive or slightly active in thin section, indicating that their original firing temperature was 850-900°C or higher.

The EN pottery seems to have been fired in a mixed/oxidised atmosphere, though there are indications for mixed/reduced conditions in some samples. With the exception of a very few samples of 'Scraped' ware, which seem to have been fired in reducing atmosphere, all MN pottery was oxidised. The LN samples were fired in reducing conditions, especially the Black Burnished and Grey wares, in order to achieve the desirable dark surface colour. The common presence of firing clouds on EN samples and the frequent variation in core colour observed in all wares of all periods, imply variable firing conditions and possibly a short duration of firing. Grey ware is an exceptional case, since for that particular ware maximum stability and uniformity in firing conditions were achieved (Schneider *et al.*, 1991, p. 49; Letsch and Noll, 1983, p. 141; Vitelli, 1994, p. 146). Although the majority of Grey ware samples were imported to Theopetra Cave, the very few that seem to have been produced locally do not deviate from this general rule. The above observations indicate that the pottery of the Theopetra Cave was fired in open fires or pits, where pots and fuel were in close contact and firing conditions were not entirely controlled.

Discussion

It has been argued that the transition to the Neolithic in Greece should be considered not as an event, but rather as a socially embedded process, taking place in social space and time (Kotsakis, 2003). This contradicts with previous simplistic models that perceive the Neolithic as being either an indigenous evolving or a migrational phenomenon. These models have viewed Thessalian pottery as a local invention or a product of Anatolian diffusion, without taking into account that material culture is produced by the agency of people in constructing identities, relations of ownership, control and power (Kotsakis, 2005, 2008). In this context, the early ceramic material from the Mesolithic to Neolithic transition at Theopetra Cave clearly has much to a lot to contribute to the discussion.

The early ceramic material from the cave demonstrates all the characteristics of a fully developed pottery technology, with all stages of manufacture from raw material selection, forming techniques, surface treatment and firing. Whilst there is absence of unambiguous evidence for experimentation, the use of coarse/unrefined clays deriving from a limited number of clay sources in the vicinity of the cave, the minimal raw material processing, the lack of control over firing conditions and temperatures, the absence of elaborate decoration and the limited repertoire in shapes, suggest that ceramic technology was more improvised in the Early Neolithic to later periods, during which time pottery production appears to be more standardised (Kyparissi-Apostolika, 2000b, p. 184). These comments also account for the few sherds unearthed from Mesolithic contexts, where the parallel presence of unbaked clay masses might also represent stages of pottery experimentation (Kyparissi-Apostolika, 2000c, p. 136-137).

In the Mesolithic, the Cave of Theopetra probably accommodated a group of foragers, moving within the wider region around the cave. In the beginning of the Neolithic this group may have interacted with other groups, possibly farmers, resulting in a conversion of social identity that was expressed through material culture (Kotsakis, 2000, p. 66-68). The introduction and development of a new technology, such as pottery manufacture, could have played a significant role in this process. Moreover, even if this technology was a borrowed, the ceramic evidence suggests that it was fully adjusted to local conditions, signaling the initiation of a detectable pottery tradition.

Through the course of time, the technological 'routes' followed by early potters at Theopetra Cave during pottery manufacture may have been gradually modified into more solid and less diverse choices. It is of interest that one fabric, THFG1, represents almost 70% of the samples studied and was in use throughout the Neolithic for the manufacture of many wares and types of vessels. The consistent exploitation of similar or the same raw material sources from the vicinity of the cave, the use of the same forming technique and the stability in the choices concerning surface treatment and firing, all point to the shaping of a local tradition from EN onwards. However this tradition should not be considered as static, since gradual modification of some stages of the manufacturing process such as raw material choice is observed, where a tendency for finer clay pastes is attested for the later phases. In other words, technological practice (Lemonnier, 1993; van der Leeuw, 1993; Sillar and Tite, 2000) is one way through which "society is made durable" (Latour, 1993, p. 379). It is through the unfolding of this tradition, that is of the tacitly and routinely performed manufacturing procedure, that technological activity becomes a means for expressing, materialising and negotiating social identities and cultural entities (Dobres, 2000, p. 136-141; Gosselain, 2000, p. 190; Dietler and Herbich, 1998, p. 244-248).

It is exactly for this reason that the study of the pottery from Theopetra Cave is so intriguing; in a cultural landscape dominated by tells, argued to be permanently inhabited and acting as markers of symbolic relations with the ancestors and of collective identity expressed through temporal and spatial continuity (Kotsakis, 1999, p. 73-74), cave habitation is usually viewed as something discontinuous and periodic. In terms of organisation of pottery production, the picture obtained from the Theopetra pottery assemblage, in particular for MN and LN, does not differ significantly from the one we have from other Neolithic settlements in Thessaly (Pentedeka, 2008, p. 209). Even if permanent habitation is not the case for Theopetra Cave, the group, or groups, that occupied the site seem to have considered its technological tradition as a rooted facet of its identity, being therefore more difficult to modify (Gosselain, 2000, p. 208-210). In any case, this rather solid potting tradition does not hinder contacts and loans between the inhabitants of Theopetra Cave and the rest of Thessaly, as expressed by imported pottery and local adoption of wares and decorative styles widespread in the whole of the Thessalian plain (Pentedeka, 2008, p. 205-210). The degree to which Theopetra Cave was involved in these intraregional contacts cannot be easily discerned by pottery analysis alone, but should be approached by a larger comparative and contextual study of all aspects of its material culture.

Acknowledgements

This project was funded through a research grant from the Institute for Aegean Prehistory (INSTAP). Sampling permits were issue by the Greek Ministry of Culture. The authors would like to thank Dr. Nina Kyparissi-Apostolika for entrusting them with the ceramic analysis of the Theopetra material, Dr. Evangelia Kiriatzi, Director of the Fitch Laboratory (British School at Athens), for supporting this project in many ways, including permission to use the Laboratory's infrastructure and Prof. Kostas Kotsakis, who prompted us to study pottery from Neolithic Thessaly with the aid of archaeological science, frequently giving shape to our ideas. Many thanks are also owed to Fotis Ifantidis for his technical expertise on setting up the images.

References

Andreou, S., Fotiadis, M., and Kotsakis, K. 2001. Review of Aegean Prehistory V: the Neolithic and Bronze Age of Northern Greece. In: Cullen, T. (Ed.), *Aegean Prehistory, A Review*. American Journal of Archaeology Supplement 1, Archaeological Institute of America, Boston: 259–327.

Bullock, P., Federoff, N., Jongerius, A., Stoops, G., and T. Tursina 1985. *Handbook for Soil Thin Section Description*. Wolverhampton: Wayne Research.

Caputo, R. 1990. *Geological and Structural Study of the Recent and Active Brittle Deformation of the Neogene-Quaternary Basins of Thessaly (Central Greece)*. Unpublished doctoral dissertation, Aristotle University of Thessaloniki.

Daszkiewicz, M. and Schneider, G. 2001. Klassifizierung von Keramik durch Nachbrennen von Scherben. *Zeitschrift für Schweizerische Archäologie und Kunstgeschichte*, 58: 25-31.

Dietler, M. and Herbich, I. 1998. *Habitus*, Techniques, Style: An Integrated Approach to the Social Understanding of Material Culture and Boundaries. In: Stark, M.T. (Ed.) *The Archaeology of Social Boundaries*. Smithsonian Institution Press, Washington and London: 232-263.

Dobres, M.-A. 2000. *Technology and Social Agency*. Blackwell, Oxford/Malden.

Facorellis, Y. and Maniatis, Y. 2000. Evidence for 50000 years of human activity in the cave of Theopetra by ^{14}C. In: Kyparissi-Apostolika, N. (Ed.) *Theopetra Cave: Twelve years of excavation and research 1987-1998. Proceedings of the International Conference, Trikala, 6-7 November 1998*. Greek Ministry of Culture, Athens: 53-68.

Facorellis, Y., Kyparissi-Apostolika, N. and Maniatis, Y. 2001. The cave of Theopetra, Kalambaka. Radiocarbon evidence for 50000 years of human presence. *Radiocarbon*, 43: 1029-1048.

Ferrière, J. 1982. *Paléogéographies et tectoniques superposes dans les Hellénides internes: les massifs de l'Orthrys et du Pelion (Grèce continentale)*. Société Géologique du Nord Publication No 8. Société Géologique du Nord, Villeneuve d'Ascq, Cedex.

Gosselain, O.P. 2000. Materializing Identities: An African Perspective. *Journal of Archaeological Method and Theory*, 7: 187-217.

Katsarou, S. 2000. Monochrome wares as indicator of the process of choice - The case of Theopetra Cave. In: Kyparissi-Apostolika, N. (Ed.) *Theopetra Cave: Twelve years of excavation and research 1987-1998. Proceedings of the International Conference, Trikala, 6-7 November 1998*. Greek Ministry of Culture, Athens: 234-262.

Kaznesi, A. in press. Approaches to the painted pottery from Theopetra Cave. In: Kyparissi-Apostolika, N. (Ed.) *Theopetra Cave: The Neolithic Period.*

Kiriatzi, E. 2000. *Keramiki Technologia kai Paragogi. I Keramiki tis Ysteris Epochis tou Chalkou apo tin Toumba Thessalonikis.* Unpublished doctoral dissertation, Aristotle University of Thessaloniki.

Kotsakis, K. 1983. *Keramiki technologia kai keramiki diaforopoiisi: provlimata tis graptis keramikis tis Mesis Neolithikis epohis tou Sesklou.* Unpublished doctoral dissertation, Aristotle University of Thessaloniki.

Kotsakis, K. 1999. What Tells Can Tell: Social Space and Settlement in the Greek Neolithic. In: Halstead, P. (Ed.) *Neolithic Society in Greece.* Sheffield Academic Press, Sheffield: 66-76.

Kotsakis, K. 2000. Mesolithic to Neolithic in Greece. Continuity, discontinuity or change of course?. *Documenta Praehistorica*, XXVII: 63-73.

Kotsakis, K. 2003. From the Neolithic side: The Mesolithic/Neolithic interface in Greece. In: Galanidou, N. and Perlès, C. (Eds.) *The Greek Mesolithic: Problems and Perspectives.* British School at Athens Studies, 10, British School at Athens: 217-221.

Kotsakis, K. 2005. Across the border: unstable dwellings and fluid landscapes in the earliest Neolithic of Greece. In: Bailey, D., Whittle A. and Cummings V. (Eds.) *(Un)settling the Neolithic.* Oxbow Books, Oxford: 8-15.

Kotsakis, K. 2008. A sea of agency: Crete in the Context of the Earliest Neolithic in Greece. In: Isaakidou, V. and Tomkins, P. (Eds.) *Escaping the Labyrinth: Cretan Neolithic in Context.* Oxbow Books, Oxford: 49-72.

Kyparissi-Apostolika, N. 1999. The Palaeolithic deposits of Theopetra Cave in Thessaly, Greece. In: Bailey, G., Adam, E., Panagopoulou, E., Perlès, C. and Zachos, K. (Eds.) *The Palaeolithic Archaeology of Greece and adjacent areas. Proceedings of the ICOPAG Conference, Ioannina 1994.* British School at Athens Studies, 3, British School at Athens: 232-239.

Kyparissi-Apostolika, N. 2000a. The excavations in Theopetra Cave 1987-1998. In: Kyparissi-Apostolika, N. (Ed.) *Theopetra Cave: Twelve years of excavation and research 1987-1998. Proceedings of the International Conference, Trikala, 6-7 November 1998.* Greek Ministry of Culture, Athens: 17-36.

Kyparissi-Apostolika, N. 2000b. The Neolithic period in Theopetra Cave. In: Kyparissi-Apostolika, N. (Ed.) *Theopetra Cave: Twelve years of excavation and research 1987-1998. Proceedings of the International Conference, Trikala, 6-7 November 1998.* Greek Ministry of Culture, Athens: 181-234.

Kyparissi-Apostolika, N. 2000c. The Mesolithic/Neolithic transition in Greece as indicated by the data in Theopetra Cave. *Documenta Praehistorica*, XXVII: 133-140.

Kyparissi-Apostolika, N. 2003. The Mesolithic in Theopetra Cave: new date on a debated period of Greek prehistory. In: Galanidou, N. and Perlès, C. (Eds.) *The Greek Mesolithic: Problems and Perspectives*. British School at Athens Studies, 10, British School at Athens: 189-198.

Kyparissi-Apostolika, N. (Ed.) in press. *Theopetra Cave: The Neolithic Period*.

Latour, B. 1993. Ethnography of a "High-tech" Case: About Aramis. In: Lemonnier, P. (Ed.) *Technological Choices: Transformations in Material Culture Since the Neolithic*. Routledge, London and New York: 372-398.

Lemonnier, P. 1993. Introduction. In: Lemonnier, P. (Ed.) *Technological Choices: Transformations in Material Culture Since the Neolithic*. Routledge, London and New York: 1-35.

Letsch, J. and Noll, W. 1983. Mineralogie und Technik der frühen Keramiken Thessaliens. *Neues Jahrbuch Mineralogische Abhandlungen*, 147: 109-146.

Manolis, S., Aiello, L., Henessy, R. and Kyparissi-Apostolika, N. 2000. The Middle Palaeolithic footprints from Theopetra Cave (Thessaly, Greece). In: Kyparissi-Apostolika, N. (Ed.) *Theopetra Cave: Twelve years of excavation and research 1987-1998. Proceedings of the International Conference, Trikala, 6-7 November 1998*. Greek Ministry of Culture, Athens: 81-86.

Pentedeka, A. 2002. *Technologia kai koinoniki tautotita: I erithri keramiki sto Neolithiko Sesklo*. Unpublished Masters dissertation, Aristotle University of Thessaloniki.

Pentedeka, A. 2008. *Diktia andallagis tis keramikis kata ti Mesi kai Neoteri Neolithiki sti Thessalia*. Unpublished doctoral dissertation, Aristotle University of Thessaloniki.

Pentedeka, A. and Dimoula, A., in press. Pots in a cave? The petrographic analysis of the ceramic material from Neolithic Theopetra. In: Kyparissi-Apostolika, N. (Ed.) *Theopetra Cave: The Neolithic Period*.

Savoyat, E. and Lalechos, N. 1972. *Geological Map of Greece, Kalambaka Sheet, scale 1:50.000*. Institute of Geology and Mineral Exploration, Athens.

Schneider, G., Knoll, H., Gallis, K. and Demoule, J.-P. 1991. Transition entre les cultures néolithiques de Sesklo et de Dimini: recherches minéralogiques, chimiques et technologiques sur les céramiques et les argiles. *Bulletin de Correspondance Hellénique*, 115: 1-64.

Sillar, B. and Tite, M.S. 2000. The Challenge of "Technological Choices" for Materials Science Approaches in Archaeology. *Archaeometry*, 42: 2-20.

van der Leeuw, S. E. 1993. Giving The Potter A Choice: Conceptual Aspects of Pottery Techniques. In: Lemonnier, P. (Editor) *Technological Choices: Transformations in Material Culture Since the Neolithic*. Routledge, London and New York: 238-288.

Vitelli, K.D. 1994. Experimental Approaches to Thessalian Neolithic Ceramics: Grey Ware and Ceramic Colour. In: Decourt, J-C., Helly, B. and Gallis, K. (Eds.) *La Thessalie: Quinze années de recherches archéologiques, 1975-1990: Bilans et perspectives. Actes du Congrès International, Lyon, 17-22 Avril 1990*. Ministry of Culture, Athens: 143-148.

Whitbread, I.K. 2001. Ceramic Petrology, Clay Geochemistry and Ceramic Production-from Technology to the Mind of the Potter. In: Brothwell, D.R and Pollard, A.M. (Eds.) *Handbook of Archaeological Sciences*. Wiley, London: 449-459.

Whitbread, I.K. 1995. *Greek Transport Amphorae. A petrological and Archaeological Study*. Fitch Laboratory Occasional Paper, 4, British School at Athens.

Whitbread, I.K. 1989. A Proposal for the Systematic Description of Thin Sections. Towards the Study of Ancient Ceramic Technology. In: Maniatis, Y. (Ed.) *Archaeometry: Proceedings of the 25th International Symposium*. Elsevier, Amsterdam: 127-138.

Whitbread, I.K. 1986. The Characterisation of Argillaceous Inclusions in Ceramic Thin Sections. *Archaeometry*, 28: 79-88.

Zisi, N. 2005. *Stadiaki thermiki katastrofi kristallikon domon kai epakolouthes antidraseis se polisistasiaka argilouha ilika. Theoritiki kai peiramatiki proseggisi*. Unpublished doctoral dissertation, Aristotle University of Thessaloniki.

FINE-GRAINED MIDDLE BRONZE AGE POLYCHROME WARE FROM CRETE: COMBINING PETROGRAPHIC & MICROSTRUCTURAL ANALYSIS

Edward W. Faber

Department of Archaeology, University of Nottingham, UK
(Edward.Faber@nottingham.ac.uk)

Peter M. Day

Department of Archaeology, University of Sheffield, UK

Vassilis Kilikoglou

Institute of Materials Science, N.C.S.R. 'Demokritos', Athens, Greece

Introduction

Petrographic and chemical compositional analyses are established techniques for addressing the production and consumption of Minoan ceramics on Crete, yet both of these approaches may have drawbacks when applied to fine-grained pottery. Due to geological similarities between the raw materials exploited across Crete neither petrography nor chemical analysis may be able to differentiate clearly between the products of different sources, and both may rely on archaeological information to clarify any distinctions. It would be preferable to use other analytical information rather than archaeological information to separate different sources, and this paper suggests a technique that may help with this is microstructural analysis of the clay micromass using scanning electron microscopy (SEM).

Several distinct regional petrographic fabrics have been identified within Minoan ceramics, whose characterisation often relies on the identification of diagnostic inclusions amongst the coarse-grained fabrics, for example the Mirabello region (Betancourt, 1984; Whitelaw *et al.*, 1997; Poursat and Knappett, 2005, p. 24-26), the South-Coast region (Whitelaw *et al.*, 1997; Poursat and Knappett, 2005, p. 21-24), the Mesara region (Wilson and Day, 1994; Shaw *et al.*, 2001, p. 115-120) and the Palaikastro region (Day, 1991, p. 142). Petrographic analysis of fine-grained fabrics may be perceived as less informative than studying coarser-grained fabrics because they do not contain as many identifiable inclusions and thus it may be more difficult to assess provenance. In the case of fine-grained Minoan pottery this can become more complicated since many are made from calcareous Neogene clays that occur throughout the island and vary little in their petrographic composition (Figure 1) (Day *et al.*, 1999, p. 1028; Hein *et al.*, 2004). Despite these drawbacks, some petrographic studies of Minoan fine-grained fabrics have been used successfully to address pottery production and consumption, especially those ceramics that have been associated with coarse-grained fabrics by other archaeological means (Day and Wilson, 1998, p. 355; Day *et al.*, 1999, 2006).

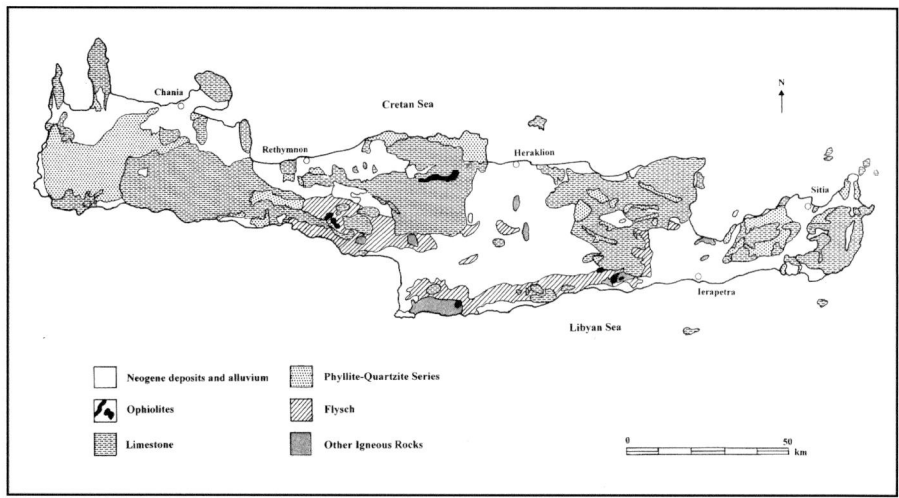

Figure 1. Geological map of Crete. Based on Creutzberg *et al.* (1977) and Thomson *et al.* (1998).

As an alternative method for investigating both fine- and coarse-grained ceramics, much work has been undertaken using chemical analysis to differentiate between production groups on Crete. This approach led to the identification of three broad chemical compositional zones on Crete, namely West Crete, Central Crete and East Crete (Jones 1986, p. 256-258). However, it can be difficult to distinguish between products of these areas since the compositions of the three control groups appear to overlap. Day *et al.* (1999) and Hein *et al.* (2004) have shown that this overlap may be due to the use of the geochemically similar but geographically different deposits of Neogene clays that occur across Crete (Figure 1). Recent work has shown that variability in chemical compositional reflects both ceramic provenance and technology, but when faced with chemical data it is hard to separate these two influences without referring to other information (Day *et al.*, 1999).

This paper addresses the use of petrographic analysis on fine-grained Middle Minoan polychrome ware from Crete and introduces the potential of SEM microstructural analysis of the fabric as an effective complementary analytical technique due to its direct compatibility with petrographic studies of the clay micromass. When interpreted together, the results of these two approaches can further our understanding of the regional differences in the production of polychrome ware on Crete and of the distribution of the polychrome ware ceramics.

Polychrome Ware on Crete

Middle Minoan polychrome ware, frequently referred to as 'Kamares Ware', is a high quality ceramic that often has complex decorative motifs and a restricted distribution, being found mainly in the palace complexes or in sites of special significance, such as the Kamares Cave (Walberg, 1983, 1987a). The polychrome decorative effect is found on a multitude of vessel shapes, including hemispherical cups, bridge-spouted jars, jugs and jars, pithoi and fruitstands (Walberg, 1983, p. 4-11, 1987a, p. 14-19; Betancourt, 1985, p. 97-100; MacGillivray, 1998, p. 65-90). There are many examples of vessels with similar decorative schemes that are thought to form ceramic sets (MacGillivray, 1987, p. 274; Van de Moortel, 1997, p. 323, 345-346; Knappett, 2005), which would have been used together for the pouring and drinking of liquids and appear to indicate the performance of ritual consumption or celebration in and around the palaces (Day and Wilson, 1998; Knappett, 2005, p. 133-166). Polychrome ware first appears in the Prepalatial period, during the MM IA ceramic phase (c. 2050-2000 BC to 1925-1900 BC) (Manning, 1995, p. 217), but the majority of examples of this ware derive from Protopalatial period deposits, spanning the MM IB-MM IIB ceramic phases (c.1925-1900 BC to c.1750-1720 BC) (Manning, 1995, p. 217).

Polychrome ware has a black-slipped surface, on top of which decoration was applied consisting of abstract and naturalistic motifs, using two or more over-slip pigments, normally in white, red, orange and occasionally purple, to create the polychrome effect. The use of two or more over-slip pigments, including at least one made from an iron-rich material, distinguishes the polychrome ware from other Middle Minoan wares, such as East Cretan White-on-Dark ware (Betancourt, 1984). Polychrome ware was made from both fine-grained and coarse-grained fabrics, with the former generally used for smaller pots and the latter for larger vessels.

There is a strong argument for identifying regionalism in the production of polychrome ware purely on stylistic grounds, particularly within the broad divisions of Central Crete, East-Central Crete (Malia-Lasithi) and East Crete. The distinctions can be less clear on a smaller scale, potentially leading to an uncertainty in the interpretation of the stylistic divisions within regions, especially within Central Crete. Some authors (e.g., Walberg, 1983, p. 90-109, 150-153, 1987a, p. 12, 1987b, p. 281-284; Betancourt, 1990, p. 27-41; Van de Moortel, 1997, p. 634-636) suggest that there is a clear division between the products of North-Central Crete and South-Central Crete, with any examples exhibiting close similarity seen as the result of periods of closer contact between the two regions. In contrast, other authors (e.g., MacGillivray, 1987; Day and Wilson, 1998) have suggested that in some cases the close similarity between the polychrome ware of North-Central Crete and South-Central Crete may have been due to Knossos and Phaistos both consuming products from a shared source.

Sample	Stylistic group	Fabric	Vitrification	Sample	Stylistic group	Fabric	Vitrification
MON9811	CC	CCf	NA	KNO95289	CC	CCf *	3
MON9817	CC	CCf	NA	MPY9905	CC / ECC	PK calc	U
MON9831	CC	CCf	NA	KNO95264	EC(AFS)	PK calc	4
MPY9920	CC	CCf	NA	MPY9902	EC	PK calc	4
MPY9921	CC / ECC	CCf	NA	PK0001	EC	PK calc	4
MPY9924	CC	CCf	NA	PK0003	EC	PK calc	4
PHA9815	CC	CCf	NA	PK0004	EC	PK calc	4
PHA9816	CC	CCf	NA	PK0007	EC	PK calc	4
PHA9819	CC	CCf	NA	PK0009	EC	PK calc	4
MAL0001	M (Tartan Style)	CCf	1	PK0011	EC	PK calc	4
MAL0002	M (Trichrome)	CCf	1	PK0012	EC	PK calc	4
MAL0003	M (Trichrome)	CCf	1	PK0013	EC	PK calc	4
MAL0004	M (Trichrome)	CCf	1	PK0014	EC	PK calc	4
MAL0005	M (Tartan Style)	CCf	1	PK0015	EC	PK calc	4
MAL0018	M	CCf	1	PK0016	EC	PK calc	4
MAL0019	M	CCf	1	PK0019	EC	PK calc	4
MAL0020	M	CCf	1	PK0021	EC	PK calc	4
MAL0021	M	CCf	1	PK0022	EC	PK calc	4
MAL0022	M	CCf	1	PK0023	EC	PK calc	4
MAL0023	M	CCf	1	PK0025	EC	PK calc	4
MAL0024	M	CCf	1	PK0026	EC	PK calc	4
MAL0027	M	CCf	1	PK0028	EC	PK calc	4
MAL0030	M	CCf	1	MPY9925	EC	PK non	4
KNO0102	CC	CCf	2	PK0002	EC	PK non	4
KNO95261	CC	CCf	3	PK0010	EC	PK non	4
KNO95285	CC	CCf	3	PK0020	EC	PK non	4
KNO95286	CC	CCf	3	PK0029	EC?	PK non	4
KNO95288	CC	CCf	3	KNO9415	CC?	vf	NA
KNO95290	CC	CCf	3	KNO9416	CC	vf	NA
KNO95294	CC	CCf	3	KNO9417	CC	vf	NA
KNO95295	CC	CCf	3	MON9816	CC	vf	NA
KNO95296	CC	CCf	3	MON9823	CC	vf	NA
KNO95297	CC	CCf	3	MPY9923	CC	vf	NA
KNO95305	CC	CCf	3	GAL0001	CC	vf	2
KNO95319	CC	CCf	3	KNO0106	CC	vf	2
KNO95333	CC	CCf	3	KNO94A37	CC	vf	2
KNO95337	CC	CCf	3	KNO94A38	CC	vf	2
KNO95343	CC	CCf	3	KNO95191	CC / ECC	vf	2
KNO95345	CC	CCf	3	KNO95263	CC	vf	2
KNO95388	CC	CCf	3	KNO95293	CC	vf	2
KNO95392	CC	CCf	3	KNO95390	CC	vf	2
KNO95393	CC	CCf	3	KNO95391	CC	vf	2
MAL0006	CC / ECC	CCf	3	KNO95394	CC	vf	2
MAL0011	CC	CCf	3	KNO95405	CC	vf	2
MPY9908	CC	CCf	3	KNO95406	CC	vf	2
MPY9912	CC / ECC	CCf	3	MAL0010	CC / ECC	vf	2
MPY9913	CC / ECC	CCf	3	MAL0012	CC / ECC	vf	2
MPY9926	CC	CCf	3	MPY9901	CC / ECC	vf	2
MPY9930	CC	CCf	3	MPY9903	CC / ECC	vf	2
PHA9801	CC	CCf	3	MPY9904	CC / ECC	vf	2
PHA9808	CC	CCf	3	MPY9906	CC	vf	2
PHA9811	CC	CCf	3	MPY9907	CC / ECC	vf	2
PHA9818	CC	CCf	3	MPY9910	CC / ECC	vf	2
PHA9827	CC	CCf	3	MPY9911	CC	vf	2
PHA9830	CC	CCf	3	MPY9914	CC / ECC	vf	2
PHA9832	CC	CCf	3	MPY9915	CC	vf	2
PHA9835	CC	CCf	3	MPY9916	CC / ECC	vf	2
PHA9840	CC	CCf	3	MPY9917	CC	vf	2
PHA9843	CC	CCf	3	MPY9918	CC / ECC	vf	2
PHA9861	CC	CCf	3	MPY9919	CC / ECC	vf	2
PK0008	CC	CCf	3	MPY9922	CC / ECC	vf	2
PK0017	CC	CCf	3	PK0030	CC	vf	2
PK0027	EC?	CCf	3	MAL0013	CC / ECC	vf *	2
MAL0007	CC	CCf *	U	MAL0015	CC / ECC	vf *	2
PHA9845	CC	CCf *	NA	MPY9909	CC / ECC	vf *	2
MAL0029	M	CCf *	1				

Table 1. Comparision between stylistic groups, petrographic fabrics and vitrification microstructural groups of the Middle Bronze Age fine-grained polychrome ware ceramics analysed in this paper. CC = Central Cretan, ECC = East Central Cretan, M = Maliote, EC = East Cretan, EC(AFS) = East Cretan Alternating Floral Style, vf = Very fine-grained fabric, vf * = coarser subset of very fine-grained fabric, CCf = Central Cretan fine-grained fabric, CCf * = coarser subset of fine-grained fabric, PK calc = Palaikastro calcareous fabric, PK non = Palaikastro non-calcareous fabric, NA = not examined.

Sampling and Analytical Methodology

It was within this context that 208 examples of Middle Minoan polychrome ware were sampled from seven sites across Central, East-Central and East Crete: Knossos, Phaistos, Malia, Palaikastro, Myrtos Pyrgos, Monastiraki and Galatas (Figure 2), to investigate their provenance and the technology used in their production. The sampling criteria were based on a study of the shapes, decoration, typology and hand specimen fabrics of polychrome ware ceramics from securely dated Middle Minoan deposits at each site rather than based on assumptions concerning local or non-local origin (Faber, 2007). The samples were analysed using an integrated approach of petrographic, chemical (neutron activation analysis) and microstructural techniques (SEM). The results discussed in this paper are based on the thin section and SEM analyses undertaken, and fuller discussions of the analytical work, including the chemical compositional results, are reported elsewhere (Faber *et al.*, 2002; Day *et al.*, 2006; Faber, 2007).

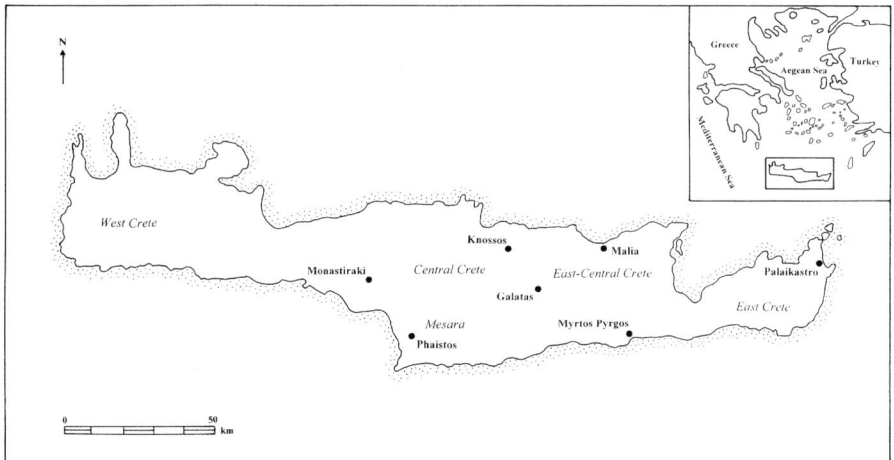

Figure 2. Location of the sites from which ceramics were sampled for this project. The approximate location of the main regional designations of Crete discussed in the text are shown in italics.

The thin-sections were classified into petrographic fabric groups and described using a modified version of the system proposed by Whitbread (1995, p. 379-388). The groups were formed and characterised according to criteria of quantity, shape, size and composition of the mineral and rock inclusions, as well as the texture, colour and optical activity of the clay micromass. The use of these criteria provides information on aspects of the provenance and technology of the pottery, reflecting both the geological environments from which the raw materials originated and the potters' choices during raw material selection and preparation as well as of the techniques used during production of the vessels.

Fresh fracture specimens of the polychrome ware samples were examined using SEM to assess the morphological features of the microstructure and the patterns of vitrification, using as a basis the work of Maniatis and Tite (1975, 1981) and Tite and Maniatis (1975a, 1975b). With this information it is possible to estimate the equivalent firing temperature of the vessel and to infer information about raw material selection and firing procedures (Tite and Maniatis, 1975b; Kilikoglou, 1994).

Fine-grained Fabrics within Middle Minoan Polychrome Ware

Although a variety of fine- and coarse-grained fabrics were identified during the petrographic analysis of this study, the majority of polychrome ware vessels were constructed using fine-grained clay pastes. Most of the samples were classified into four petrographic fabrics, with the remaining vessels assigned to fabrics that only contained a small number of samples or a single sample (Faber, 2007). In order to provide adequate data with which to address the problems of characterising fine-grained polychrome ware ceramics the following discussion will refer only to those fine-grained fabrics identified in this study as that have a sufficient number of members (Table 1).

Palaikastro fine-grained fabrics

The petrographic analysis of the samples selected in this study has revealed the presence of two related fine-grained fabrics that originate in the Palaikastro region of East Crete: a common fine-grained calcareous Palaikastro fabric and a rarer fine-grained non-calcareous Palaikastro fabric (Table 1). There is a great deal of similarity between these fabrics since their characteristic inclusions are essentially the same, but they are differentiated primarily on the proportion of calcite in the clay matrix. The fine-grained calcareous Palaikastro fabric contains small sub-angular to sub-rounded aplastic inclusions of monocrystalline and polycrystalline quartz, biotite mica, alkali feldspar and iron oxide concentrations, with sub-angular to sub-rounded rock fragments of grey phyllite, sandstone, biotite schist, and siltstone, and textural concentration features (Figure 3A-C). The grey phyllite is characteristic of the outcrops of the 'Phyllite-Quartzite Series' that occur in the vicinity of Palaikastro (Figure 1) (IGSR, 1959). The clay matrix is most notable for being optically inactive in all samples and for its mottled colouration. These ceramics are very similar to the fine-grained calcareous fabric identified in Neopalatial ceramics (Day, 1991, Palaikastro fabric 1, p. 142). The mottling of the clay matrix appears to indicate clay mixing, and may be the result of the deliberate blending of calcareous and non-calcareous clay to alter its working properties. Both these clay types are present in the probable source area in the marine calcareous deposits of the Toplou Formation and the terrigeneous-clastic non-calcareous clay deposits of the Kastri Formation (Gradstein, 1973, p. 531-539). The fine-grained, non-calcareous Palaikastro fabric contains small sub-angular to sub-rounded aplastic inclusions of monocrystalline and polycrystalline quartz, biotite mica, alkali feldspar and iron oxide concentrations, with sub-angular to sub-rounded rock fragments of the same grey phyllite, chert, sandstone, biotite schist, and siltstone, as well as textural concentration features (Figure 3D). The matrix is also notable for being optically inactive in all samples. In this case, the fabric would appear to derive from the deposits of the fluvial terrigeneous-clastic non-calcareous clay in the area surrounding Palaikastro (Gradstein, 1973, p. 531-534, 564) and the outcrops of the Phyllite-Quartzite Series that occur in the vicinity (IGSR, 1959).

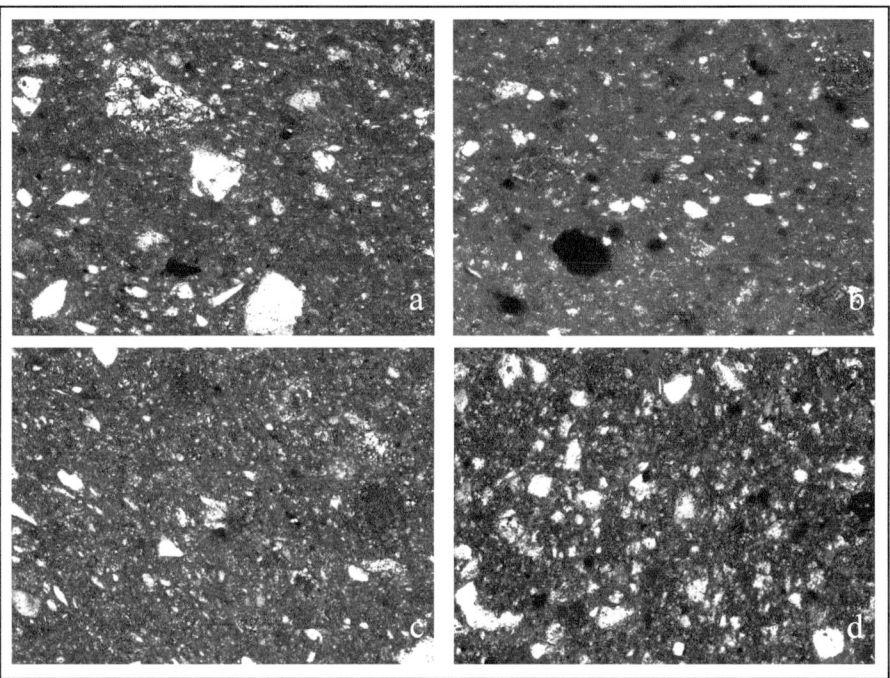

Figure 3. Photomicrographs of Middle Bronze Age fine-grained polychrome ware ceramics analysed in this paper. a-c) Palaikastro fine-grained calcareous fabric samples MPY9905, KNO95264 and PK0003, d) Palaikastro fine-grained non-calcareous fabric sample PK0029. All images taken in crossed polars. Image width = 0.7 mm.

Unsourced very fine-grained calcareous fabric

One of the largest fine-grained fabrics identified during the petrographic examination is defined by its distinctive very fine-grained texture and general lack of characteristic inclusions, which may reflect the degree of refinement undertaken during the preparation of the clay paste. This fabric is made from a calcareous clay which exhibits frequent large textural concentration features (Figure 4A), but which contains very few aplastic inclusions that, when present, consist of small sub-rounded to sub-angular monocrystalline or polycrystalline quartz and biotite mica (Figure 4B-D). The frequent textural concentration features within the calcareous fabric are probably the result of heterogeneity in this source of Neogene clay (Day, 1995, p. 159-161; Day *et al.*, 2006, p. 41). Due to the general lack of distinctive aplastic inclusions it is not possible to assign a provenance to most members of this fabric on purely petrographic grounds. Since the ceramics in this group occur at sites that are relatively far apart, such as Palaikastro and Monastiraki (Figure 2), the question is whether their apparent resemblance is due to a shared geographical provenance, rather than the use of similar raw materials in different locations.

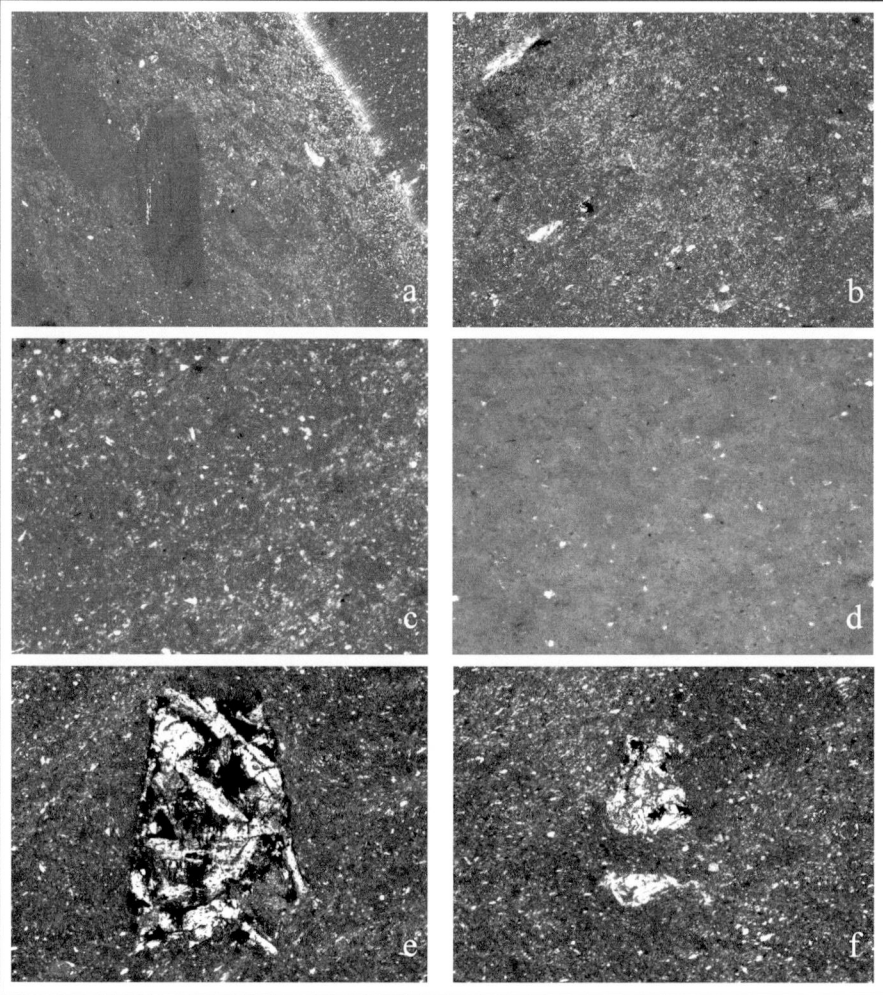

Figure 4. Photomicrographs of Middle Bronze Age fine-grained polychrome ware ceramics analysed in this paper. a) Unsourced very fine-grained calcareous fabric sample KNO95405 with large textural concentration features that are characteristic of this fabric, b-d) Unsourced very fine-grained calcareous fabric samples GAL0001, MAL0010 and PK0030, e) Unsourced very fine-grained calcareous fabric sample MPY9909 with inclusion of altered basalt, f) Unsourced very fine-grained calcareous fabric sample MPY9909 with inclusion of serpentinite. All images taken in crossed polars. Image width = 0.7 mm, except a = 2.9 mm.

A small subgroup of this fabric contains rare, sub-rounded to sub-angular inclusions of basic igneous rock and serpentinite (Figure 4E,F). The combination of these inclusions suggests that they could derive from the ophiolite series and flysch mélange, and suggests a possible link with the Early Minoan South-Coast Fabrics, which contain a variety of rounded sand or angular rock fragments largely of basaltic and serpentiferous rocks (Whitelaw et al., 1997; Poursat and Knappett, 2005, p. 21-24). However, since the ophiolite series and flysch mélange occur throughout Central and East-Central Crete, these inclusions are only indicative of a source within Central or East-Central Crete (Figure 1).

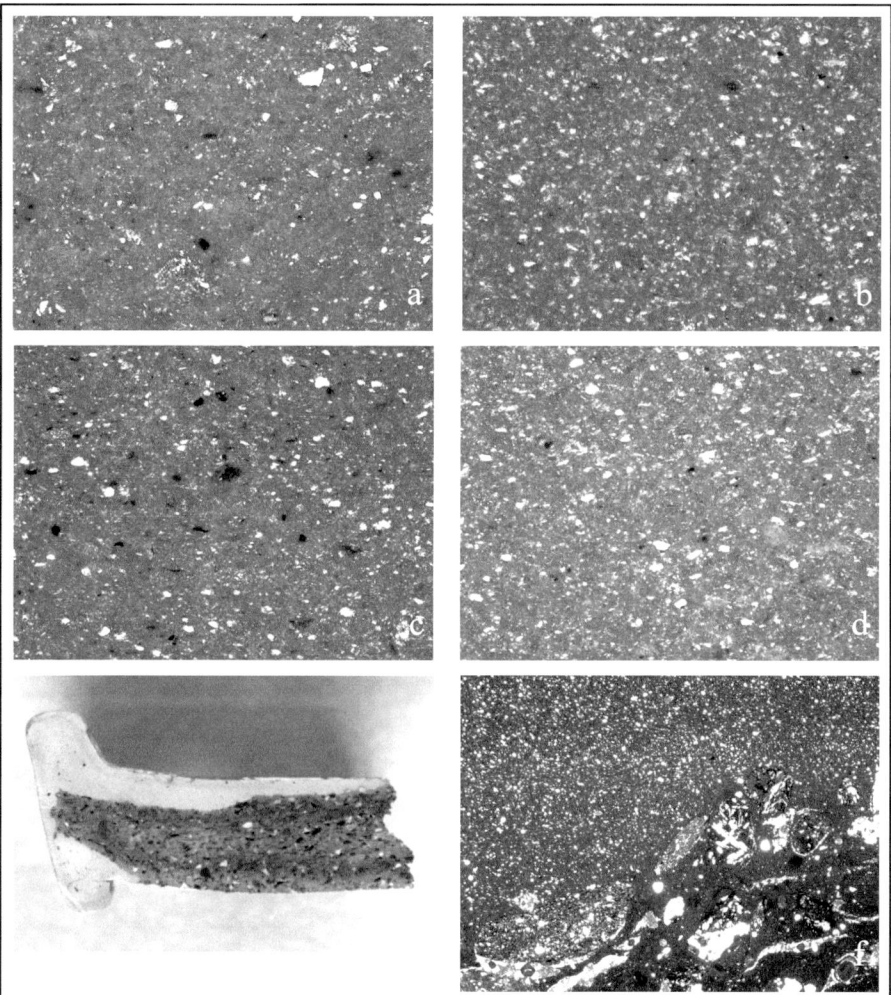

Figure 5. Photomicrographs of Middle Bronze Age fine-grained polychrome ware ceramics analysed in this paper. a-d) Central Cretan fine-grained calcareous fabric samples PHA9840, MAL0024, MPY9924 and MON9811, e) Cross-section of Central Cretan fine-grained calcareous fabric sample PHA9815 illustrating the layering technique, f) Sample PHA9815 showing the Central Cretan fine-grained calcareous fabric and a Mesara coarse-grained calcareous fabric. Samples in a, c, d and f are typologically Central Cretan, while the sample in b is typologically Maliote. All images taken in crossed polars, except e. Image width = 0.7 mm, except e = 36 mm, f = 2.9 mm.

Central Cretan fine-grained calcareous fabric

The Central Cretan fine-grained calcareous fabric is the most common fine-grained fabric used for the production of Middle Minoan polychrome pottery analysed in this study (Table 1). It is composed of a calcareous clay that contains some small aplastic inclusions of quartz, feldspar, biotite mica, as well as rare small igneous and metamorphic rock fragments, plus textural concentration features (Figure 5A-D). Although the samples analysed form a coherent petrographic fabric it is not distinctive enough to make secure suggestions concerning provenance as the clay matrix derives from Neogene clay that occurs across the island and the fine inclusions are undiagnostic (Figure 1) (Hein et al., 2004).

The ceramics made from the Central Cretan fine-grained calcareous fabric include samples that are typologically distinct (Table 1) and therefore the fabric may not represent the products of a single production group. Within this fabric the typologically Maliote ceramics from Quartier Mu, including those described as Tartan Style, appear petrographically similar to the typologically Central Cretan ceramics found at sites such as Phaistos, Knossos, Myrtos Pyrgos and Monastiraki (Figure 5A-D). This petrographic resemblance appears to result from the use of similar Neogene clays that exhibit a great deal of homogeneity across the island (Day et al., 1999; Hein et al., 2004). It is possible to assign a definite Mesara provenance to some of these samples as there are several vessels that have been made using a layering technique, which combines a coarse-grained calcareous fabric and the Central Cretan fine-grained calcareous fabric (Figure 5E,F) (Day et al., 2006, p. 46-48; Faber, 2007). Although some of the other ceramics made from the Central Cretan fine-grained calcareous fabric appear similar to Early Minoan fine-grained calcareous fabric groups 7 and 8 described from the Mesara region of Central Crete by Wilson and Day (1994, p. 61-62), it is not possible to differentiate the typologically Maliote and typologically Central-Cretan samples purely on petrographic criteria, other than those made from layered fabrics.

Microstructural Analysis

Although some of the fine-grained fabrics can be clearly characterised by petrography, others have proved more difficult. Ceramic petrography in isolation has been shown as unable to distinguish between samples of the different typologies made from the Central Cretan fine-grained calcareous fabric, except in the exceptional and unusual circumstances of the layered fabrics. Thus other information is required and evidence is presented here to suggest that microstructural analysis of the clay micromass may aid the characterisation of fine-grained fabrics used for making polychrome ware.

Research on the kinetics of clay vitrification during firing (Tite and Maniatis, 1975a, 1975b; Maniatis and Tite, 1975, 1981) has demonstrated that predictable morphological changes will occur in the level of vitrification observed in clay exposed to a specific combination of heating rate, firing duration and maximum temperature. These changes are affected by the proportion of 'active' calcium contained within the clay paste, and thus calcareous clays (>5% Ca content) will behave differently from low or non-calcareous clays (<5% Ca content). The most distinctive microstructure is that of calcareous clay fired between 850-1050°C, where the level of vitrification remains essentially unchanged across the temperature range and exhibits a stable network of smooth filaments forming a cellular structure (Tite and Maniatis 1975a: 122).

Examination of the vitrification microstructures exhibited by Middle Minoan polychrome ware in this study revealed four regular patterns of microstructural development across the different stages of vitrification (Table 2), each possibly representing the use of a different clay paste. Not all of the polychrome ware ceramics exhibit one of these patterns and therefore there may be other patterns that have not as yet been identified (Table 1).

The first microstructural pattern is typified by the presence of very fine filaments of glass that form a network, which appears to be quite disjointed and does not form a smooth, cellular microstructure (Figure 6A,B). Refiring experiments on ceramics from this group show that the vitrification microstructure continues to exhibit the disjointed filaments even when fired to 1100°C, although they have formed slightly thicker filaments than were apparent at lower temperatures (Table 2). The microstructure of these samples may suggest that the clay paste contained a significant proportion of well-distributed very fine-grained calcite, the dissociation of which has caused the disruption observed in the glass filaments.

The second microstructural pattern consists of a sequence of samples that exhibit fine filaments of glass with a more uniform size range (Table 2), which form an almost perfect example of the smooth, cellular structure that develops in calcareous ceramics fired at between 850-1050°C (Figure 6C,D). At temperatures above 1050°C the filaments coalesce and eventually form a continuous vitrified surface at around 1080°C. The presence of fine filaments in samples fired between 850-1050°C suggests that the calcite in this clay paste was fine-grained and well distributed throughout, but neither as fine-grained nor in as large a quantity as in the first microstructural pattern described.

The third microstructural pattern consists of a sequence of calcareous ceramics displaying a smooth, cellular structure of fine glass filaments but with evidence of occasional disjointed filaments (Figure 6E,F). The prevalence of disjointed filaments separates the samples of this vitrification pattern from those of the second vitrification pattern. The samples fired at temperatures above 1050°C exhibit the characteristic thicker filaments that are formed at higher temperatures and eventually form a continuous vitrified surface above 1080°C. Again, the fine filaments are probably the result of well-distributed fine-grained calcite in the clay paste, but the filaments are thicker than in the first two patterns described.

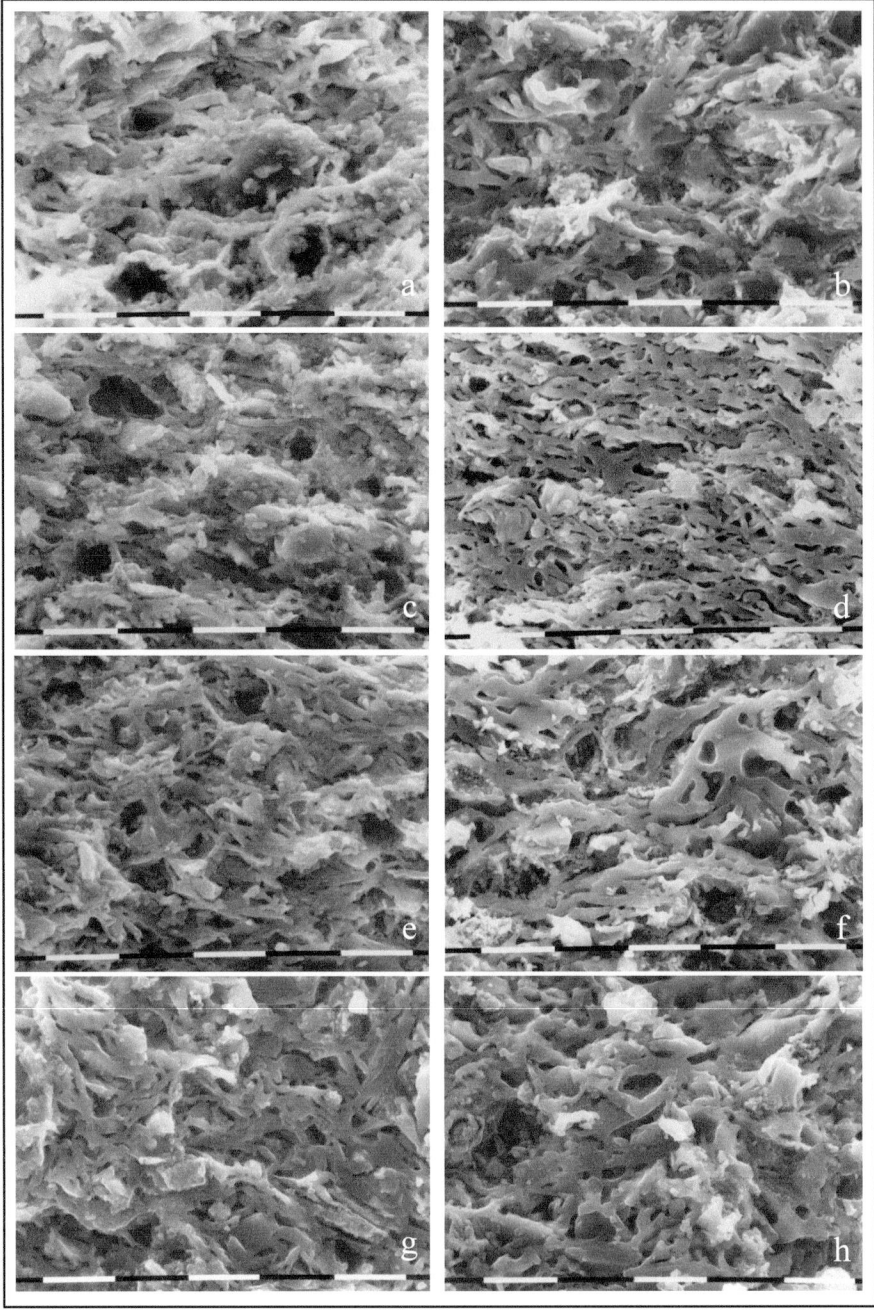

Figure 6: Scanning electron micrographs of vitrification microstructure patterns identified in Middle Bronze Age fine-grained polychrome ware ceramics analysed in this paper, with equivalent firing temperature estimates. a,b) Samples MAL0019 (850-950°C) and MAL0021 (>1080°C) with first pattern, c,d) Samples MPY9919 (800-950°C) and KNO95293 (1000-1080°C) with second pattern, e,f) Samples MPY9912 (800-950°C) and KNO95295 (1050-1080°C) with third pattern, g,h) Samples PK0019 (950-1050°C) and KNO95264 (1050-1080°C) with fourth pattern. Scale bar units = 10 μm.

Vitrification pattern	Calcareous content	Thickness of vitrified filaments	
		when closest to a smooth cellular microstructure	when forming thicker structures
1	moderate to high	0.7-1.9μm (850-950°C)	1-4.3μm (>1080°C)
2	low to moderate	1-2μm (800-950°C)	1-5.1μm (1000-1080°C)
3	low to moderate	0.7-3.1μm (800-950°C)	1.7-5.9μm (1050-1080°C)
4	non to moderate	0.9-4.3μm (900-1050°C)	1.4-6.8μm (>1050°C)

Table 2. Comparison between the thickness of the vitrified clay filaments for each of the four vitrification microstrutural patterns identified in the Middle Bronze Age fine-grained polychrome ware ceramics analysed in this paper. Equivalent firing temperature estimates for each of the stages are given.

In contrast to the other vitrification patterns identified in this study, the fourth type displays a significantly different microstructure. The ceramics in this group exhibit thicker glass filaments, which include globular areas up to twice the thickness of the main part of the filament and more rounded ends to the filaments in all vitrification stages of the sequence (Figure 6G,H). These globular areas and rounded ends continue to be present in samples even when the filaments have begun to coalesce into thicker structures. The vitrification structures demonstrated by the calcareous ceramics of this group may suggest that the calcite grains in the clay paste were coarser and less well distributed than in the other calcareous pastes observed. Since there are non-calcareous samples that also exhibit this pattern, it is possible to suggest that the various patterns described here reflect differences between the clays other than just the calcareous content.

Discussion

A comparison between the petrographic fabrics, typology and microstructural patterns exhibited by the polychrome ware fine-grained ceramics analysed in this study (Table 1) demonstrates that in most cases the microstructural patterns observed in the vitrification microstructure of the clay pastes correspond closely to the petrographic fabric classification. The microstructural patterns appear to reflect the different regional production groups of polychrome ware. Both the fine-grained calcareous Palaikastro fabric and the fine-grained non-calcareous Palaikastro fabric exhibit the fourth type of vitrification microstructure. The unsourced very fine-grained calcareous fabric exhibits the second vitrification pattern, which helps to confirm the shared geographical origin of these samples, but does not resolve its source. Finally, the Central Cretan fine-grained calcareous fabric exhibits both the first and the third vitrification microstructure patterns, which correspond to samples that originate from Malia and those that originate from the Mesara respectively. This distinction could not be made on petrographic evidence alone. Thus, in comparison to the petrographic analyses, the microstructural patterns observed in the vitrification microstructure of the body clay may provide a clearer means by which to differentiate between the ceramics of different production groups that exploited Neogene clays on Crete. In particular, the vitrification microstructure helps to distinguish between samples of polychrome ware originating from Central and East-Central Crete, which have been shown to be difficult to separate chemically (Jones, 1986, p. 256-258).

It is remarkable that such regular microstructural patterns have developed in several ranges of ceramics. Each microstructural pattern contains ceramics made during the different stages of the Middle Minoan period, which suggests that specific raw materials and firing techniques were chosen and used over several generations. A similar (or indeed greater) continuity has previously been demonstrated for ceramic traditions in the Mesara (Day et al., 2006). In part, the continuity in production techniques may reflect the necessary technical choices for producing the characteristic polychrome decoration, which requires specific firing conditions to form the combination of a vitrified, reduced slip and non-vitrified, oxidised pigments (Noll et al., 1971, 1975; Noll 1982; Faber et al., 2002; Faber 2007). Such continuity in behaviour might support the wide use of calcareous clays since the broad temperature range (850-1050°C) at which the stable smooth, cellular network of filaments developed enables the reliable repetition of one aspect of the firing process.

Conclusions

The evidence from the different petrographic fabrics identified in the polychrome ware samples in this study highlights the variation in the level of information that may be recovered from fine-grained fabrics. There is little question that the Palaikastro fabrics provide clear evidence for a positive provenance determination, due to the distinctive small grey phyllite inclusions and the characteristics of the clay matrix, but in contrast the provenance of other fine fabrics is less clear. Even when there may be difficulties in attributing provenance, the petrographic analysis of the fine-grained polychrome ware ceramics can reveal much about the technological practices used. The consistency of these fabrics over such a long time is suggestive of traditions of production that reinforce the deliberate selection and manipulation, including refinement, of specific raw materials by the potters. There may also be evidence for clay mixing within the patchiness of the clay micromass of the Palaikastro calcareous fabric that might indicate an incomplete mixing of the raw materials. The Central Cretan fabric provides an example in which typological evidence suggests that it represents production in at least two places, but this cannot be detected petrographically. In this instance, ceramic petrography could not distinguish between the centres of production as similar raw materials were used at both production centres. This strongly suggests that petrographic analysis of fine-grained ceramics should be supported by other analytical techniques or additional archaeological information.

This study indicates that microstructural analysis of the clay, including the morphology and patterns of vitrification, can complement petrography in differentiating between samples from multiple sources. Using this approach it was possible to differentiate between the Palaikastro fabrics, the unsourced very fine-grained calcareous fabric and the Central Cretan fine-grained calcareous fabric, thereby reinforcing the petrographic results while at the same time demonstrating the efficacy of this technique. In the case of the Central Cretan fine-grained calcareous fabric it has also been able to differentiate between samples from different typological groups that were previously were classified together by petrography. Thus microstructural analysis may help to distinguish between samples made from different sources of raw materials including, but not restricted to, samples made from geologically similar but geographically different raw materials. In the case of the fine-grained Middle Minoan polychrome ware, the microstructural patterns observed in the various stages of the vitrification microstructure of the body would appear to be the clearest technological indicator for identifying the products from different locations.

Acknowledgements

We are indebted to the following for permission to study and sample material referred to in this paper: G. Cadogan, F. Carinici, M.S.F. Hood, A. Kanta, V. La Rosa, C.F. Macdonald, J.A. MacGillivray, N. Momigliano, J.-C. Poursat, G. Rethemiotakis, L.H. Sackett, the Council of the British School at Athens, the 23rd Ephorate of Prehistoric and Classical Antiquities and the Conservation Directorate of the Greek Ministry of Culture, the 24th Ephorate of Prehistoric and Classical Antiquities and the Conservation Directorate of the Greek Ministry of Culture, and the 25th Ephorate of Prehistoric and Classical Antiquities and the Conservation Directorate of the Greek Ministry of Culture. The research presented here forms part of a doctoral thesis completed at the University of Sheffield that was supported by the EC TMR Research Network 'GEOPRO' (Integrating Geochemical and Mineralogical Techniques: A new approach to raw materials and archaeological ceramic provenance) (Grant number ERB FMRX CT 98-0165), a Wingate Scholarship from the Harold Hyam Wingate Foundation; and an educational grant from the Charles Henry Foyle Trust. The geological map in Figure 1 and the location map in Figure 2 were produced by David Taylor.

References

Betancourt, P.P. (Ed.) 1984. *East Cretan White-on-Dark Ware*. The University Museum, Pennsylvania.

Betancourt, P.P. 1985. *The History of Minoan Pottery*. Princeton University Press, Princeton.

Betancourt, P.P. 1990. *Kommos II: The Final Neolithic through Middle Minoan III Pottery*. Princeton University Press, Princeton.

Creutzberg, N., Drooger, C.W. and Meulenkamp, J.E. 1977. *General Geological Map of Greece. Crete Island, 1:200,000 map*. Institute of Geology and Mining Exploration, Athens.

Day, P.M. 1991. *A Petrographic Approach to the Study of Pottery in Neopalatial East of Crete*. Unpublished doctoral thesis, Cambridge University.

Day, P.M. 1995. Pottery production and consumption in the Sitia Bay area during the New Palace period. In: Tsipopoulou, M. and Vagnetti, L. (Eds.) *Achladia: Scavi e ricerche della Missione Greco-Italiana in Creta Orientale (1991-1993)*. Gruppo Editoriale Internazionale, Roma: 149-175.

Day, P.M. and Wilson, D.E. 1998. Consuming power: Kamares Ware in Protopalatial Crete. *Antiquity*, 72: 350-358.

Day, P.M., Relaki, M. and Faber, E.W. 2006. Pottery making and social reproduction in Bronze Age Mesara, Crete. In: Wiener, M. H., Warner, J. L., Polonsky, J. and Hayes, E. E. (Eds.) *Pottery and Society: The Impact of Recent Studies in Minoan Pottery. Gold Medal Colloquium in Honor of Philip P. Betancourt. 104th Annual Meeting of the Archaeological Institute of America, New Orleans, Louisiana, 5 January 2003*. Archaeological Institute of America, Boston: 22-72.

Day, P.M., Kiriatzi, E., Tsolakidou, A. and Kilikoglou, V. 1999. Group therapy in Crete: a comparison between the analyses by NAA and thin-section petrography of Early Minoan Pottery. *Journal of Archaeological Science*, 26: 1025-1036.

Faber, E.W. 2007. *Middle Minoan Polychrome Pottery: an Integrated Microstructural, Geochemical and Mineralogical Investigation of its Production Technology and Provenance*. Unpublished doctoral thesis, University of Sheffield.

Faber, E.W., Kilikoglou, V., Day, P.M. and Wilson, D.E. 2002. A technological study of Middle Minoan polychrome pottery from Knossos, Crete. In: Kilikoglou, V., Hein, A. and Maniatis, Y. (Eds.) *Modern Trends in Scientific Studies on Ancient Ceramics. Papers Presented at the 5th European Meeting on Ancient Ceramics. Athens 1999*. BAR International Series 1011. Archaeopress, Oxford: 129-141.

Gradstein, F.M. 1973. The Neogene and Quaternary deposits in the Sitia district of Eastern Crete. *Annales Géologiques des Pays Helléniques*, 25: 527-572.

Hein, A., Day, P.M., Quinn, P.S. and Kilikoglou, V. 2004. Geochemical diversity of Neogene clay deposits in Crete and its implications for Provenance studies of Minoan pottery. *Archaeometry*, 46: 357-384.

IGSR. 1959. *Geological Map of Greece: Siteia. 1:50,000*. Institute for Geology and Subsurface Research, Athens.

Jones, R.E. 1986. *Greek and Cypriot Pottery: a Review of Scientific Studies*. Fitch Laboratory Occasional Paper 1. The British School at Athens, Athens.

Kilikoglou, V. 1994. Scanning Electron Microscopy. In: Wilson, D.E. and Day, P.M. *Ceramic regionalism in prepalatial Crete: the Mesara imports in EM I to EM IIA Knossos*. Annual of the British School at Athens, 89: 70-75.

Knappett, C. 2005. *Thinking Through Material Culture: an interdisciplinary perspective*. University of Pennsylvania Press: Philadelphia.

MacGillivray, J.A. 1987. Pottery workshops and the Old Palaces in Crete. In: Hägg, R. and Marinatos, N. (Eds.) *The Function of the Minoan Palaces*. Skrifter Utgivna av Svenska Institutet i Athen, Stockholm: 273-279.

MacGillivray, J.A. 1998. *Knossos: Pottery Groups of the Old Palace Period*. British School at Athens Studies 5. British School at Athens, London.

Maniatis, Y. and Tite, M.S. 1975. A scanning electron microscope examination of the bloating of fired clays. *Transactions and Journal of the British Ceramic Society*, 74: 229-232.

Maniatis, Y. and Tite, M.S. 1981. Technological examination of Neolithic-Bronze Age pottery from central and southeast Europe and from the Near East. *Journal of Archaeological Science,*, 8: 59-76.

Manning, S.W. 1995. *The Absolute Chronology of the Aegean Bronze Age: Archaeology, Radiocarbon and History. Monographs in Mediterranean Archaeology 1*. Sheffield Academic Press, Sheffield.

Noll, W. 1982. Mineralogie und technik der keramiken Altkretas. *Neues Jahrbuch fur Mineralogie Abhandlungen*, 142: 150-199.

Noll, W., Holm, R. and Born, L. 1971. Chemie und technik altkretischer vasenmalerei vom Kamares-Typ. *Naturwissenschaften*, 58: 615-618.

Noll, W., Holm, R. and Born, L. 1975. Painting of ancient ceramics. *Angewandte Chemie International Edition*, 14: 602-613.

Poursat, J.-C. and Knappett, C. 2005. *La Poterie du Minoen Moyen II: Production et Utilisation. Études Crétoises 33*. École Française d'Athènes, Paris.

Shaw, J.W., Van der Moortel, A., Day, P.M. and Kilikoglou, V. (Eds.) 2001. *A LM IA Ceramic Kiln in South-Central Crete: Function and Pottery Production. Hesperia Supplement 30*. American School of Classical Studies at Athens: 111-155.

Thomson, S.N., Stöckhert, B. and Brix, M.R., 1998. Thermochronology of the high-pressure metamorphic rocks of Crete, Greece: Implications for the speed of tectonic processes. *Geology*, 26: 259-262.

Tite, M.S. and Maniatis, Y. 1975a. Examination of ancient pottery using the scanning electron microscope. *Nature,* 257: 122-123.

Tite, M.S. and Maniatis, Y. 1975b. Scanning electron microscopy of fired calcareous clays. *Transactions and Journal of the British Ceramic Society,* 74: 19-22.

Van de Moortel, A. 1997. *The Transition from the Protopalatial to the Neopalatial Society in South-Central Crete: a Ceramic Perspective.* Unpublished doctoral thesis, Bryn Mawr College.

Walberg, G. 1983. *Provincial Middle Minoan Pottery.* Verlag Philipp von Zabern, Mainz am Rhein.

Walberg, G. 1987a. *Kamares: A Study of the Character of Palatial Middle Minoan Pottery. Studies in Mediterranean Archaeology and Literature Pocket Book. 49.* Paul Åströms Förlag, Göteborg.

Walberg, G. 1987b. Palatial and provincial workshops in the Middle Minoan period. In: Hägg, R. and Marinatos, N. (Eds.) *The Function of the Minoan Palaces.* Skrifter Utgivna av Svenska Institutet i Athen, Stockholm: 281-284.

Whitbread, I. K. 1995. *Greek Transport Amphorae: a Petrological and Archaeological Study.* Fitch Laboratory Occasional Paper 4, Athens.

Whitelaw, T., Day, P. M., Kiriatzi, E., Kilikoglou, V. and Wilson, D. E. 1997. Ceramic Traditions at EMIIB Myrtos, Fournou Korifi. In: Laffineur, R. and Betancourt, P.P. (Eds.) *TEXNH: Craftsmen, Craftswomen and Craftsmanship in the Aegean Bronze Age. Proceedings of the 6th International Aegean Conference. Philadelphia , Temple University, 18-21 April 1996. Aegaeum 16.* Université de Liège, Liège: 265-74.

Wilson, D.E. and Day, P.M. 1994. Ceramic regionalism in Prepalatial Central Crete: the Mesara imports at EM I to EM IIA Knossos. *Annual of the British School at Athens,* 89: 1-87.

POTTERY TECHNOLOGY & REGIONAL EXCHANGE IN EARLY IRON AGE CRETE

Marie-Claude Boileau

Fitch Laboratory, British School at Athens, Greece
(fitchlab@bsa.ac.uk)

Anna Lucia D'Agata

Istituto di Studi sulle Civiltà dell'Egeo e del Vicino Oriente, CNR, Roma, Italy

James Whitley

School of History and Archaeology, Cardiff University, UK

Introduction

Interest in the Early Iron Age of Crete is rapidly expanding and focuses upon the social and political developments that took place between the end of the Bronze Age and 'the emergence of the polis' from the 8th century BC onwards (D'Agata, 1999, 2001a, 2001b; Deger-Jalkotzy and Lemos, 2006; Nowicki, 2000; Whitley, 1991, 1997, 2001, 2004). Early Iron Age ceramic scholarship has been dominated by the study of painted fine wares, particularly those that have turned up in burials. Scientific analyses, in the form of intensive application of geochemical and petrographic analyses, have by contrast been directed towards prehistory, leaving the early periods of the 1st millennium BC somewhat neglected. As such, previous analytical work on central Cretan Early Iron Age pottery has been minimal and consists mainly of chemical analyses (Jones, 1986; Liddy, 1996; Tomlinson and Kilikoglou, 1998), and very rare petrographic studies, such as that on fine-grained fabrics from Eleutherna (Nodarou, 2008). This paper presents the first petrographic results of a project investigating one of the least known periods in the history of Crete, the Early Iron Age. These results have been obtained through the combination of ceramic petrography, neutron activation analysis and detailed typo-chronological study.

One aim of our project is to transfer the rich analytical knowledge of Cretan Bronze Age pottery to the subsequent periods. To this end, a large-scale analytical study of coarse wares from central Crete was designed to investigate pottery technology and regional exchange networks. Comparison between the two different communities of Knossos and Sybrita allowed us to shed some light on the extent to which there was basic continuity in the production and consumption of coarse and cooking wares during this period and the degree to which production was geared to local consumption. Moreover, the degree to which Knossos or Sybrita was importing coarse wares either from within Crete or from elsewhere in the Aegean and the Mediterranean was assessed. The project was successful in characterising the raw materials used in the production of coarse utilitarian pottery at both sites and confirmed a local provenance for the majority of the stylistically assigned local wares. It also was possible to assess the amount of non-local coarse wares and identify their origin of manufacture, as well as to investigate clay paste technology, highlighting diachronic and synchronic changes in pottery production and consumption from the 12th to the 7th centuries BC.

Materials and Methods

The regional and chronological scope of the study called for a comparative approach based on ceramic assemblages showing uninterrupted occupation during the Early Iron Age. The material selected for this analytical study comes from two sites located in north-central and west-central Crete: Knossos, a major centre with continuity of occupation, and Thronos Kephala (Sybrita), a settlement established in Late Minoan III C times (Figure 1). Excavations conducted at Knossos by the British School at Athens have brought to light Early Iron Age deposits belonging to both funerary and domestic contexts. While our knowledge of the early Greek town is very limited mainly because it was severely disturbed by later occupation, excavations at various locations have produced at least one stratified deposit corresponding to each of the Early Iron Age periods (Coldstream, 1972, 2000, 2001). Complete vessels from the tombs of the North Cemetery provide a ceramic repertoire quite different in terms of shape, style and fabric from the Knossos town pottery which is clearly of domestic character, mainly used for cooking and storing activities (Coldstream, 2001, p. 21). Published deposits from different areas were selected for sampling, including the Southwest Houses (Coldstream and Macdonald, 1997), the Geometric Well (Coldstream, 2000, p. 279-284), Little Palace North (Hatzaki et al., 2008), Villa Dionysus (Coldstream and Hatzaki, 2003), and tombs from the North Cemetery (Coldstream and Catling, 1996; Coldstream, 2001). Local coarse and cooking pots were chosen, as well as semi-fine wares assigned stylistically to a local production. These samples include all morphological classes and date from the Sub-Minoan to the Orientalizing periods as established by Brock (1957).

In west-central Crete, Greek-Italian excavations at the site of Thronos Kephala, generally identified with the Classical polis of Sybrita, have uncovered a settlement, which was continuously occupied from the 12th to the 7th centuries BC (D'Agata 1999, 2001a, 2001b, 2003, 2007, in press). The site is located on the summit of Kephala hill on the southwest slopes of Psiloritis massif. Thronos Kephala was founded in Late Minoan III C Early, just after the collapse of the Late Bronze Age political entities, and suffered a severe destruction in the course of the 7th century BC. The excavated area consists of three sectors: the north and the south plateau, and a central area close to the hilltop where more than 40 pits containing ceramic material of domestic nature, including fine, coarse and cooking wares have been unearthed. The remains of the settlement include, on the north plateau, Buildings 1, 2, 3, and the large Building B1; on the south plateau, Building A1. The sherds selected for analysis come from these ritual and domestic contexts. They include all morphological classes, and date from the Late Minoan III C to the Early Orientalizing periods. Coarse to semi-fine grained wares considered of local manufacture based on style were carefully selected.

Based on the coarse-grained nature of the majority of the sampled ceramics, thin section petrography was chosen as the main analytical technique. Petrographic analysis of 414 thin sections was carried out according to the methodology described by Whitbread (1986, 1989, 1995). Samples were grouped into fabrics based on the mineralogy of the non-plastic inclusions, grain-size distribution, colour and optical activity of the groundmass. A second phase of the project on the geochemistry of 12 clays and 203 samples of semi-coarse to semi-fine wares by neutron activation analysis is currently underway.

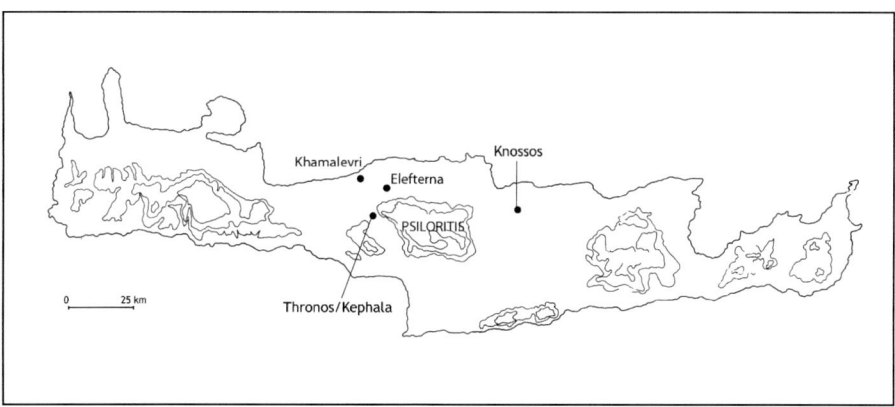

Figure 1. Map of Crete with the Early Iron Age sites analysed in this study.

With regards to provenance, 'local' attribution of fabrics groups was based on available references on the geology of central Crete and on published petrographic data. The geology around Knossos is essentially characterised by Neogene limestones, marls and clays, while metamorphic outcrops of the phyllite-quartzite series are found at the southern and western edges of the Heraklio basin (IGME, 1996). On the basis of published petrographic studies of Knossian ceramics, it is possible to identify fabrics whose origin of production lies in north-central Crete, in and around the Heraklio basin (Day, 1988, 1995, 1997; Day and Wilson, 1998; Day et al., 1999; Jones, 1986; Quinn and Day, 2007; Riley, 1983; Tomkins and Day, 2001; Tomkins et al., 2004; Wilson and Day, 1994, 1999). It is however still problematic to locate, for example, the clay deposits which yield the low-grade metamorphic inclusions found in the well documented fabrics at Knossos (Wilson and Day, 1999, p. 38). Such metamorphic outcrops do occur on the western and southern edges of the Heraklio basin, but not in the immediate vicinity of the site. The term 'broadly local' has been used as the non-plastic inclusions are consistent with the geology of north-central Crete and similar fabrics have been found consistently in the Knossos assemblage since the Early Neolithic onwards, but corresponding geological sources are located at some distance from the site (IGME, 1996).

Sybrita is located in an area mainly characterised by Neogene continental deposits of conglomerates, sandstones, clays and outcrops of the Phyllite-Quartzite series with shale, phyllites, quartz-phyllites, and quartzite immediately to the north of the site. Alluvial deposits of calcareous and phyllitic composition are also found just south of the site (IGME 1985, 1991). Geological samples collected in the main drainages in west-central Crete, such as the Potamos, Amari and Aghios Vasileios valleys, has provided additional information on regional and local raw material sources.

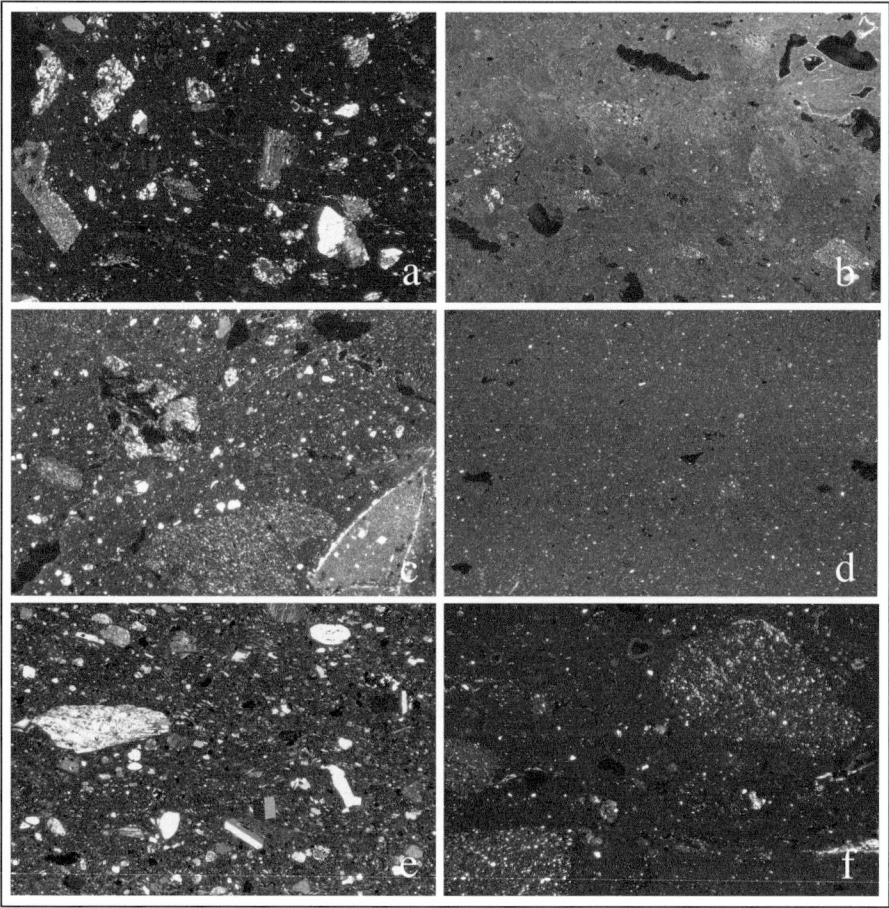

Figure 2. Thin section micrographs of Iron Age pottery from Knossos. a) Knossos Fabric Group 1, low-grade metamorphic (sample KN137), b) Knossos Fabric Group 2, siltstone and sandstone (sample KN188), c) Knossos Fabric Group 3, mix of metamorphic, sedimentary and altered igneous rock fragments (sample KN146), d) Knossos Fabric Group 6, fine-grained (sample KN 84), e) Knossos Fabric Group 4, red micaceous fabric (sample KN92), f) Knossos Fabric Group 5, grey siltstone in red clay base (sample KN155). All images taken in crossed polars. Image width = 5.9 mm, except d = 4.6 mm.

Results

The petrographic examination of the 214 Early Iron Age pottery samples from Knossos permitted the identification of seven fabric groups and 23 two-sample fabrics and loners. Knossos Fabric Group 1 is characterised by poorly sorted sand-sized inclusions of metaquartzite, quartz-mica schist, phyllite, mono- and polycrystalline quartz and few to rare sandstone, siltstone, chert, epidote, serpentinite, altered volcanic rocks (mainly devitrified basalt), and calcimudstone inclusions (Figure 2A). The clay base is orange and well fired as the groundmass is optically inactive to weakly active. Knossos Fabric Group 2 is very similar to Fabric Group 1 in terms of firing, mineralogy of non-plastics and clay base but differs in the higher amount of large siltstone and sandstone inclusions (Figure 2B). Knossos Fabric Group 3 differs from the two previous groups by its abundant fine fraction composed of quartz, chert, mica, epidote and micrite, and its red-firing base clay (Figure 2C). The coarse fraction is characterised by phyllites, quartzite, altered igneous, sedimentary rock fragments and calcimudstone. Altogether, these three groups represent the coarse-grained 'broadly' local fabrics at Early Iron Age Knossos used to manufacture large to medium-size vessels. Knossos Fabric Group 6 is the main 'broadly' local fine-grained group and is characterised by the rare presence of low-grade metamorphics, quartz, siltstone and or red textural concentration features set in a very fine and often mottled groundmass (Figure 2D).

Knossos Fabric Group 4, characterised by a coarse fraction of muscovite mica inclusions, iron oxides, unaltered simple twinned feldspars, rounded monocrystalline quartz, muscovite mica shimmer aggregates set in a bright, red-firing clay base (Figure 2E), does not fit with the geology of Crete. The petrographic characteristics point to an off-island origin of production, most probably from the Cyclades where red micaceous fabrics are common, even though we cannot entirely exclude an east Cretan origin (see Poulou-Padadimitriou and Nodarou, 2007, p. 757). Knossos Fabric Group 7 exhibits similar non-plastic inclusions as Fabric Group 4 but has abundant brown phyllites and a lower amount of mica laths and aggregates. A Cycladic origin of production is highly probable in this case also. Another non-local fabric is Knossos Fabric Group 5, which is characterised by large rounded inclusions of grey siltstone set in a very fine-grained groundmass (Figure 2F). Even though we cannot entirely exclude a local production near Knossos, such fabrics are found in south-central Crete (Wilson and Day, 1999). Loner KN 178, characterised by large inclusions of very altered igneous rock fragments, grey siltstone inclusions and vegetal temper set in a bright red and very fine groundmass (Figure 3A) could be an import to Knossos from east Crete (Day, 1995). Other imports have been identified in the petrographic loners. For example KN 32 is believed to come from the Vrysinas area based on the coarse sand-sized quartzite and phyllite inclusions set in a bright red low-fired groundmass (Kordatzaki, 2007). Lastly, KN 24, a Cycladic import, is characterised by a calcareous-rich clay base with mica laths and rounded quartz inclusions in the coarse fraction.

The 200 pottery thin sections from Sybrita were classified into seven fabric groups and 14 single-sample fabrics. The majority of which are characterised by mineral and rock inclusions consistent with the local rock formations of the Phyllite-Quartzite Series and alluvium and Neogene deposits (IGME 1985, 1991).

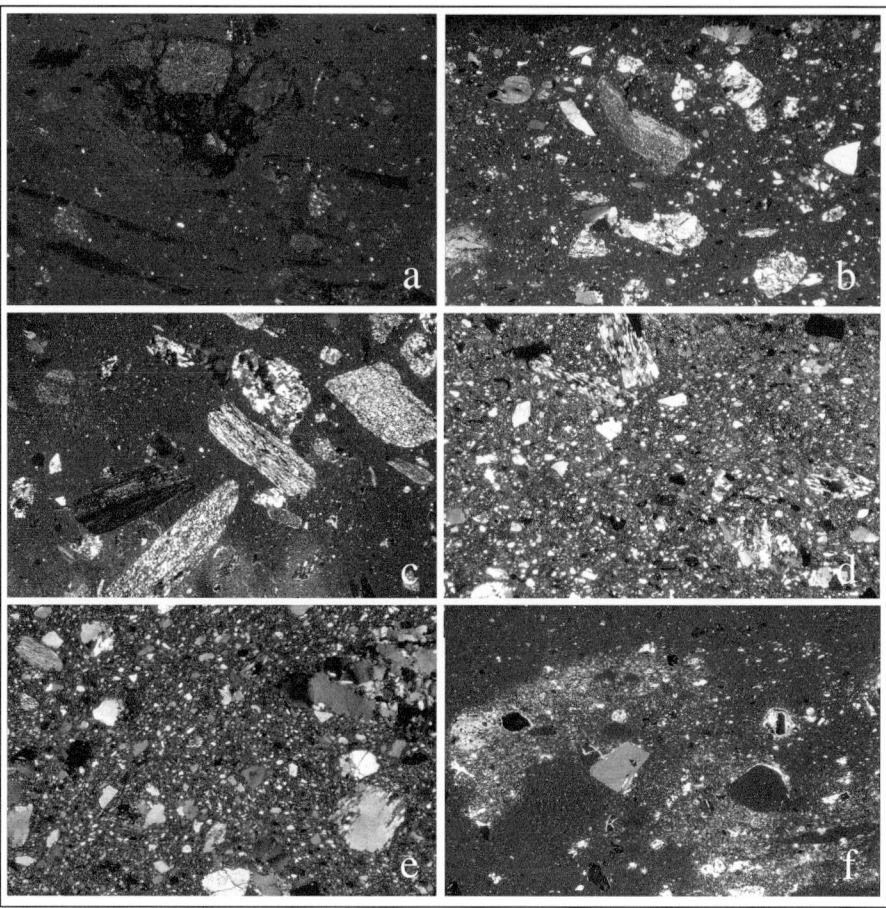

Figure 3. Thin section micrographs of Iron Age pottery from Knossos and Sybrita. a) Knossos loner from the Mesara (sample KN178), b) Sybrita Fabric Group 1a, low-grade metamorphic (sample SY111), c) Sybrita Fabric Group 1b, with light grey core (sample SY4), d) Sybrita Fabric Group 2, red-firing with abundant silicate fine fraction (sample SY132), e) Red metamorphic clay sampled near Sybrita (sample SGS 8), f) Sybrita Fabric Group 5, red TCFs in calcareous Neogene clay (sample SY74). All images taken in crossed polars. Image width a, d = 5.9 mm, c, f = 7.4 mm, b = 9 mm, e = 4.6 mm.

Sybrita Fabric Groups 1a and 1b are characterised by low-grade metamorphic inclusions set in an orange-firing base clay (Figure 3B). The coarse fraction consists of dominant inclusions of phyllite, quartz-phyllites, quartzite, and common to rare mono- and polycrystalline quartz, feldspar, chert, schist, and calcimudstone (micrite and sparite). Samples in Group 1b exhibit a finer clay paste with a distinctive grey core and thin orange margins, which is the results of a fast firing with incomplete oxidisation of the clay walls (Figure 3C; Figure 5C). Sybrita Fabric Group 2 varies from Group 1 in a number of respects, including its red-firing clay base, abundant quartz-rich fine fraction and higher concentration of quartzite and rarity of phyllites within the coarse-fraction (Figure 3D). This fabric is directly comparable to the red metamorphic clays sampled in the vicinity of Sybrita (Figure 3E). Sybrita Fabric Group 5 is significantly different from the other local fabrics with its fine calcareous Neogene clay base and red textural concentration features. Non-plastic inclusions consist of monocrystalline quartz, micrite, quartzite, microfossils (foraminifera), clay pellets, phyllite, and chert (Figure 3F). Such clays (and fabrics) are found in many places on Crete (Hein *et al.*, 2004b) and without a diagnostic coarse fraction, it is very difficult to suggest an area of production. There is no reason for it to be non-local as such clays are found near Sybrita. It points however to the use of different raw materials, if not to the products of a different local workshop. Finally, Sybrita Fabric Group 6 corresponds to the finer version of Group 1 with a scarce coarse-fraction of low-grade metamorphics and calcimudstones and a fine fraction composed of quartz, biotite, opaques and rare textural concentration features (Figure 4A).

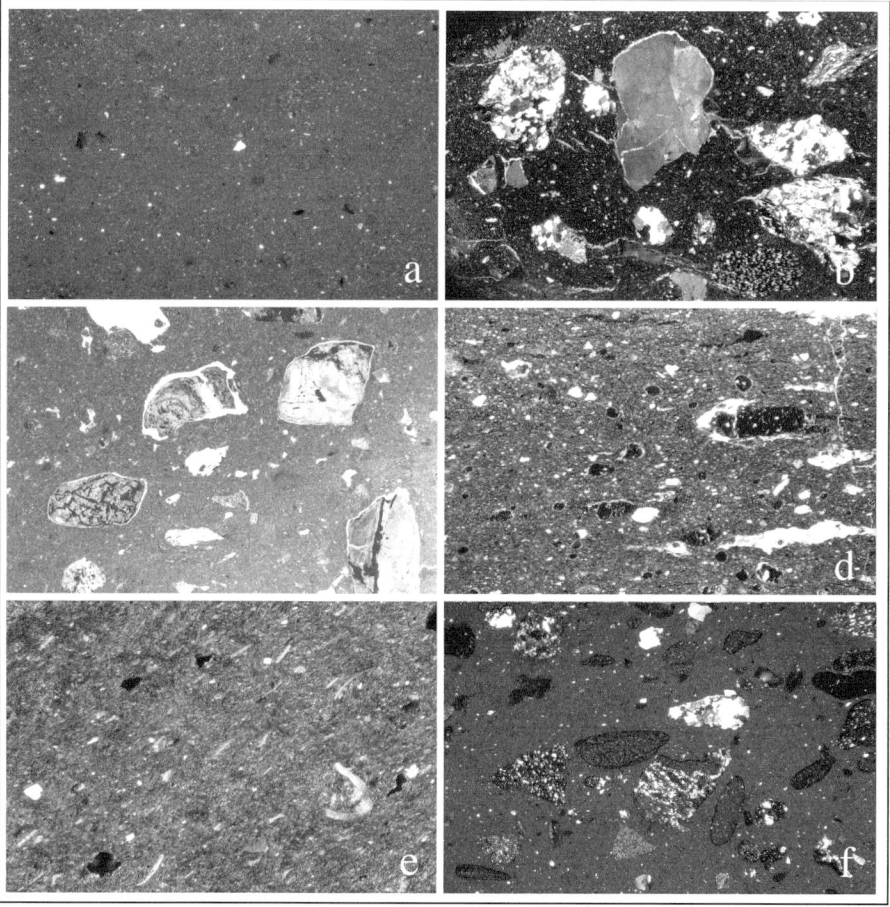

Figure 4. Thin section micrographs of Iron Age pottery from Sybrita. a) Sybrita Fabric Group 6, fine-grained (sample SY10), b) Sybrita Fabric Group 3, muscovite schist (sample SY126), c) Sybrita loner, serpentinite (sample SY60), d) Sybrita Fabric Group 7, black TCFs (sample SY118), e) Sybrita loner, shell-rich (sample SY92), f) Sybrita loner, water-worn phyllites (sample SY67). All images taken in crossed polars except c, d. Image width b, c = 5.9 mm, a = 3.0 mm, d = 3.7 mm, e = 42.5 mm, f = 7.4 mm.

Beyond these groups, a number of non-local fabrics have also been identified in the Sybrita assemblage. Sybrita Fabric Groups 3 and 4, and loner SY 60, characterised by mica-schist inclusions including some with epidote (Figure 4B), garnet-schist or predominated by large serpentinite inclusions (Figure 4C) could originate from one or more workshops located in the Aghios Vasileios valley where ophiolitic and schist/greenstone rock formations are found; or even elsewhere as ophiolitic sequences appear in other areas of Crete, namely in the south-central part of the island (IGME, 1984) where similar Bronze Age fabrics have been identified (Belfiore *et al.*, 2007; Buxeda i Garrigós *et al.*, 2001; Myers and Betancourt, 1990; Shaw *et al.*, 2001; Wilson and Day, 1994). Fabric Group 7, characterised by black textural concentration features, or clay pellets, (Figure 4D), and sample SY 92, a shell-rich calcareous fine-grained fabric (Figure 4E), are considered imports from Eleutherna where similar Early Iron Age fabrics have been identified (Nodarou, 2008). Sample SY 11 has a fine calcareous fabric with sponge spicules. It compares well with the LM III C fabrics of Khamalevri, near Rethymnon (Moody *et al.* 2003, p. 97; Nodarou, in press). A bright red coarse-grained fabric rich in quartzite matches the local Late Bronze Age fabrics of Vrysinas, south of Rethymnon (Kordatzaki, 2007) as well as the red metamorphic clays we sampled in that area. A similar sample was identified in the Knossos assemblage. Another loner, SY 67, is characterised by a well-fired calcareous clay base tempered with water-worn grey siltstone and phyllitic inclusions (Figure 4F). It is related petrographically to fabrics from the western Mesara in south-central Crete (Buxeda i Garrigós *et al*, 2001; Belfiore *et al.*, 2007; Myers and Betancourt, 1990; Shaw *et al.*, 2001; Wilson and Day, 1994, p. 53). Other samples, characterised by a mix of sedimentary and metamorphic rock fragments, are similar to fabrics from north-central Crete but do not match the Knossian fabric groups analysed in this study.

Discussion

In terms of clay paste technology, our analysis has shown that specific clay recipes were used to produce particular vessel forms: coarse wares, cooking pot wares and finer-grained wares have distinct clay paste technology. This feature is most striking in the Sybrita assemblage, particularly between cooking pots and storage/transport vessels. Cooking pot fabrics have an abundant quartz-rich fine fraction and were made using red, iron-rich, metamorphic clays. Sybrita Fabric Group 2, the cooking pot fabric, directly matches the red metamorphic clays sampled around the site. It suggests that potters deliberately used already coarse raw materials without adding or removing non-plastic inclusions and that these iron-rich clays high in silicate were adequate for cooking pots.

The clay paste technology of coarse-grained fabrics of large and medium-sized vessels used for storage or transport, i.e. pithoi, amphorae, craters, basins, shows that potters reworked the constituent raw materials. The strong bimodality of the grain size distribution as seen by the coarse sand set in a fine clay base rarely occurs naturally. It is believed that potters levigated clays and added as temper sand derived from the phyllite series. Evidence for clay mixing was also observed in some samples, suggesting that potters probably mixed calcareous Neogene clays with iron-rich metamorphic ones. Clay mixing is a practice that has already identified in Cretan pottery production (Day *et al.*, 1999; Hein *et al.*, 2004a).

The finer-grained fabrics have higher calcareous content as attested by buff, pink or light orange refiring colours. No clear evidence of levigation was noticed, but the very fine clay pastes, strongly suggests that potters deliberately removed coarse inclusions. These fine-grained fabrics were mainly used in the production of smaller vessels, such as hydriae, skyphoi, and aryballoi. At Knossos, however, some amphorae and craters, especially those excavated in the Geometric Well and in the North Cemetery, were manufactured with fine-grained fabrics. This suggests that potters were skilled at manufacturing large vessels using very fine-grained pastes.

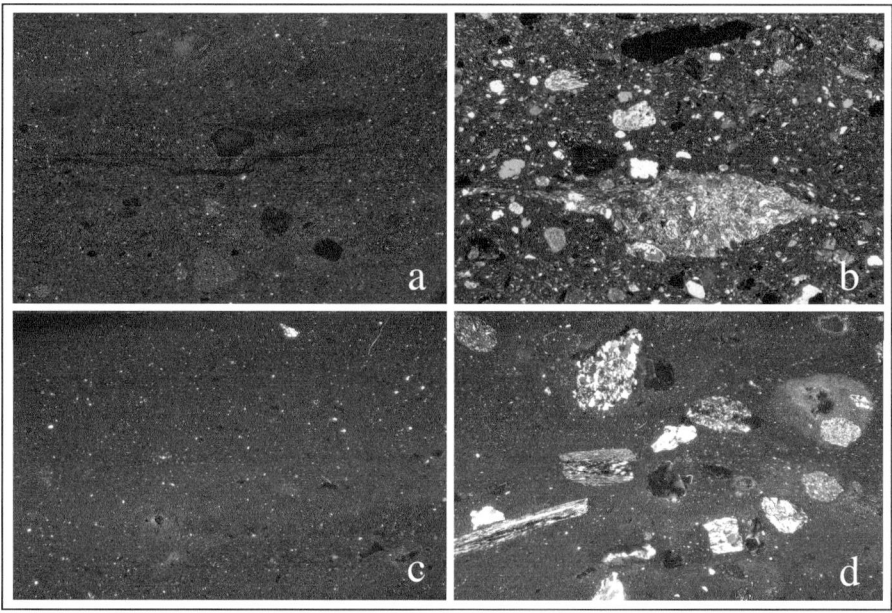

Figure 5. Thin section micrographs of Iron Age pottery from Sybrita. a) Ribbon of iron rich clay (sample KN33), b) Porphyroclast of micaceous clay in red metamorphic groundmass (sample KN139), c) very fine fabric of upper part of SY190, d) coarse fabric of lower part of sample SY190. All images taken in crossed polars. Image width = 5.9 mm.

The mixing of different clays is evidenced by streaks, clay 'porphyroclasts' and ribbons of different colour or granulometry within the groundmass (Figure 5A). For example, Knossos Fabric group 4 (red micaceous fabric) has large porphyroclasts of highly micaceous yellowish clay incompletely mixed with the surrounding red clay base are visible (Figure 5B). As discussed in the previous paragraphs, the techniques of levigation and tempering clays were also used, the latter mainly evidenced by coarse sand-sized non-plastic inclusions set in a very fine groundmass and exhibiting a strong bimodal grain-size distribution. Additionally, the presence of burnt out organic voids, seen mainly in the pithoi samples, is another evidence for tempering. Lastly, one sample form the Sybrita assemblage shows that different clay bases were used for different parts of the same vessel: the upper part (neck and shoulder) has a very fine-grained fabric (Figure 5C), as opposed to a coarsely tempered fabric for the lower body (Figure 5D). Similar features have been identified on Bronze Age ceramics from the Mesara in south-central Crete (Day *et al.*, 2006; Wilson and Day, 1994, p. 61).

Our analysis has shown that there is a good degree of continuity in the raw materials exploited and in clay paste preparation throughout the periods of the Early Iron Age for the main local fabric groups. Within these groups however, small diachronic changes are visible in the clay preparation: there are more calcimudstone inclusions set in a orange groundmass as well as a higher amount of fine-fraction detrital minerals in the samples from the later periods, while in the earlier period fabrics the inclusions are coarser-grained and set in a fine and more iron-rich clay base. Low fired fabrics, attested by an optically active groundmass are also more common in the later periods.

The results of the study have also highlighted a number of changes in pottery production and consumption patterns, namely the appearance and disappearance of certain fabric groups. For example, the Knossos red micaceous cooking pots are not present at the site in the Subminoan period but start to appear in the Protogeometric B period and become common in the Geometric period. At Sybrita, Fabric Group 1b, the main local group with the different firing technology, prevails in the Protogeometric period. This could reflect changes in the organisation of pottery production, and more specifically in the firing procedure and/or the type of kiln use. Furthermore, the fine calcareous fabric was also produced mainly in the earlier periods at Sybrita. Other fabrics appear only later, such as the imports from Eleuftherna, which date to the Geometric period.

Conclusions

Overall then the picture is one of considerable continuity in patterns of production and consumption during the Early Iron Age north-central Crete. While Cretan communities continue to produce for what they could consume locally, networks of exchange for coarse utilitarian pots (or their contents) appear to have existed as intra- and extra-regional imports were identified at both sites. Distinct patterns in the production and consumption of coarse wares for both cooking and storage, or transport are evident at Sybrita and Knossos and, in general, persist over the whole of the Early Iron Age. Moreover, each site has a distinct pattern of imports from other parts of Crete. The variety of fabrics, local and otherwise, to be found among the ceramic material of Sybrita, prompts reflections on the role played by that centre in the regional trade dynamics within the island. No off-island importation has been firmly ascertained, which is in fact hardly surprising given the inland position of the site. The presence of material from other parts of the island is, however, of great interest. Research is still in the initial stage, and we have yet, therefore, to process the data available in order to be able to advance a well-grounded interpretation. Nevertheless, there can be little doubt that one of the main reasons for the outward projection observed at the site is to be seen in the special position occupied by the hill of Thronos Kephala within the variegated landscape hemmed in between the Psiloritis mountain to the east and mount Kedros to the south. Situated at the northern end of the Amari valley, and to the immediate north of the village of Thronos, the hill of Thronos Kephala stands at a natural crossroads upon which converge valleys and landscapes from various parts of the island

Within this general picture of continuity however, changes take place in the Protogeometric phase when transformations and innovations can be discerned within the ceramic repertoire. At Sybrita, while Late Minoan III C and Subminoan pottery is remarkably similar from a stylistic point of view, it is in the Protogeometric that changes happen and the ceramic repertoire is clearly transformed. At Knossos, petrographic analysis determined that the coarse micaceous fabric was an import whose origin of production may be off-island, perhaps from the Cyclades. This was unexpected, as based on quantity and function of the pots, this group was believed to have been produced on site (Coldstream, 2001, p. 61). Our results call for a re-evaluation of the consumption patterns and trade networks for this utilitarian class of pottery. The introduction of the micaceous fabric coincides with a number of other signs of greater external contact during that latter part of the 9th BC, namely: new styles of metalwork in bronze and gold, with oriental affinities (Brock 1957, p. 198-199; Boardman 1961, p. 134-137; Boardman 1967, p. 63-64); and the introduction of the Protogeometric B style itself on Knossian fine wares (Brock 1957, p. 143; Coldstream, 1998; Kaiser, 2005), often described as Greece's first true 'Orientalizing' style. Overall then the data from Sybrita allows greater precision in our understanding of trade and interaction within Crete than we can yet achieve for Knossos. But the data from Knossos sheds new light on wider Mediterranean connectivity.

Acknowledgements

We are grateful to the Greek Ministry of Culture, the Greek Archaeological Service, especially the Ephoria of Heraklio and the Ephoria of Chania, as well as to the Greek Institute of Geology and Mineral Exploration for permission to sample pottery and collect geological samples. We are indebted to the Institute for Aegean Prehistory and the Fitch Laboratory, British School at Athens for financial assistance. For permission to sample published and unpublished material we thank the British School at Athens, ICEVO Roma, E. Hatzaki, C. MacDonald, A. Prent and the late Prof. J. N. Coldstream. Finally, we are most grateful to the following colleagues for sharing their knowledge of Cretan fabrics and geological material: E. Kiriatzi, P. Day, E. Nodarou and P. Quinn.

References

Belfiore, C. M., Day, P. M., Hein, A., Kilikoglou, V., La Rosa, V., Mazzoleni, P. and Pezzino, A. 2007. Petrographic and chemical characterization of pottery production of the Late Minoan I kiln at Haghia Triada, Crete. *Archaeometry,* 49: 621-653.

Boardman, J. 1961. *The Cretan Collection in Oxford: The Dictaean Cave and Iron Age Crete*. Clarendon, Oxford.

Boardman, J. 1967. The Khaniale Tekke tombs II. *Annual of the British School at Athens,* 62: 57-75.

Brock, J.K. 1957. *Fortetsa: Early Greek Tombs Near Knossos.* British School at Athens, Supplementary Volume 2, Cambridge University Press, Cambridge.

Buxeda i Garrigós, J., Kilikoglou, V. and Day, P. M. 2001. Chemical and mineralogical alterations of ceramics from a Late Bronze Age kiln at Kommos, Crete: the effect on the formation of a reference group. *Archaeometry,* 43: 349-371.

Coldstream, J.N. 1972. Knossos 1951-61: Protogeometric and Geometric pottery from the Town. *Annual of the British School at Athens,* 67: 63-98.

Coldstream, J.N. 1998. Minos Redivivus: some nostalgic Knossians of the ninth century BC. In: Cavanagh, W.G., Curtis, M., Coldstream. J.N. and Johnston, A.W. (Eds.) *Post-Minoan Crete: Proceedings of the First Colloquium.* British School at Athens Studies Series, 2, British School at Athens, London: 58-61.

Coldstream, J.N. 2000. Evan's Greek Finds: The Early Greek Town of Knossos, and it's encroachment on the borders of the Minoan palace. *Annual of the British School at Athens,* 95: 259-299.

Coldstream, J.N. 2001. The Early Greek Period: Subminoan to Late Orientalizing. In: Coldstream, J. N., Eiring, L.J. and Forster, G. (Eds.) *Knossos Pottery Handbook. Greek and Roman.* British School at Athens Studies, 7, British School at Athens, London: 21-76.

Coldstream, J.N. and Catling, H.W. (Eds.) 1996. *Knossos North Cemetery. Early Greek Tombs.* British School at Athens Supplementary Volume, 28, British School at Athens, London.

Coldstream, J.N. and Hatzaki, E.M. 2003. Knossos: Early Greek Occupation under the Roman Villa Dionysos. *Annual of the British School at Athens*, 98: 279-306.

Coldstream, J.N. and Macdonald, C.F. 1997. Knossos: Area of the South-West Houses, Early Hellenic Occupation. *Annual of the British School at Athens,* 92: 191-245.

D'Agata, A.L. 1999. Defining a Pattern of Continuity during the Dark Age in Central-Western Crete: Ceramic Evidence from the Settlement of Thronos/Kephala (Ancient Sybrita). *Studi Micenei ed Egeo-Anatolici,* 41: 181-218.

D'Agata, A.L. 2001a (1997-2000). Ritual and Rubbish in Dark Age Crete: The Settlement of Thronos Kephala (ancient Sybrita) and the pre-Classical Roots of a Greek city. *Aegean Archaeology,* 4: 45-59.

D'Agata, A.L. 2001b. Public versus Domestic? A Geometric Monumental Building at Thronos Kephala Amariou. *8th International Congress of Cretan Studies, Herakleion September 9-14 1996*, A1, Herakleion: 327-339.

D'Agata, A.L. 2003. LM IIIC-SM Pottery Sequence at Thronos Kephala (ancient Sybrita) and Its Connections with the Greek Mainland. In: Deger-Jalkotzy, S. and Zavadil, M. (Eds.) *LH IIIC Chronology and Synchronisms. Proceedings of the International Workshop held at the Austrian Academy of Sciences at Vienna, May 7th and 8th, 2001*. Österreichischen Akademie der Wissenschaften, Vienna: 23-35.

D'Agata, A.L. 2007. Evolutionary Paradigms and Late Minoan III. On a Definition of Late Minoan IIIC Middle. In: Deger-Jalkotzy, S., Zavadil, M. (Eds.) *Chronology and Synchronisms II. LH IIIC Middle. Proceedings of the International Workshop held at the Austrian Academy of Sciences at Vienna, October 29th and 30th, 2004*. Österreichischen Akademie der Wissenschaften, Vienna: 89-118.

D'Agata, A.L. in press. Subminoan: A Neglected Phase of the Cretan Pottery Sequence. *Studi Micenei ed Egeo-Anatolici*.

Day, P.M. 1988. The production and distribution of storage jars in Neopalatial Crete. In: French, E.B. and Wardle, K.A. (Eds.) *Problems in Greek Prehistory*. Bristol Classical Press, Bristol: 499-507.

Day, P.M. 1995. Pottery production and consumption in the Sitia Bay area during the New Palace period. In: Tsipopoulou, M. and Vagnetti, L. (Eds.) *Achladia: Scavi e Ricerche della Missione Greco-Italiana in Creta Orientale (1991-1993)*. Gruppo Editoriale Internazionale, Rome: 149-173.

Day, P.M. 1997. Ceramic exchange between towns and outlying settlements in Neopalatial East Crete. In: Hagg, R. (Ed.) *The Function of the "Minoan Villa". Proceedings of the 8th International Symposium at the Swedish Institute, Athens*. Paul Åströms Förlag, Stockholm: 219-228.

Day, P.M. and Wilson, D.E. 1998. Consuming Power: Kamares Ware in Protopalational Knossos. *Antiquity, 72*: 350-58.

Day, P.M., Kiriatzi, E., Tsolakidou, A. and Kilikoglou, V. 1999. Group therapy in Crete: a comparison between analyses by NAA and thin section petrography of Early Minoan pottery. *Journal of Archaeological Science, 26*: 1025-1036.

Day, P.M., Relaki, M. and Faber, E. W. 2006. Pottery making and social reproduction in Bronze Age Mesara, Crete. In: Wiener, M.H., Warner, J.L., Polonsky, J. and Hayes, E. E. (Eds.) *Pottery and Society: The Impact of Recent Studies in Minoan Pottery*. Gold Medal Colloquium in Honor of Philip P. Betancourt. 104th Annual Meeting of the Archaeological Institute of America, New Orleans, Louisiana, 5 January 2003. Archaeological Institute of America, Boston: 22-72.

Deger-Jalkotzy, S. and Lemos, I. (Eds.) 2006. *Ancient Greece from the Mycenaean Palaces to the Age of Homer*. Edinburgh Leventis Studies, 3, Edinburgh University Press, Edinburgh.

Hatzaki, E., Prent, M., Coldstream J.N., Evely, D. and Livarda, A. 2008. Knossos the Little Palace North Project, Part One: the early Greek periods, *Annual of the British School at Athens*: 235-289.

Hein, A., Day, P.M., Cau Ontiveros, M.A. and Kilikoglou, V. 2004a. Red clays from Central and Eastern Crete: geochemical and mineralogical properties in view of provenance studies on ancient ceramics. *Applied Clay Science, 24*: 245-255.

Hein, A., Day, P.M., Quinn, P.S. and Kilikoglou, V. 2004b. The geochemical diversity of Neogene clay deposits in Crete and its implications for provenance studies of Minoan pottery. *Archaeometry*, 46: 357-384.

IGME. 1984 *Geological Map of Greece: Timbakion. 1:50 000*. Institute of Geology and Mineral Exploration.

IGME. 1985 *Geological Map of Greece: Melambes. 1:50 000*. Institute of Geology and Mineral Exploration.

IGME. 1991. *Geological Map of Greece: Perama. 1:50 000*. Institute of Geology and Mineral Exploration.

IGME. 1996 *Geological Map of Greece: Heraklion. 1:50 000*. Institute of Geology and Mineral Exploration.

Jones, R.E. 1986. *Greek and Cypriot Pottery. A Review of Scientific Studies*. Fitch Laboratory Occasional Paper, 1, British School at Athens, Athens.

Kaiser, I. 2005. Protogeometric B – Minoan and Oriental influences on a Cretan pottery style of the second half of the 9th century BC. In: Πεπραγμένα Θ' Διεθνούς Κρητολογικού Συνεδρίου: Τόμος Α5: Αρχαία Ελληνική καί Ρωμαϊκή Περίοδος,, Heraklion: 63-70.

Kordatzaki, G. 2007. Κεραμική από το ιερό κορυφής του Βρύσινα. Ένα σύνθετο τεχνοσύστημα παραγωγής και χρήσης κατά την $2^{η}$ χιλιετία π.Χ. Unpublished Ph.D. thesis, University of Crete, Rethymno.

Liddy, D.J. 1996. A chemical study of decorated Iron Age pottery from the Knossos North Cemetery. In: Coldstream, J. N. and Catling, H. W. (Eds.) *Knossos North Cemetery. Early Greek Tombs*. British School at Athens Supplementary Volume, 28, British School at Athens, London.: 465-516.

Moody, J., Robinson, H.L., Francis, J., Nixon, L. and Wilson, L. 2003. Ceramic fabric analysis and survey archaeology: the Sphakia Survey. *Annual of the British School at Athens*, 98: 37-105.

Myers, G.H. and Betancourt, P.P. 1990. The Fabrics at Kommos. In: Betancourt, P.P. (Ed.) *Kommos II. The Final Neolithic through Middle Minoan III Pottery*. Princeton University Press, Princeton: 3-13.

Nodarou, E. 2008. Petrographic Analysis of Selected Early Iron Age Pottery from Eleutherna. In: Kotsonas, A. *The Archaeology of Tomb A1K1 of Orthi Petra in Eleutherna: The Early Iron Age Pottery*. University of Crete: 345-362.

Nodarou, E. in press. Το ΥΜ ΙΙΙΒ/Γ κεραμικό σύνολο από το Χαμαλεύρι Ρεθύμνου: προκαταρκτικά αποτελέσματα από την πετρογραφική ανάλυση. Proceedings of the 10th Cretological Congress (Chania 2006).

Nowicki, K. 2000. *Defensible Sites in Crete c.1200 – 800 B.C. (LMIIIB through Early Geometric)*. Aegaeum, 21, Université de Liège, Liège.

Poulou-Papadimitriou, N. and Nodarou, E. 2007. La céramique protobyzantine de Pseira : la production locale et les importations, étude typologique et pétrographique. In: Bonifay, M. and Tréglia, J.-C. (Eds.) *LRCW 2 Late Roman Coarse Wares, Cooking Wares and Amphorae in the Mediterranean. Archaeology and Archaeometry.* Vol. II BAR International Series 1662, II : 755-766.

Quinn, P.S. and Day, P.M. 2007. Calcareous Microfossils in Bronze Age Aegean Ceramics: Illuminating Technology and Provenance. *Archaeometry,* 49: 775-793.

Riley, J. A. 1983. The contribution of ceramic petrology to our understanding of Minoan society. In: Krzyszkowska, O. and Nixon, L. (Eds.) *Minoan society. Proceedings of the Cambridge Colloquium 1981*. Bristol, Bristol Classical Press: 283-292.

Shaw, J.W., Van der Moortel, A. Day, P.M. and Kilikoglou, V. (Eds.) 2001. *A LM IA Ceramic Kiln in South-Central Crete: Function and Pottery Production*. Hesperia Supplement, 30, American School of Classical Studies at Athens.

Tomkins, P. and Day, P.M. 2001. Production and exchange of the earliest ceramic vessels in the Aegean: a view from Early Neolithic Knossos, Crete. *Antiquity,* 75: 259-260.

Tomkins, P., Day, P.M. and Kilikoglou, V. 2004. Knossos and the earlier Neolithic landscape of the Herakleion basin. In: Cadogan, G., Hatzaki, E. and Vasilakis, A. (Eds.) *Knossos: Palace, City, State*. British School at Athens Studies, 12, British School at Athens, London: 51-59.

Tomlinson, J.E. and Kilikoglou, V. 1998. Neutron activation analysis of pottery from the Early Orientalizing kiln at Knossos. *Annual of the British School at Athens*, 93: 385-388.

Whitbread, I.K. 1986. The Characterization of Argillaceous Inclusions in Ceramic Thin Sections. *Archaeometry,* 28: 79-88.

Whitbread, I.K. 1989. A Proposal for the Systematic Description of Thin Sections Towards the Study of Ancient Ceramic Technology. In Maniatis, Y. (Ed.) *Archaeometry. Proceedings of the 25th International Symposium*. Amsterdam: 127-138.

Whitbread, I.K. 1995. *Greek Transport Amphorae: A Petrological and Archaeological Study*. Fitch Laboratory Occasional Paper, 4, British School at Athens.

Whitley, J. 1991. *Style and Society in Dark Age Greece*. Cambridge University Press, Cambridge.

Whitley, J. 1997. Cretan laws and Cretan literacy. *American Journal of Archaeology,* 101: 635-661.

Whitley, J. 2001. *The Archaeology of Ancient Greece*. Cambridge University Press, Cambridge.

Whitley, J. 2004. Style Wars: towards an explanation of Cretan exceptionalism. In: Cadogan, G., Hatzaki, E. and Vassilakis, A. (Eds.) *Knossos: Palace, City, State*. British School at Athens Studies, 12, British School at Athens, London: 433-442.

Wilson, D.E. and Day, P.M. 1994. Ceramic regionalism in Prepalatial Central Crete: the Mesara imports at EM I to EM II A Knossos. *Annual of the British School at Athens*, 89: 1-87.

Wilson, D.E. and Day, P.M. 1999. EM II Ware groups at Knossos: The 1907-1908 South Front tests. *Annual of the British School at Athens*, 94: 1-62.

THE MOVEMENT OF MIDDLE BRONZE AGE TRANSPORT JARS
A PROVENANCE STUDY BASED ON PETROGRAPHIC AND CHEMICAL ANALYSIS OF CANAANITE JARS FROM MEMPHIS, EGYPT

Mary Ownby

Department of Archaeology, University of Cambridge, UK
(mfo22@cam.ac.uk)

Janine Bourriau

McDonald Institute for Archaeological Research, University of Cambridge, UK

Introduction

The discovery at the site of Memphis, Egypt of transport vessels, referred to as Canaanite jars, dating to the Middle Bronze Age (MBA, 2000-1550 BC) provided a unique opportunity to examine the contacts between Egypt and the Levant during this period. These vessels appear to derive from several areas of the Levant based on their form and fabric. However, the MBA Canaanite jars at Memphis represented only one aspect of trade to Egypt, and their relationship to the numerous Canaanite jars found at the site of Tell el-Dab'a in the Eastern Nile Delta was also of interest. This is because during the period in which the MBA Canaanite jars reached Memphis, c. 1750-1550 BC, the town was controlled by a group of Levantine rulers called the Hyksos, who were based at the site of Tell el-Dab'a. While the textual sources provide this information, archaeologically only the Canaanite jars present possible evidence for a relationship between the Egyptians at Memphis and the mixed Egyptian-Levantine population at Tell el-Dab'a. Additionally, there is little evidence south of Memphis for contact with Tell el-Dab'a beyond scattered examples of Canaanite jars, scarabs, and vessels of Tell el-Yahudiyeh ware, most of which were probably produced in Egypt (Arnold *et al.*, 1993; Ben-Tor, 2007; Kaplan, 1980). Therefore, the provenance of the MBA Canaanite jars from Memphis was important in order to understand the nature of trade relations within Egypt between the 'foreign' population at Tell el-Dab'a and the native population at Memphis. Due to the cultural differences between Memphis and Tell el-Dab'a (Bader, 2007) there was the distinct possibility the two corpora did not originate from the same production areas in the Levant.

The site of Tell el-Dab'a has revealed a unique set of material culture with ties to both Egypt and the Levant (Bietak, 1996). From the excavations, it appears that beginning in the first half of the 2nd Millennium BC peoples from the Levant began to settle in great numbers in the Eastern Delta and over time the population of immigrants grew. The site of Tell el-Dab'a became an important centre with a port situated on a branch of the Nile, giving easy access to the trading networks of the Eastern Mediterranean. While no longer the capital of Egypt, the city of Memphis continued to flourish during this period. Excavations by the Egyptian Exploration Society revealed a workshop area with lower-middle class housing (Giddy and Jeffreys, 1991; Jeffreys and Giddy, 1993).

However, the settlement strata at Memphis cover only the period from c. 1750-1550 BC, while the strata at Tell el-Dab'a date from c. 1950-1550 BC. In addition, the excavated area at Memphis was small, so that vertically and horizontally Memphis corresponds to only a part of the site of Tell el-Dab'a. This caveat needs to be kept in mind for the comparisons of the MBA Canaanite jars from the two sites.

Canaanite jars are a type of transport vessel made from many different clays and tempering materials (Grace, 1956) (Figure 1). The vessels are predominantly wheel-made, however, the bases can be moulded (Rye, 1981, p. 134-137). Their form varied considerably, particularly over time, suggesting that several areas of manufacture existed and a gradual evolution of shape occurred (Amiran, 1969). Residue analysis on Late Bronze Age (LBA, 1550-1150 BC) Canaanite jars by Serpico *et al.* (2003), suggests that they were used to transport wine, oil, and resin around the Eastern Mediterranean. This study also attempted to analyze residues from LBA Canaanite jars from Memphis, but was unsuccessful due to poor preservation in the damp soil of the Nile Valley.

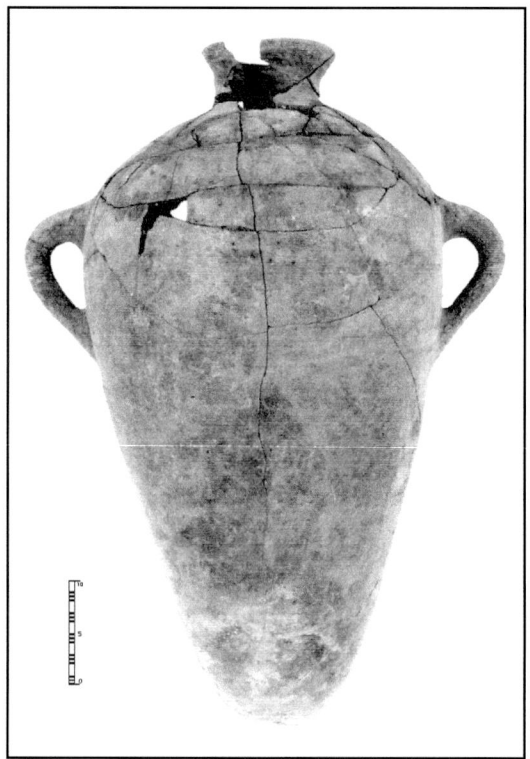

Figure 1. Image of MBA Canaanite jar (from Aston, 2004, plate LXXVb).

The preliminary results of the current petrographic and chemical analyses of the MBA Canaanite jars from Memphis suggest several possible sources for the provenance of these ceramics. Comparison with the Canaanite jars from Tell Dab'a revealed similarities between the two corpora, but also some notable differences. Thus, it appears that cultural factors may have played a role in the types of jars that were imported to the sites, and that trade between Egypt and the Levant in the Middle Bronze Age was more complicated than previously thought.

Previous Compositional Analyses of Canaanite Jars from Tell el-Dab'a

The first compositional study of Canaanite Jars from Tell el-Dab'a was the neutron activation analysis (NAA) of McGovern (2000). These results were compared to NAA data on pottery from various sites in Egypt and the Levant to determine where the Tell el-Dab'a material might have originated. The results indicated that most of the vessels were manufactured in southern Palestine. McGovern (2000) suggested that the Levantine peoples living at Tell el-Dab'a were from this area and maintained trade contacts with their former homeland. However, the samples within the comparative database had not been firmly established as local productions, and therefore could not assist in determining the provenance of the ceramics at Tell el-Dab'a (Goren, 2003). Additionally, the database contained little comparative material from the northern Levant, so it is perhaps unsurprising that most of the data fell into groups containing the southern Palestinian pottery. Due to these issues, Cohen-Weinberger and Goren (2004) conducted a petrographic study to establish with certainty the provenance of the imported pottery at Tell el-Dab'a and to reconstruct the trade relationships between this site and the Levant. This study identified 11 petrographic groups originating from the coast of Syria south to the coast and inland regions of Palestine. Over 70% of the imported vessels were suspected to have originated from the northern Levant.

Methodology

For the current study, fifty-six sherds from the Memphis corpus of MBA Canaanite jars were selected for study out of an assemblage of 1000 sherds. The sampled sherds encompassed the entire range of fabrics identified in the hand specimen fabric classification system. Standard procedures were employed to produce the thin sections, which were thoroughly examined with a petrographic microscope. The minerals were identified and their size, shape, and frequency were recorded, along with the overall frequency of inclusions, sorting, the optical activity of the fabric, and information on the nature of the clay (Whitbread, 1989; Whitbread, 1995). Geological maps from Lebanon and Israel were consulted to locate regions where all of the mineralogical components and clay types could be found (Bartov, 1994; Dan *et al.*, 1975; Dubertret, 1945; Dubertret, 1949; Dubertret and Weztel, 1945; Ilaiwi, 1985; Sneh *et al.*, 1998a,b; Wetzel, 1945). Additionally, thin sections of modern beach sand from several Israeli sites were examined to characterize the coastal sediment at these localities. This was important as studies of Canaanite jars (e.g. Cohen-Weinberger and Goren, 2004; Smith *et al.*, 2004) have highlighted the use of coastal sand as temper.

In order to further identify examples of locally available raw materials, thin sections were examined from the Late Bronze Age Canaanite jars from the Uluburun shipwreck and the Late Bronze Age Amarna Letters from the Near East that were found in Egypt at Amarna (Goren *et al.*, 2004). The LBA Canaanite jars, while not contemporaneous, are a good comparative source particularly for illustrating the materials along the Carmel Coast of Israel. The Amarna Letters are not ideal comparative material, since they are made from light coloured fine clay with few inclusions that was selected for the purpose of writing and the tablets are generally not well fired. Nevertheless, the ability to link these tablets with specific sites through the texts meant that their petrographic features could characterize available resources within a particular area.

A total of 217 previously prepared thin sections of imported material from Tell el-Dab'a were examined at the Austrian Academy of Sciences in Vienna. Only forty-three of the examined thin sections had also been analyzed by Cohen-Weinberger and Goren (2004). Additionally, thin sections representative of some of the petrographic groups were not present. A research trip to Tell el-Dab'a allowed for a more comprehensive comparison of the Memphis MBA Canaanite jars to those found at this site to more clearly understand the connections between the two corpora.

Chemical compositional data from the Memphite MBA Canaanite jars was obtained through Inductively Coupled Plasma Atomic Emission Spectrometry (ICP-AES) and Inductively Coupled Plasma Mass Spectrometry (ICP-MS). This technique was chosen as it provides comparative data with a similar level of sensitivity, accuracy and precision to the existing NAA data from a selection of the MBA Canaanite jars from Memphis (Porat *et al.*, 1991; Tsolakidou and Kilikoglou, 2002; Al-Dayel, 1995). The samples were taken from the sherds following the procedures used for the NAA study. For analysis, the material was digested by lithium metaborate fusion, which is thought to produce the best results for ceramic material (Tsolakidou *et al.*, 2002). ICP-AES analysis gave elemental data for the major elements (Ca, Al, Fe, K, Mg, Mn, Na, P, Si, Ti) and the minor elements (Ba, Cr, Ni, Sr, V, Zn). ICP-MS data were acquired for the following elements: Ce, Co, Cr, Cs, Dy, Eu, Hf, La, Lu, Rb, Sc, Sm, Ta, Th, U, and Yb. These elements were selected intentionally to match with the elemental data available from the previous NAA of Memphis MBA Canaanite jars.

The compositional data were analyzed statistically by Principal Components Analysis (PCA), Hierarchical Cluster Analysis (HCA), K-means Cluster Analysis (KCA), and Discriminant Analysis (DA) after transformation of the raw data into base 10 logarithms. These statistical tests employ different methods for examining the variability and structure in the data, and were used together to provide a more rigorous analysis of the data (Shennan, 1997). The petrographic group assignments for the samples were tested through DA whereby each sample was assigned to a group, and the chemical similarity in that group was compared to the compositional data from the other groups. In order to combine the ICP and NAA data, the results for the individual elements from samples with both types of data were plotted against each other (Ashton and Hughes, 2005). The equation for the slope of the best-fit line passing through the points was used to adjust the ICP data to be more similar to the NAA data. Then the combined data were transformed into base 10 logarithms before being examined by the above statistical tests.

Results

The petrographic data and the chemical data both suggested that four major compositional groups exist within the MBA Canaanite jar samples (Figure 2). However, the attempt to combine the ICP and NAA data into one compositional set proved problematic. This was determined based on the misclassification of samples from the same sherd and the lack of congruency with the petrographic results. Therefore, the two data sets were kept separate for the statistical analysis (Figure 3).

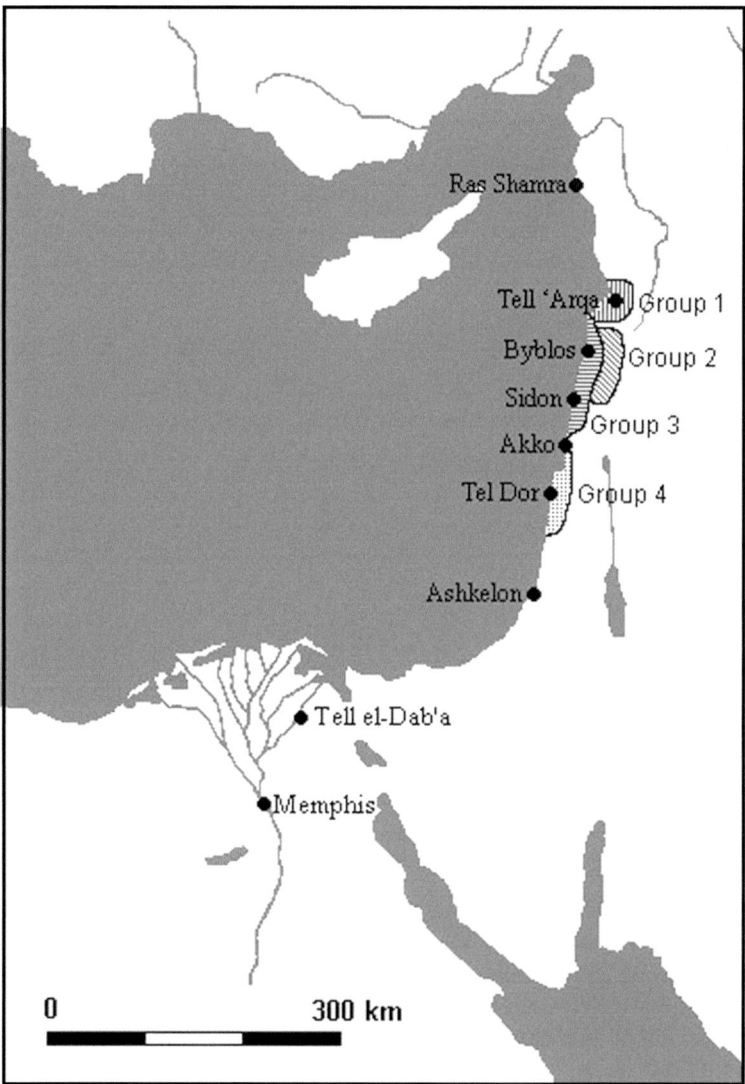

Figure 2. Map of Memphite MBA Canaanite Jar source locations interpreted in this study.

The first group, Group 1, consists of seven samples made from a Neogene marl clay and poorly sorted inclusions of chalk, micritic and sparry limestone, chert, chalcedony, geode quartz, opaques, and alkali olivine basalt (Figure 4A,B). Some of the basalt is altered to iddingsite and minerals deriving from the basalts such as plagioclase and pyroxenes also exist in the matrix. Deposits of Neogene clay are found in the Akkar Plain and near Tripoli (Beydoun, 1977). The geode quartz, chert, and chalcedony are to be found in Santonian-Campanian or Eocene and Cenomanian-Turonian deposits (Beydoun, 1977). Both the type of clay and inclusions suggest a provenance near the Lebanon Mountains, and probably along the Akkar Plain where the Nahr el-Kebir river brings in Pliocene basalt material from further inland. The lack of bioclasts characteristic of coastal sediments further implies an inland source. Important MBA sites in this area are Tell 'Arqa and Tell Kazel (Thalmann, 2006; Badre et al., 1994). Examination of the LBA Amarna tablets sent from this area, however, revealed that the basalts were not similar in their crystalline structure. Therefore, for this group the general designation of Akkar Plain would be appropriate, although a precise provenance has yet to be identified (Figure 2). While none of the thin sections from Tell el-Dab'a contained similar basalt inclusions, the hand specimen examination of Canaanite jar fabrics with basalt fragments did suggest some of the Tell el-Dab'a samples were similar to the Memphite samples in this group.

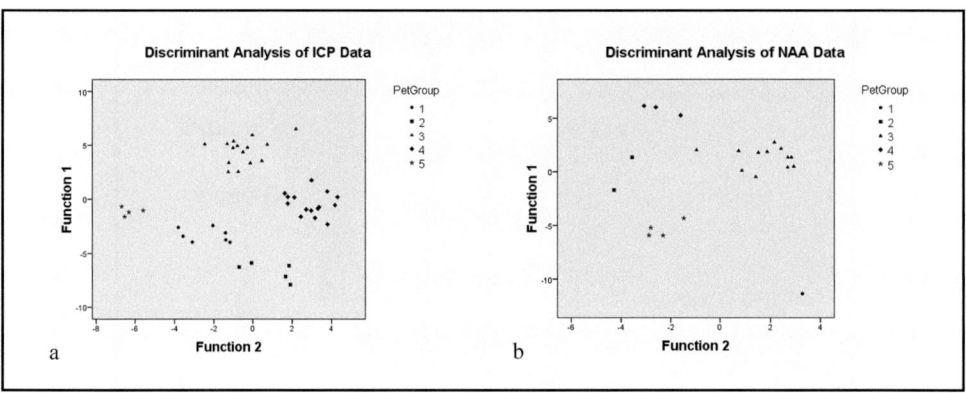

Figure 3. Chemical compositional groups of MBA Canaanite Jars from Memphis as determined by discriminant analysis of a) ICP and b) NAA data.

The six samples within Group 2 contain chert, chalcedony, geode quartz, iron-stained chalk, micritic and sparry limestone, biotite and/or muscovite mica, opaques, and very rare quartz (Figure 4C,D). The inclusions are poorly sorted and none of the samples contain typical coastal sediments. The clay appears to be an iron rich rendzina, a soil that develops on the limestone outcrops throughout the Levant due to the Mediterranean climate (Wieder and Adan-Bayewitz, 2002, p. 397-398). This soil is found within the chalk exposures of the Eocene deposits that also contain chert. The sedimentary inclusions once again relate to the Santonian-Campanian or Eocene, and Cenomanian-Turonian deposits in the Lebanese mountains. Therefore, the location of manufacture of these vessels is likely to be inland Lebanon, possibly near Byblos (Figure 2) (Dunand, 1954). From the LBA comparative material, no exact parallels were seen, although the inclusions of chert, chalcedony, and geode quartz were seen in samples deriving from Lebanon. Examination of MBA Canaanite jar sherds from Tell el-Dab'a found several nearly identical matches to the Memphite samples in this group, although none of the thin sections studied revealed this link.

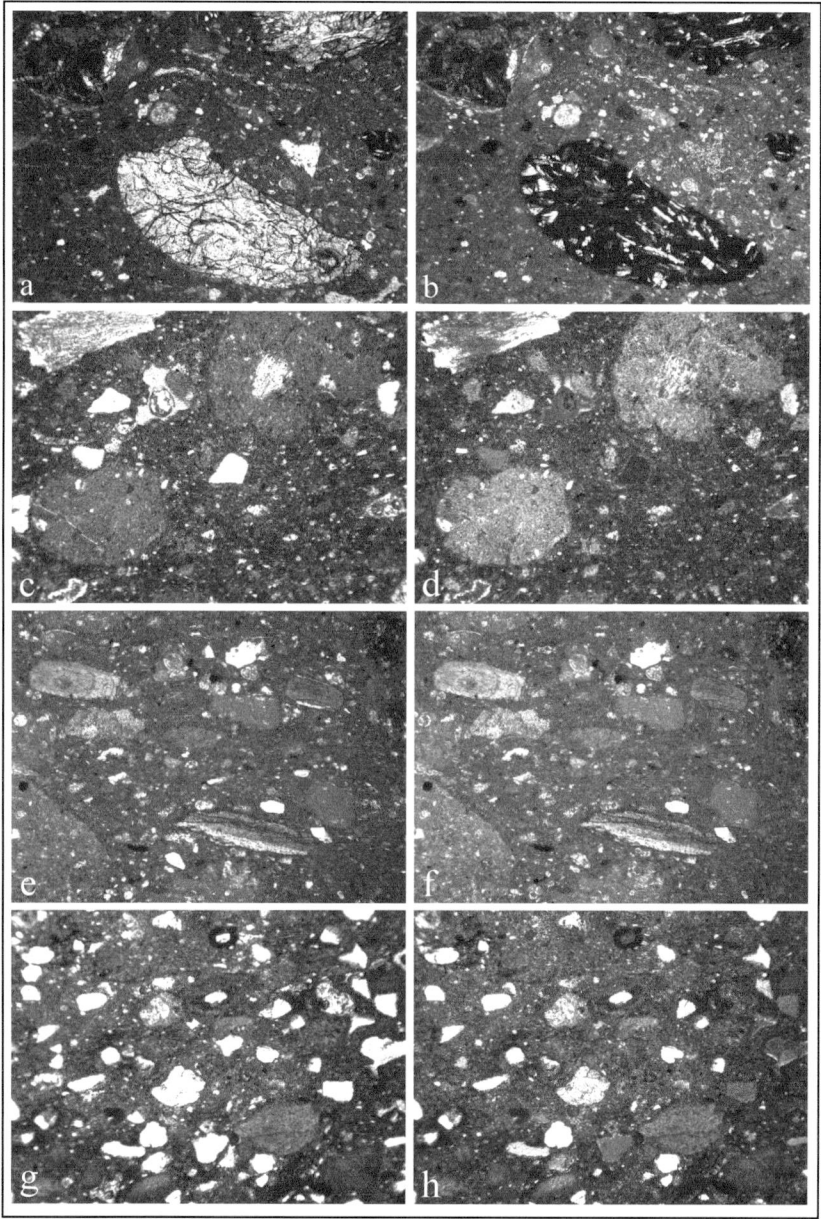

Figure 4. Photomicrographs of MBA Canaanite Jars from Memphis. a,b) Group 1 in PPL and XP showing large hypocrystalline basalt fragments, c,d) Group 2 in PPL and XPL showing iron-stained chalk inclusions, e,f) Group 3 in PPL and XP showing clast of the alga *Amphiroa* in upper left, g,h) Group 4 in PPL and XP showing high proportion of quartz grains. Image width = 2.75 mm.

The twenty-one heterogeneous samples in Group 3 are manufactured of similar materials, namely rendzina clay and bioclastic coastal sand, but they differ in the quantities of these components (Figure 4E,F). The poorly sorted inclusions consist of bioclastic coastal sand that is dominated by fragments of coralline algae of the genus *Amphiroa*. This fossil is often found in Quarternary beach deposits of Pleistocene and younger age (Sivan, 1996). An inland sedimentary raw material source is suggested by the chert, chalcedony, geode quartz, and micritic limestone and chalk found in the samples. The lack of quartz in the beach sand and the sedimentary inclusions indicate that the production of these ceramics took place along the Lebanese coast. Coastal sand in this area has very little quartz and contains prevalent bioclasts, and the sedimentary outcrops are close to the coast in several areas, increasing the likelihood of finding coastal sands with both chert and chalk inclusions. The probable location of manufactured for these samples is therefore the coastal area between Akko and Sidon, though some may have been produced near Beirut or Tripoli where small amounts of coastal sand can also occur (Figure 2). Archaeological sites with MBA levels in this area are Byblos, Beirut, and Sidon (Dunand, 1954; Badre, 1997; Doumet-Serhal, 2004). The lack of kiln sites excavated along the coast of Lebanon impedes a more exact provenance assignment. While the clay matrix of Group 3 did not match the LBA comparanda, the inclusions were similar and substantiate the assignment of this group to the coast of Lebanon. Comparison with the Tell el-Dab'a thin section samples from the Lebanese coast supported the provenance assignment for the Memphite material and confirmed the visual similarities between the samples. However, the Tell el-Dab'a samples from this area can be produced from several different clay types, while the Memphite Canaanite jars appear to be made from geologically similar clays.

The thirteen Group 4 samples also contain inclusions indicative of a coastal provenance (Figure 4G,H). In this case, the coastal sand temper consists of almost equal proportions of bioclasts and quartz grains that are fairly sorted. Clasts of the alga *Amphiroa* are rarely present and various amounts of feldspars, opaques, chalk, and micritic and sparry limestone can be seen, while chert and epidote inclusions are occasionally present. The clay is indicative of a rendzina soil, but with some examples having *terra rossa* added. *Terra rossa* is an iron-rich clay that forms on limestone and dolomite outcrops and is thought to have been commonly added by the Levantine potters to other clays (Wieder and Adan-Bayewitz, 2002, p. 395, 406; Bentor, 1996). The lack of Amphrioa and increase in quartz places the location of manufacture of these samples further south than Group 3 and perhaps in northern Palestine along the Carmel Coast (Figure 2). In this area, quartz grains, which originate from the Nile River and are deposited along the Palestinian coast due to the Mediterranean currents, are less common and there is an increase in the bioclasts (Rohrlich and Goldsmith, 1984). One sample from this group contained a single inclusion of unaltered holocrystalline basalt. This basalt fragment most likely derives from the Upper Cretaceous volcanics that outcrop along the Carmel Ridge. The fresh nature of this basalt fragment might differentiate it from the more altered basalt inclusions found in the Jezreel Valley (Sivan, 1996). Several important MBA sites are known along the northern coast of Palestine, namely Tel Akko and Tel Nami (Artzy, 1993; Dothan, 1993).

There are a few petrographically unique samples that contain similar amounts of quartz and bioclasts, signifying that they may have originated on the northern coast of Palestine. One such sample appears to have been made from Hamra, a type of soil is found only within the Palestinian coastal plain deposits from Ashdod northwards (Gvirtzman *et al.*, 1998). The LBA Canaanite jars from the Uluburun shipwreck produced in this area are related to the Memphite samples due to the presence of coastal sand, although they may have different base clays. Within the MBA Canaanite jars from Tell el-Dab'a, a few samples appear to derive from this region, but most of the examples originate from southern Palestine and are made entirely of Hamra soil (Cohen-Weinberger and Goren, 2004, p. 77). In the sherd comparisons between the Tell el-Dab'a and Memphite material, some matches were found, but it was still evident that the Tell el-Dab'a samples exhibited greater diversity in the raw materials utilized and therefore their provenance.

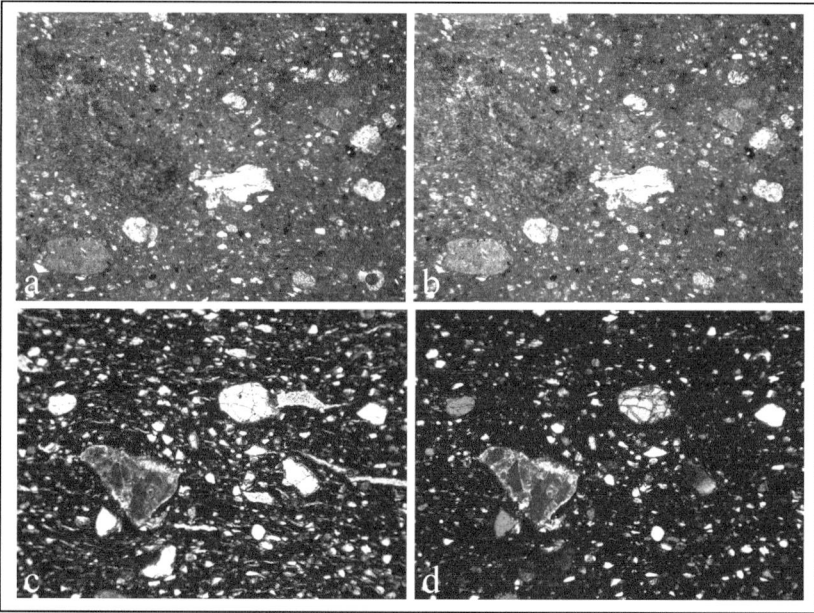

Figure 5. Photomicrographs of MBA Canaanite Jars from Memphis. a,b) Sample Ow 7 in PPL and XP showing large limestone fragment, c,d) Sample Ow 48 in PPL and XPL showing fragment of kurkar in lower left. Image width = 2.75 mm.

In addition to the main compositional groups, nine other samples were petrographically and chemically distinct. Two of these samples may represent Egyptian fabrics composed of calcareous marl clays. Sample Ow 7 was made of a microfossiliferous marl and contained poorly sorted limestone and travertine inclusions, suggestive of an inland, probably Levantine source (Figure 5A,B). One sample, Ow 48, consists of a loess soil with limestone, coarse quartz, and kurkar (Figure 5C,D). Kurkar is a calcareous, cemented quartzitic sandstone found predominantly along the coast of Palestine (Gvirtzman *et al.*, 1998). Additionally, the distribution of loess soil in southern Israel, close to the coast in this area, is suggestive of the provenance of this sample (Yaalon and Dan, 1974). The amount of quartz in the coastal sand may link the

Memphite example to the site of Ashkelon or other important MBA sites nearby (Stager, 1993). This sample is related to Tell el-Dab'a Group K, coming from the Shephelah region, in which Ashkelon is a major site (Cohen-Weinberger and Goren, 2004, p. 79, 81).

Finally, as is the case in most provenance studies of pottery, five samples could not be assigned to a specific location of production. This was mainly due to a lack of diagnostic inclusions and/or the employment of materials that are widely spread throughout the Levant. These samples are thus assigned to the general Levantine area (Group 5).

The above comparison between the Tell el-Dab'a and Memphite MBA Canaanite jars revealed that certain fabrics were present at both sites. However, some fabrics at Tell el-Dab'a were not present in the Memphis material (Cohen-Weinberger and Goren, 2004). This seems to be due to the rarity of these fabrics and probable chronological changes in the trade they represent. One unique fabric at Tell el-Dab'a is Group A, believed to originate from the northern coast of Syria and having a serpentine-rich clay containing volcanic rock fragments, limestone, and rare radiolarian chert (Cohen-Weinberger and Goren, 2004, p. 71-73). None of the Memphite Canaanite jars contain these inclusions, nor do the thin sections appear to be similar to the Group A examples. A further group not seen in the Memphite MBA Canaanite jars was Group F, characterized by rendzina soil with chalk and various volcanic tuff inclusions (Cohen-Weinberger and Goren, 2004, p. 76-77). The volcanic tuffs indicate a provenance in the Mount Carmel region of Palestine. While this fabric was not examined in thin section, none of the Memphite samples contain tuff.

Several Tell el-Dab'a fabrics not seen in the Memphite material were used exclusively to manufacture Canaanite jars. This includes the very rare fabric in Group C produced from an ocher-yellow clay containing dark red hematite particles, calcite crystals, and micritic calcitic bodies (Cohen-Weinberger and Goren, 2004, p. 74-75). The likely provenance is the Lebanese coast near Byblos. Although thin sections were not available for comparison, the general description of this fabric does not match any of the Memphite material. Another fabric employed exclusively for Canaanite jars is Group H (Cohen-Weinberger and Goren, 2004, p. 78). The samples were produced in inland Palestine and are made of *terra rossa* with quartz and/or calcareous sand inclusions and sometimes plant remains. This description does not match any of the Memphite samples. Finally, Group I, also from inland Palestine, features samples with a characteristic dolomitic marl clay and calcareous sand inclusions (Cohen-Weinberger and Goren, 2004, p. 78-79). While a thin section of this fabric was not seen, a sherd of this type was examined and proved that none of the Memphite samples are from this area.

Discussion

The provenance interpretation of the MBA Canaanite jars from Memphis in this study suggests that the coast of Lebanon was the primary region of production. In all likelihood, this relates to the political and economic importance of sites in this area and to the products, such as resin, that originate in the region. The second major area for the production of the Memphite Canaanite jars appears to be the northern coast of Palestine. Overall, the variety and geographic extent of the potential sources of the fabrics substantiates the view that a network of sites along the Levantine coast participated in trading goods to Egypt.

Comparison between the MBA Canaanite jars from Tell el-Dab'a and Memphis has revealed some interesting compositional similarities and differences. The examination of the Tell el-Dab'a thin sections and sherds that are believed to have been manufactured along the coast of Lebanon revealed that they were quite similar to the Memphis samples from this area. Furthermore, the identification of sherds similar to Memphite Groups 1, 2 and 4 confirmed that Canaanite jars manufactured in the Akkar Plain, in inland Lebanon, and along the northern coast of Palestine occur at both sites. In the case of the more unusual petrographic compositions at Memphis, it proved difficult to find exact matches, although only a small proportion of the total Tell el-Dab'a material was available for examination and further parallels may exist. Nevertheless, the similarities that were identified are significant enough to suggest that both the Memphite and Tell el-Dab'a MBA Canaanite jars were imported from similar production locations. Furthermore, the majority of Canaanite jars from both sites came from areas on the Lebanese Coast.

While the Memphite jars originate predominantly from northern Lebanon to northern Palestine, those from Tell el-Dabᶜa come from coastal Syria to southern Palestine. Therefore, Memphis did not receive examples of jars from all of the various production locations throughout the Levant, and instead appears to have acquired mostly jars from the more prolific production centers. The explanation for this discrepancy is probably to be found in the nature of the contact between the populations at Memphis and Tell el-Dab'a, and the stark dissimilarity between the two sites. Tell el-Dab'a was an expansive settlement with inhabitants from a range of socio-economic backgrounds and a sizeable portion of the site has been investigated. From the limited area of Memphis that has been excavated, the site was characterized by small houses and grain silos of the artisan class. The character of the Memphite settlement may suggest that the Canaanite jars were not received with their contents intact, but rather are evidence of the reuse of large and useful storage jars. Within the Egyptian ceramic repertoire, no large handled vessels existed for the purpose of storage of goods on a mass scale. Furthermore, Tell el-Dab'a probably consisted of a multi-ethnic population that may have specifically imported certain vessels to provide familiar goods from their homeland. Perhaps these rare types, including the fabrics utilized exclusively for the manufacture of Canaanite jars, were not as likely to be recycled and/or traded to the Egyptians at Memphis.

Conclusions

This study has confirmed that the MBA Canaanite jars from Memphis were similar to those found at Tell el-Dab'a, and that the jars from both sites were predominantly manufactured in the north-central Levant, undoubtedly at important Middle Bronze Age cities. Given the cultural differences between Memphis and Tell el-Dab'a, and the lack of evidence to suggest Hyksos control of Memphis, the similarities in the proposed origin of the MBA Canaanite jars from both sites is significant. While confirmation that the Memphite vessels came to the site via Tell el-Dab'a will be difficult to acquire, there is good evidence to suggest this possibility or to at least confirm that the Egyptians at Memphis acquired Levantine materials from sites that were also supplying Tell el-Dab'a; although reuse of the jars at Memphis must still be seriously considered. The petrographic analysis, in conjunction with macroscopic examination, has improved our understanding of connections between the two sites during a complex period in Egypt, and between the Levant and Egypt during the Middle Bronze Age. Additional research will focus on refining the provenance of the MBA Canaanite jars from Memphis and comparison of this material to the LBA Canaanite jars also found at Memphis. The latter research will assess the changes in importation of Canaanite jars from the Levant to Egypt between the Middle and Late Bronze Age.

Acknowledgments

The research presented in this paper forms part of the doctoral study of M. Ownby being conducted at the University of Cambridge under the supervision of Dr. Charles French. The thin sections were produced in the McBurney Geoarchaeological Laboratory, University of Cambridge. Assistance with the petrography was provided by Dr. Laurence Smith and Dr. Judith Bunbury. Consultation of the thin sections and sherd material from Tell el-Dab'a was made possible through Prof. Manfred Bietak, Dr. Irene Forstner-Müller, and Dr. Karin Kopetzky. Prof. Yuval Goren most generously assisted in examining the thin sections and made his vast comparative collection available for M. Ownby to study. The trip to work in the Laboratory for Comparative Microarchaeology at Tel Aviv University, Israel was made possible by a grant from the Anglo-Israel Archaeological Society. The ICP-AES analysis was conducted at the University of London Royal Holloway Department of Geology with the assistance of Dr. Emma Tomlinson and Ms. Sue Hall. The ICP-MS analysis was carried out at the University of Kingston Department of Geology under the guidance of Dr. Kym Jarvis and Dr. Benoit Disch. The work at both of these facilities was financially supported by the Natural and Environmental Research Council of the United Kingdom. Statistical advice was provided by Dr. Mike Hughes. The samples studied were exported by permission of the Supreme Council of Antiquities, Egypt.

References

Al-Dayel, O. 1995. *Characterization of Egyptian Ceramics by Neutron Activation Analysis*. Unpublished doctoral dissertation, University of Manchester.

Amiran, R. 1969. *Ancient pottery of the Holy Land from its beginnings in the Neolithic period to the end of the Iron Age*. Rutgers University Press, New Brunswick.

Arnold, D., Arnold, F. and Allen, S. 1993. Canaanite Imports at Lisht, The Middle Kingdom Capital of Egypt. *Ägypten und Levante*, V: 13-32.

Artzy, M. 1993. Tel Nami. In: Stern, E., Lewison-Gilboa, A. and Aviram, J. (Eds) *The New Encyclopaedia of Archaeological Excavations in the Holy Land, 4 vols*. Simon & Schuster, London: 1095-1098.

Ashton, S.-A. and Hughes, M. 2005. Large, Late and Local? Scientific Analysis of Pottery Types from Al Mina. In: Villing, A. (Ed.) *The Greeks in the East*. British Museum Research Publication No. 157. British Museum Press, London: 93-103.

Aston, D. 2004. *Tell el-Dab'a XII. A Corpus of Late Middle Kingdom and Second Intermediate Period Pottery*. Verlag der Österreichischen Akademie der Wissenschaften, Wien.

Bader, B. 2007. A Tale of Two Cities: First Results of a Comparison Between Avaris and Memphis. In: Bietak, M. and Czerny, E. (Eds) *The Synchronization of Civilizations in the Eastern Mediterranean in the Second Millennium B.C. II, Vol. III*. Verlag der Österreichischen Akademie der Wissenschaften, Wien: 249-267.

Badre, L. 1997. Excavations of the American University of Beirut Museum 1993-1996. *Bulletin d'Archéologie et d'Architecture Libanaises*, 2: 6-94.

Badre, L., Gubel, É., Capet, E. and Panayot, N. 1994. Tell Kazel (Syrie), Rapport préliminaire sur les 4e-8e campagnes de fouilles (1988-1992). *Syria*, 71: 259-346.

Bartov, Y. 1994. *Geological Photomap of Israel and Adjacent Areas, 1:750,000, 2nd Edition*. The Geological Survey, Jerusalem.

Ben-Tor, D. 2007. Scarabs of the Middle Kingdom: Historical and Cultural Implications. *Bulletin of the Egyptological Seminar*, 17: 14-25.

Bentor, Y. 1996. *The Clays of Israel: guidebook to the excursion*. Israel Programme for Scientific Translation, Jerusalem.

Beydoun, Z.R. 1977. The Levantine countries: the geology of Syria and Lebanon (maritime regions). In: Nairn, A.E.M., Kanes, W.H. and Stehli, F.G. (Eds) The ocean basins and margins. 4A: *The eastern Mediterranean*. Plenum Press, New York and London: 319-353.

Bietak, M. 1996. *Avaris, the Capital of the Hyksos. Recent Excavations at Tell el-Dab'a*. British Museum Press, London.

Cohen-Weinberger, A. and Goren, Y. 2004. Levantine-Egyptian Interactions During the 12th to the 15th Dynasties based on the Petrography of the Canaanite Pottery from Tell el-Dab'a. *Ägypten und Levante,* XIV: 69-100.

Dan, Y., Raz, Z., Yaalon, D.H. and Koyumdjisky, H. 1975. *Soil Map of Israel, 1:500,000*. Survey of Israel, Jerusalem.

Dothan, M. 1993. Tel Acco. In: Stern, E., Lewison-Gilboa, A. and Aviram, J. (Eds) *The New Encyclopaedia of Archaeological Excavations in the Holy Land, 4 vols.* Simon & Schuster, London: 17-23.

Doumet-Serhal, C. 2004. Excavating Sidon 1998-2003. In: Doumet-Serhal, C. (Ed.) in collaboration with A. Rabate and A. Resek *A Decade of Archaeology and History in the Lebanon*. The Lebanese British Friends of the Beirut National Museum, London: 102-123.

Dubertret, L. 1945. *Carte géologique au 50.000e Feuille de Beyrouth*. République. Libanaise, Ministére des Travaux Publics, Beyrouth.

Dubertret, L. 1949. *Carte géologique au 50.000e Feuille de Saida*. République. Libanaise, Ministére des Travaux Publics, Beyrouth.

Dubertret, L. and Weztel, R. 1945. Carte géologique au 50.000e Feuille de Batroun. République. Libanaise, Ministére des Travaux Publics, Beyrouth.

Dunand, M. 1954. *Fouilles de Byblos 1933-1938, Vol. 2*. Geuthner, Paris.

Giddy, L.L. and Jeffreys, D.G. 1991. Memphis, 1990. *Journal of Egyptian Archaeology*, 77: 1-6.

Goren, Y. 2003. Book Review: McGovern, P.E. 2000. The Foreign Relations of the "Hyksos". *Bibliotheca Orientalis*, LX/1-2: 105-110.

Goren, Y., Finkelstein, I. and Na'aman, N. 2004. *Inscribed in Clay, Provenance Study of the Amarna Tablets*. Emery and Claire Yass Publications in Archaeology of the Institute of Archaeology, Tel Aviv University, Tel Aviv.

Grace, V.R. 1956. The Canaanite Jars. In: Weinberg, S.S. (Ed.) *The Aegean and the Near East: Studies Presented to Hetty Goldman on the Occasion of Her Seventy-fifth Birthday*. J.J. Augustin, Locust Valley, NY: 80-109.

Gvirtzman, G., Netser, M. and Katsav, E. 1998. Last-Glacial to Holocene kurkar ridges, hamra soils, and dune fields in the coastal belt of central Israel. *Israel Journal of Earth Sciences*, 47: 29-46.

Ilaiwi, M. 1985. *Soil Map of Arab Countries, Soil Map of Syria and Lebanon*. The Arab Center for the Studies of Arid Zones and Dry Lands, Soil Sciences Division, Damascus.

Jeffreys, D.G. and Giddy, L.L. 1993. Looking for Memphis. *Egyptian Archaeology*, 1: 5-8.

Kaplan, M. 1980. *The Origin and Distribution of Tell el Yahudiyeh Ware*. Studies in Mediterranean Archaeology, 62, Paul Åström Vorlag, Göteborg.

McGovern, P.E. 2000. *The Foreign Relations of the "Hyksos". A neutron activation study of Middle Bronze Age pottery from the Eastern Mediterranean*. BAR International Series, 888, Archaeopress, Oxford.

Porat, N., Yellin, J., Heller-Kallai, L. and Halicz, L. 1991. Correlation between Petrography, NAA, and ICP Analyses: Application to Early Bronze Egyptian Pottery from Canaan. *Geoarchaeology*, 6: 133-149.

Rohrlich, V. and Goldsmith, V. 1984. Sediment Transport along the Southeast Mediterranean: A Geological Perspective. *Geo-Marine Letters*, 4: 99-103.

Rye, O. 1981. *Pottery Technology, Principles and Reconstruction*. Manuals on Archaeology, 4, Taraxacum, Washington.

Serpico, M., Bourriau, J., Smith, L., Goren, Y., Stern, B. and Heron, C. 2003. Commodities and Containers: A Project to Study Canaanite Amphorae Imported into Egypt during the New Kingdom. In: Bietak, M. (Ed.) *The Synchronization of Civilizations in the Eastern Mediterranean in the Second Millennium B.C. II. Proceedings of the SCIEM 2000-EuroConference, Haindorf, 2nd of May-7th of May 2001, CChEM4, Vienna*. Verlag der Österreichischen Akademie der Wissenschaften, Wien: 365-376.

Shennan, S. 1997. *Quantifying Archaeology, 2nd Edition*. Edinburgh University Press, Edinburgh.

Sivan, D. 1996. *Paleogeography of the Galilee coastal plain during the Quaternary*. Report GSI/18/96. Geological Survey of Israel, Jerusalem.

Smith, L., Bourriau, J., Goren, Y., Hughes, M. and Serpico, M. 2004. The Provenance of Canaanite Amphorae found at Memphis and Amarna in the New Kingdom: results 2000-2002. In: Bourriau, J. and Phillips, J. (Eds) *Invention and Innovation: The Social Context of Technological Change 2, Egypt, the Aegean and the Near East, 1650-1150 BC. Proceedings of a conference held at the McDonald Institute for Archaeological Research, Cambridge, 4-6 September 2002*. Oxbow Books, Oxford: 55-77.

Sneh, A., Bartov, Y. and Rosensaft, M. 1998a. *Geological Map of Israel 1:200,000, Sheet 1*. State of Israel Ministry of National Infrastructures Geological Survey of Israel, Jerusalem.

Sneh, A., Bartov, Y. and Rosensaft, M. 1998b. *Geological Map of Israel 1:200,000, Sheet 2*. State of Israel Ministry of National Infrastructures Geological Survey of Israel, Jerusalem.

Stager, L.E. 1993. Ashkelon. In: Stern, E., Lewison-Gilboa, A. and Aviram, J. (Eds) *The New Encyclopaedia of Archaeological Excavations in the Holy Land, 4 vols*. Simon & Schuster, New York: 103-112.

Thalmann, J.-P. 2006. *Tell Arqa-1. Les niveaux de l'âge du Bronze*. Institut Français du Proche-Orient, Beyrouth.

Tsolakidou, A. and Kilikoglou, V. 2002. Comparative analysis of ancient ceramics by neutron activation analysis, inductively coupled plasma-optical-emission spectrometry, inductively coupled plasma-mass spectrometry, and X-ray fluorescence. *Analytical and Bioanalytical Chemistry*, 374: 566-572.

Tsolakidou, A., Buxeda i Garrigos, J. and Kilikoglou, V. 2002. Assessment of dissolution techniques for the analysis of ceramic samples by plasma spectrometry. *Analytica Chimica Acta*, 474: 177-188.

Wetzel, R. 1945. *Carte géologique au 50.000e Feuille de Tripoli*. République. Libanaise, Ministére des Travaux Publics, Beyrouth.

Whitbread, I. 1989. A Proposal for the Systematic Description of Thin Sections towards the Study of Ancient Ceramic Technology. In Maniatis, Y. (Ed.) *Archaeometry: Proceedings of the 25th International Symposium*. Elsevier, New York: 127-138.

Whitbread, I. 1995. *Greek transport amphorae: a petrological and archaeological study*. Fitch Laboratory Occasional Paper,4, British School at Athens.

Wieder, M. and Adan-Bayewitz, D. 2002. Soil Parent Materials and the Pottery of Roman Galilee: A Comparative Study. *Geoarchaeology*, 17: 393-415.

Yaalon, D. and Dan, J. 1974. Accumulation and distribution of loess-derived deposits in the semi-desert and desert fringe areas of Israel. *Zeitschrift für Geomorphologie Supplementbände*, 20: 91-105.

PETROGRAPHIC ANALYSIS OF EB III CERAMICS FROM TALL AL-'UMAYRI, JORDAN: A RE-EVALUATION OF LEVELS OF PRODUCTION

Stanley Klassen

Near and Middle Eastern Civilizations, University of Toronto, Canada
(stanley.klassen@utoronto.ca)

Introduction

The use of petrography to study Early Bronze Age (EBA) pottery in the Southern Levant has increased substantially in the past few decades. With some exceptions (Dessel, 2009; Benyon et al., 1986; Porat, 1989a; Rast and Schaub, 2003) relatively few such analyses include details on manufacturing technology or go beyond the general description of fabric groups and their relationship to geological formations for the determination of provenance. Although the documentation and description of petrographic fabric groups from the Southern Levant is an important process, it is equally crucial to look within the established fabric types for possible evidence of the technologies utilized by Bronze Age potters and aspects of craft production. In response to this, the present study re-evaluates the production of Early Bronze Age III (EB III) pottery from the site of Tall al-'Umayri, Jordan, using petrography. Preliminary petrographic analysis of EB III pottery from the site indicated the presence of a diverse ceramic industry (London et al., 1991). Metric analyses further demonstrated the dispersed nature of the ceramic industry at the site during the EB III period, suggesting that household production was dominant, with larger jars likely produced at a community level (Harrison, in press). Building on this, the following study incorporates data from a recent, more extensive petrographic examination of the EB III ceramics at Tall al-'Umayri, and identifies the use of specialized clays and tempers for the production of holemouth and flared rim jars. The production of other vessel types indicates a heterogeneous industry with little correlation between the use of specific types of temper and clay. These results reinforce the previous metric analyses, and emphasize the importance of combining morphology and compositional data in order to understand ceramic technology. The petrographic analysis presented here seeks to contribute to an ongoing regional study of EBA ceramic technology in the Madaba Plain region.

Archaeological Background

The site of Tall al-'Umayri is approximately 4.3 ha in size and is located on the northern edge of the Madaba Plain in Central Jordan, where the foothills extending north towards Amman begin (Figure 1). Although the team from the Madaba Plains Project (MPP) found evidence of Early Bronze Age material in all but two of the seven Fields excavated (see Herr, 1997, figure 2.1, p. 8), it is Field D that produced the most extensive and best-preserved EBA material remains to date (Daviau, 1991; Harrison,

1995, 1997, 2000a, 2000b, in press). After four seasons of excavations conducted between 1984 and 1992, six architectural phases dating to the EBA have been identified in Field D. Field Phase 4 (see Harrison 2000b, figure 5.3, p. 97) for phasing of Field D) shows the best preservation with three domestic structures dating to EB III (see Harrison, in press, figure 3. This field phase contained a large number of ceramic vessels found *in situ*, and ended with a destruction sealing the occupation level. The pottery from this field phase therefore represents an ideal assemblage for studying the EB III ceramic industry at Tall al-'Umayri.

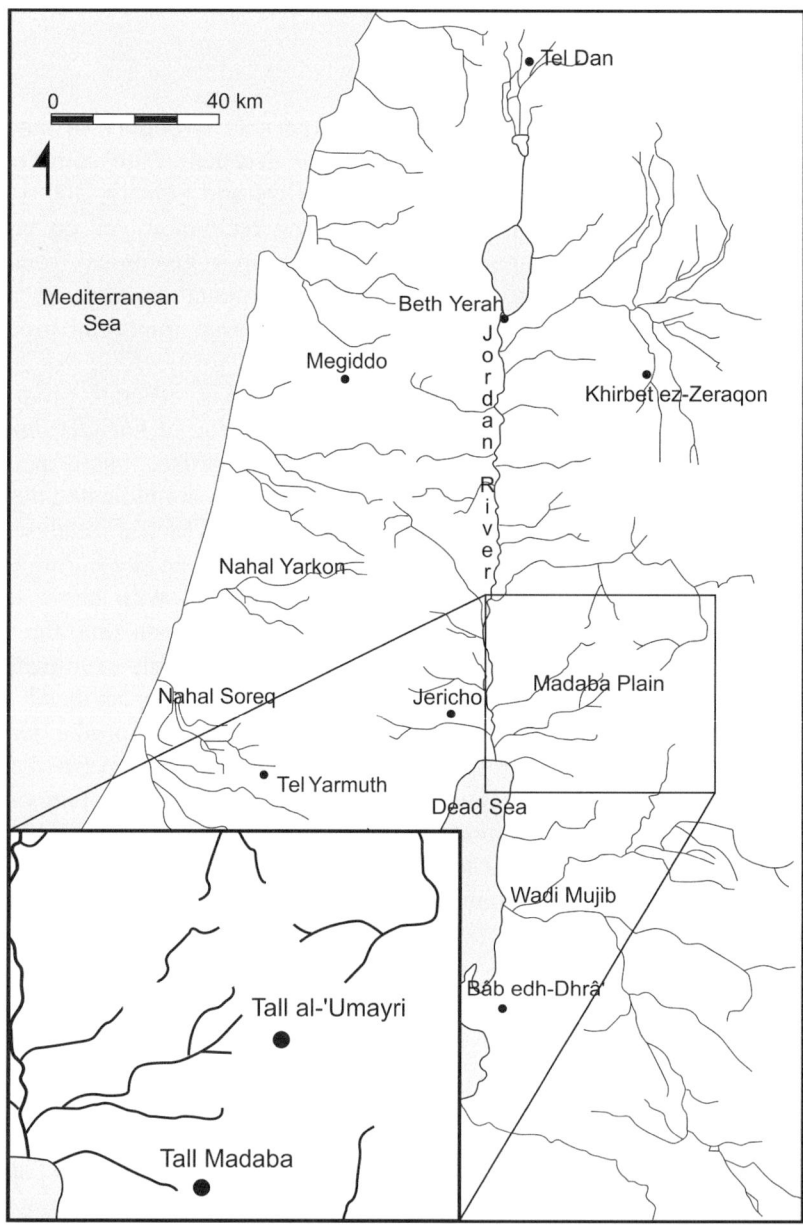

Figure 1. Map of Southern Levant with the location of Tall al-'Umayri and other EB III sites.

Research into EBA Ceramic Production at Tall al-'Umayri

A number of studies have been conducted on the EB III pottery from Field D at Tall al-'Umayri using a variety of analyses. Although most of these are preliminary in nature and focus on EB III ceramics from a number of field phases, they highlight the diversity evident in the potting industry at Tall al-'Umayri during this period. Initial reports on the organization of EBA pottery production at Tall al-'Umayri have identified at least three levels of production (Harrison, 2000a; Harrison and Savage, 2003), two at the local level (household and specialized) (London, 1991, 1995; London *et al.*, 1991), as well as non-local production. Examination of the local manufacturing technology (London 1991, 1995) indicates the ceramic industry to have been dispersed and heterogeneous with four fabrication techniques identified: hand made using pinching, moulds, coiling, and finishing on a slow turntable. It would also appear that the size of the vessel is indicative of the method of manufacturing, with craft specialists making larger jars while the smaller household vessels were likely produced either by domestic potters or by specialists. Confirmation of local pottery production is evidenced by the lower portion of a slow wheel that was recovered from excavations in Field D in 1984 (Platt and Herr, 2002).

Further research into EBA production modes was recently conducted through the analyses of the metric variability of the pottery found exclusively in Field Phase 4 within each of the three compounds in Field D (Harrison, in press). This was used to examine the 'corporate village' model of Philip (2001), by testing whether the potting industry at Tall al-'Umayri reflects household or community level production; the hypothesis being that pottery produced at the community level would show common attributes at the intra-site level with increased variability apparent within households. Harrison (in press) compared the coefficient of variation (CV) values of measurable attributes of the pottery to the CV thresholds established by Roux (2003). These stipulate that CV values of less than 3% indicate large-scale production whereas those greater than 6% are indicative of small-scale production.

Unfortunately, there were insufficient numbers of most vessel forms from Tall al-'Umayri to effectively test the corporate village model. An exception was the larger jars such as holemouth, flared rim, and necked jars. Analysis of these vessels recorded CV values that were suggestive of small-scale production at the house-compound level (Harrison, in press). However, the variability of the flared rim and holemouth jars decreased when the vessels were compared across the site, and increased slightly for the necked jars. Although the CV values are not as low as Roux's established threshold (2003, figure 8, p. 780), this decrease in CV values suggests that these vessels were likely produced at the neighbourhood or community level as opposed to the household level. The slight increase in values for the necked jars may be due to the occurrence of varying size classes within this vessel group (Roux, 2003, p. 778).

Petrography has also been applied in the analysis of the EB III potting industry at the site (London *et al.*, 1991). Although very preliminary in scope, this has reinforced the macroscopic findings of a diverse potting industry, as different fabric groups existed for each of the four vessels analyzed. London *et al.* (1991) suggested that the use of grog temper may be indicative of specialized production as it was found exclusively in the flared rim storage jar analyzed. Both calcite and fossilized shell were the preferred temper for what the authors identified as cooking pots, while a painted sherd was composed of a fine untempered carbonate rich matrix.

The petrographic data presented in the present paper is an attempt to build on the current understanding of the ceramic industry during the EB III period at Tall al-'Umayri. Thirty-six thin sections from pottery excavated from Field Phase 4 were analyzed with a variety of research goals in mind. As only four samples have been published to date (London *et al.*, 1991), a more extensive study was necessary. The expanded petrographic analysis was conducted in part, to verify the results of the preliminary study of (London *et al.*, (1991) with regards to carbonate and grog tempering, as well as the interpretation of the levels of ceramic production. Additionally, a more thorough understanding of paste composition of a larger sample size and variety of vessel forms by means of petrography is used to evaluate the recent metric study and determine if ceramic production was indeed restricted to household and community levels.

Materials and Methods

The data presented in this paper is based on thin sections from pottery excavated exclusively from Field Phase 4 in Field D. Taken from a larger petrographic study of the EB material excavated at Tall al-'Umayri, this study presents the data from two separate analyses. All sample numbers beginning with the abbreviation 'UM' were analyzed by the author (Tables 1 and 2), and data assigned to samples designated with 'P' are taken from a study at Yarmouk University by Nisar *et al.* (2000). Although perhaps limited in size, the combined number of samples used represents a cross-section of the principal vessels types identified during excavation of Field Phase 4. The repertoire appears to be largely domestic in nature with vessel types related to storage, serving, cooking, and food processing (Daviau, 1991; Harrison, 1997, 2000a; Mitchel, 1989).

Although there may appear to be some discrepancy within petrofabric groups between samples from the two different analyses (Tables 1 and 2), this is largely due to the differing methodologies used. In the analysis of the 'UM' series, the analytical procedures of Mason (2004) and Whitbread (1986, 1995) have been used as well as the "primary characteristics" of petrographic description put forth by Freestone (1991). Petrofabric groups were characterized based on the relative mineral abundance of the aplastic inclusions evident in the thin sections, including the mineralogy of the silt size fraction of the matrix. These values are expressed as percentages and were estimated using comparison charts (Terry and Chillinger, 1955). In addition, comparison charts were used to identify degrees of sorting, as well as roundness and angularity (Pettijohn *et al.*, 1987). Grain size of the inclusions was determined using the Wentworth scale (Wentworth, 1922).

Petrographic Analysis of Bronze Age Ceramics from Tall al-'Umayri, Jordan

Sample #	Vessel Type	Charcoal	Sandstone	Sed. Rock	Clay Nod.	Chert	Basic Volc.	Biotite	Muscovite	Clinopyrox	Amphibole	Plagioclase	Limestone	Shell	Opaques	Quartz	Grog	Calcite	Micrite
Petrofabric 1 - Calcareous Fossiliferous Clay with High Quartz Grog Temper																			
P405	Small Bowl	-	16(c-e)	4(d)	10(c)	-	-	-	-	-	-	-	-	-	-	-	-	-	-
P409	Small Bowl	-	8(c-e)	5(c-e)	5(c-d)	-	-	-	4(c-d)	-	-	-	-	-	7(c-d)	5(d)	2(d-e)	1(d)	-
P410	Deep Bowl	-	22(c-e)	5(c-e)	7(c)	-	-	-	-	-	-	-	-	-	4(c-e)	3(c-d)	-	-	-
P412	Deep Bowl	-	10(d-e)	10(d-e)	5(c)	-	-	-	1(e)	-	-	-	-	-	-	2(c-d)	-	-	-
P422	Platter	-	12(c-e)	4.5(c-d)	6(c)	-	-	-	-	-	-	-	-	-	2(c)	5(c-d)	tr(d)	-	-
P425	Juglet	-	16(c-e)	4(d)	10(c)	-	-	-	-	-	-	-	-	-	1(c)	5(d)	2(d-e)	-	-
Petrofabric 2 - Calcareous Fossiliferous Clay with Low Quartz																			
Petrofabric 2a - Grog Temper																			
UM 7.45	Small Bowl	5(b-e)	4(c-e)	5(d-f)	tr(a-c)	1(a-d)	-	-	-	-	-	-	-	-	-	-	-	-	-
UM 14.20	Flared Rim Jar	3(b-e)	1(b-f)	5(d-f)	1(a)	1(a-e)	-	tr(d-e)	-	-	-	-	-	-	-	-	-	-	-
UM 15.17	Platter	10(a-f)	3(b-g)	3(d-g)	1(a-b)	7(a-f)	tr(b-e)	tr(d-f)	-	tr(b)	tr(a-c)	-	-	tr(e)	tr(d)	-	-	-	-
P418	Deep Bowl	-	35(c-e)	3(c-e)	3(c)	-	-	-	5(c-d)	-	tr(a)	-	-	-	2(c)	6(c-e)	-	-	-
P419	Vat	-	10(c-e)	12(d-e)	3(c)	-	-	-	1(c-d)	-	-	-	-	-	-	-	3(d-e)	-	-
P427	Juglet	-	30(c-e)	4(d-e)	1(c)	-	-	-	-	-	-	-	-	-	-	2(c-d)	-	-	-
P430	Amphoriskos	-	20(c-e)	8(d-e)	2(c)	-	-	-	-	-	-	-	-	-	-	3(c-d)	-	-	-
P431	Amphoriskos	-	25(c-e)	4(d-e)	tr(c)	-	-	-	-	-	-	-	-	-	3(c)	3(c-d)	-	1(c-d)	2(d)
P432	Necked Jar	-	22(c-d)	2(c-d)	1(c)	-	-	-	-	-	-	-	-	-	-	6(c-d)	-	-	-
P448	Flared Rim Jar	-	25(c-e)	10(d-e)	3(c)	-	-	-	4(c-d)	-	-	-	-	-	-	7(d-e)	-	-	-
P458	Holemouth Jar	-	25(c-e)	4(c-d)	-	-	-	-	-	-	-	-	-	-	10(c-d)	-	5(d-e)	-	-
P461	Holemouth Jar	-	20(c-e)	15(d-e)	1(c)	-	-	-	-	-	-	-	-	-	-	4(c-d)	-	-	-
Petrofabric 2b - Calcite Temper																			
UM 4.15	Holemouth Jar	15(b-f)	8(a-f)	-	1(a-c)	1(a-f)	-	-	-	tr(a)	-	-	-	-	-	-	-	-	-
UM 8.39	Holemouth Jar	3(b-f)	15(b-g)	-	tr(a)	1(a-e)	-	-	tr*	-	-	-	-	-	-	-	-	-	-
P453	Holemouth Jar	-	30(c-e)	-	1(c)	-	-	-	-	-	-	-	-	-	1(c)	4(c-d)	-	-	-

Table 1. Details of petrofabric groups indicating percentages and grain size of inclusions identified in samples from Field Phase 4, Tall al-'Umayri analyzed in this study. a = coarse silt, b = very fine sand, c = fine sand, d = medium sand, e = coarse sand, f = very coarse sand, g = very fine granule.

Table 2. Details of petrofabric groups indicating percentages and grain size of inclusions identified in samples from Field Phase 4, Tall al-'Umayri analyzed in this study. a = coarse silt, b = very fine sand, c = fine sand, d = medium sand, e = coarse sand, f = very coarse sand, g = very fine granule.

Sample #	Vessel Type	Charcoal	Sandstone	Sed. Rock	Clay Nod.	Chert	Basic Volc.	Biotite	Muscovite	Clinopyrox	Amphibole	Plagioclase	Limestone	Shell	Opaques	Quartz	Grog	Calcite	Micrite
Petrofabric 2c - Shell Temper																			
UM 10.23	Holemouth Jar	3(b-g)	-	-	tr(a)	1(a-d)	15(b-g)	tr(e-f)	-	-	-	-	-	-	-	-	-	-	-
Petrofabric 2d - Untempered																			
UM 4.9	Jar	5(a-f)	1(a-d)	-	-	2(a-e)	1(c-e)	2(b-d)	tr(a)	-	-	-	-	-	-	1(e-g)	-	-	-
Petrofabric 3 - Calcareous Clay with High Quartz																			
Petrofabric 3a - Grog Temper																			
UM 7.19	Jug	3(b-f)	3(b-f)	7(e-f)	4(a-b)	1(a-g)	tr(c-e)	-	tr(a)	tr(a)	tr(a-b)	tr(a)	tr(a-b)	-	-	1(d-f)	-	-	-
P420	Platter	-	6(c-d)	8(e)	4(c)	-	-	-	-	-	-	-	-	-	-	3(c-d)	-	-	-
P442	Necked Jar	-	20(c-e)	6(c-e)	10(c)	-	-	-	-	-	-	-	-	-	-	5(c-d)	-	-	-
Petrofabric 3b - Calcite Temper																			
UM 8.24	Holemouth Jar	3(c-d)	15(a-g)	tr(f)	5(a-b)	1(a-e)	-	tr(d)	-	tr(a)	-	-	-	-	-	-	-	-	-
Petrofabric 4 - Calcareous Clay with Low Quartz																			
Petrofabric 4a - Grog Temper																			
UM 2.28	Deep Bowl	10(a-g)	tr(a-b)	7(e)	tr(a-b)	1(a-e)	1(d-f)	4(d-g)	-	-	-	-	-	tr(d)	-	4(d-e)	-	-	-
UM 12.18	Large Bowl	5(b-f)	tr(d-e)	2(e-g)	3(a-c)	4(a-b)	-	1(c-e)	-	tr(b)	tr(c)	-	-	-	1(d-f)	-	-	-	-
UM 14.24	Necked Jar	11(b-g)	1(b-f)	5(d-f)	1(a-b)	1(a-g)	1(d-f)	-	-	tr(b)	tr(a)	-	-	-	-	-	-	-	-
P426	Juglet	-	20(d-e)	7(c-e)	tr(b)	-	-	-	1(c-d)	-	-	-	-	-	1(c)	10(c-d)	-	-	-
P434	Necked Jar	-	10(d-e)	10(d-e)	-	-	-	-	tr(c-d)	-	-	-	-	-	10(c)	7(c-e)	-	-	-
P444	Flared Rim Jar	-	15(c-e)	7(c-e)	2(c)	-	-	-	1(c-d)	-	-	-	-	-	3(c-d)	-	-	-	-
P445	Flared Rim Jar	-	15(c-e)	5(d-e)	3(c)	-	-	-	tr*	-	-	-	-	-	-	4(c-d)	-	-	2(c-d)
Petrofabric 4b - Calcite Temper																			
UM 10.13	Holemouth Jar	4(a-d)	12(a-g)	-	tr(a)	1(a-e)	-	-	-	tr(a)	-	tr(a)	-	-	-	-	-	-	-
Petrofabric 5 - Calcareous Clay with Large Opaques																			
Grog Temper																			
UM 6.1	Flared Rim Jar	5(a-f)	1(c-e)	10(d-f)	tr(a-b)	5(a-g)	-	-	-	-	tr(a)	-	-	-	tr(a-d)	-	-	-	-

The petrographic data of Nisar *et al*. (2000) includes the size, shape, and abundance of identified inclusions. Although petrofabric groups were not recorded, the authors concluded that most of the sherds analyzed had an almost identical paste mineralogy with only minor variations. Inclusions smaller than 100 μm were not included in their analysis, which accounts for the lack of very fine sand to silt size inclusions in the "P" samples (Tables 1 and 2). Another discrepancy occurs in the identification of calcium carbonates. The combined percentage of calcium carbonates in the form of micrite, calcite (sparry, euhedral), limestone and bioclasts that are identified in the "UM" series, appears similar in percentage to the grouping of all calcium carbonates identified in the study of Nisar *et al*. (2000) under the term "calcite". Thus, the high percentage of calcite that often appears in the "P" series samples does not necessarily indicate calcite temper, but represents all the various forms of calcium carbonates identified.

Regional Geology

Tall al-'Umayri is located in the Highlands of Central Jordan on the northern edge of the Madaba Plain along the eastern rim of the Wadi Araba-Jordan Graben. It is bounded by the Wadi Araba-Dead Sea-Jordan River Depression to the west and the East Jordanian Limestone Plateau to the east (Bender, 1974; Bullard, 1972). The lithostratigraphy of the immediate area, as described by Schnurrenburger (1991, p. 370) is typical of the wider Madaba Plain region (Bullard, 1972; James, 1976; Lacelle, 1986; Al-Hunjul, 1995; Harrison, 1995, p. 10-13; Shawabekeh, 1998).

Deposits belonging primarily to the Upper Cretaceous periods of the Turonian to the Campanian ages underlie the region around the site. These are represented by the Upper Ajlun and Belqa Groups, composed primarily of the Wadi as-Sir Limestone formation represented predominately by limestone with alternating layers of marls, dolomitic and fossiliferous limestone, marls with chert nodules, and oyster beds (Bender, 1974, p. 78; Shawabekeh, 1998, p. 35-37), through the Amman Silicified Limestone formations made up of massive chert beds (lower strata) with varying layers of coquinal, fossiliferous, micritic, and phosphatic limestone, chert, and marl (Bender, 1974, p. 78; Al-Hunjul, 1995, p. 12-15). The Al-Hisa Phosphorite formations of the Maastrichtian age (Belqa Group), composed of alternating layers of silicified phosphorites, phosphatic, micritic, and chalky limestone, marl, and oyster beds (Bender, 1974, p. 79; Al-Hunjul, 1995, p. 15-17), appear more predominately to the south near Madaba. Lower Cretaceous deposits of the Kurnub Group outcrop to the west of Tall al-'Umayri, where the deeper wadis have cut through the softer sandstone. Outcrops of this formation are found in Transjordan extending from the lower part of the Dead Sea to the Wadi Zarqa. Medium-coarse-grained quartzose sandstone intermixed with rounded quartz pebbles and granules are typical of the lower levels of this formation. Silt and clay proportions increase at the upper reaches with ferruginous siltstone, sandstone, and plant fossils (Bender, 1974, p. 70-73; Shawabekeh, 1998, p. 7-29).

Red Mediterranean Soils (RMS) predominate around the site of Tall al-'Umayri. Although similar to the Terra Rosa soils of the Western Mediterranean area, the entire profile of the RMS contains a higher carbonate content (Bender, 1974, p. 187). Accumulation of sediments in wadi bottoms around the site, are identified primarily as colluvium with small amount of alluvium present (Schnurrenburger, 1991, p. 372-374).

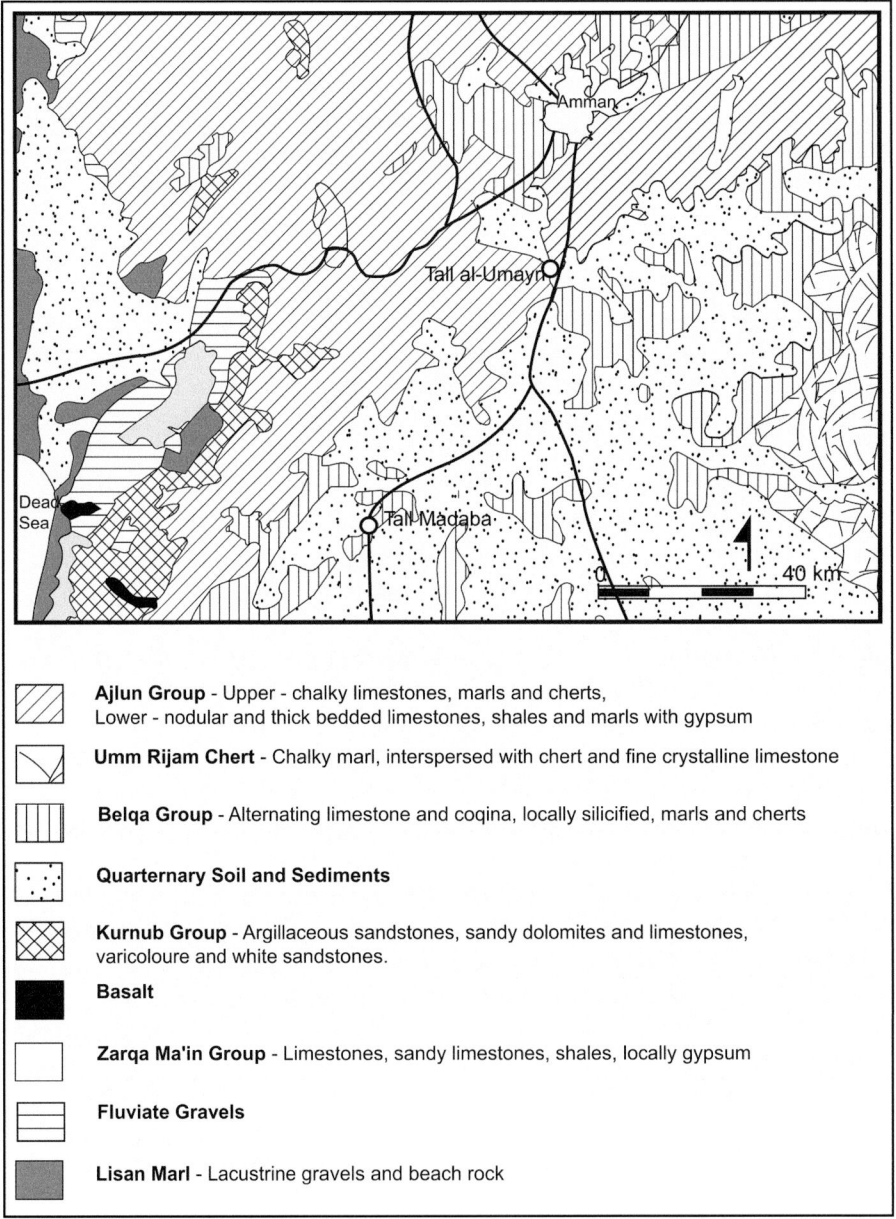

Figure 2. Geological map of the region around Tall al-'Umayri (adapted from Bender, 1974 and Bullard, 1972).

Petrofabric Descriptions

Analysis of the petrographic data from the Field Phase 4 samples has identified five distinct petrofabric groups. As described above, the petrofabric groups were characterized based on the relative mineral abundance of the aplastic inclusions evident in the entire thin section, including the mineralogy of the silt size fraction of the matrix. The major distinction between the petro-fabric groups is the occurrence of either a high or low percentage of fine quartz and the presence or absence of calcareous microfossils in the form of foraminifera and ostracods within the matrix. Although using the presence-absence of microfossils alone as a method of characterization is questioned, using these types of bioclasts together with other inclusions can produce meaningful petrographic groupings (Quinn and Day, 2007). Sub-groups within each petrofabric group are defined by tempering agents, or the lack thereof, which indicate technological modifications by the potter (Porat, 1989a, p. 25).

Brief descriptions of each of the petrofabric groups identified can be found below, with details of overall percentages and sizes of inclusions are given in Tables 1 and 2. Photomicrographs of the petrofabric groups are illustrated in Figures 3 and 4. Unfortunately no photomicrographs of the Yarmouk study were included in the report of Nisar *et al.* (2000).

Petrofabric group 1

This petrofabric group is characterized by the presence of a relatively high percentage of fine monocrystalline quartz within a calcareous, fossiliferous clay matrix (Table 1). All six samples within this group are grog tempered and were analyzed by Nisar *et al.* (2000). Although the dominant feature of this group appears to be calcite, this is less a reflection of calcite tempering than a precise identification of the form of calcium carbonates identified. Primary features include coarse sub-angular grog and fine sub-round to sub-angular quartz. Other inclusions evident are clay nodules, chert, sedimentary rock, plagioclase and sandstone. The occurrence of plagioclase in sample P409 is exceptional for this group and may indicate the addition of wadi sand based on the medium sand sized inclusions of quartz that occur as well. The high percentage of chert may also be indicative of this. Clay nodules are also a feature that is consistent within petrofabric group 1.

Petrofabric group 2

This petrofabric group is composed of calcareous fossiliferous clay and is identified by the presence of a low percentage of monocrystalline quartz within the matrix. Seventeen samples occur in this group, with twelve samples being tempered by grog, three tempered by calcite, one by shell, and one sample that is un-tempered (Tables 1 and 2).

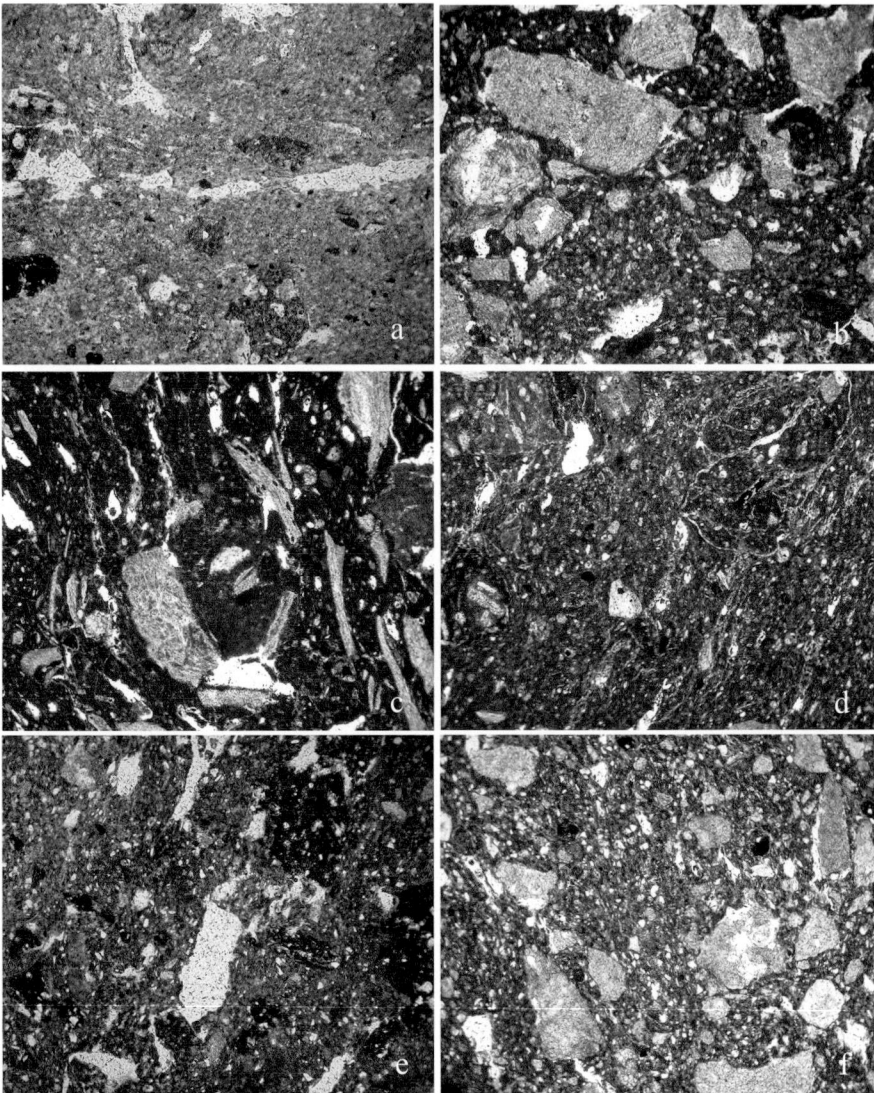

Figure 3. Photomicrographs of petrofabric groups of EB III ceramics from Tall al-'Umayri identified in this study. a) Sample UM 14.20 of PF 2a showing grog temper, b) Sample UM 4.15 of PF 2b showing micritic euhedral and subhedral calcite temper, c) Sample UM 10.23 of PF 2c showing shell temper with coquina limestone (centre and upper right), d) Sample UM 4.9 of PF 2d showing naturally occurring inclusions, e) Sample UM 7.19 of PF 3a showing grog temper, f) sample UM 8.24 of PF 3b showing calcite temper in the form of sparry euhedral, and subhedral inclusions. All images taken in plane polarised light. Image width = 3.7 mm.

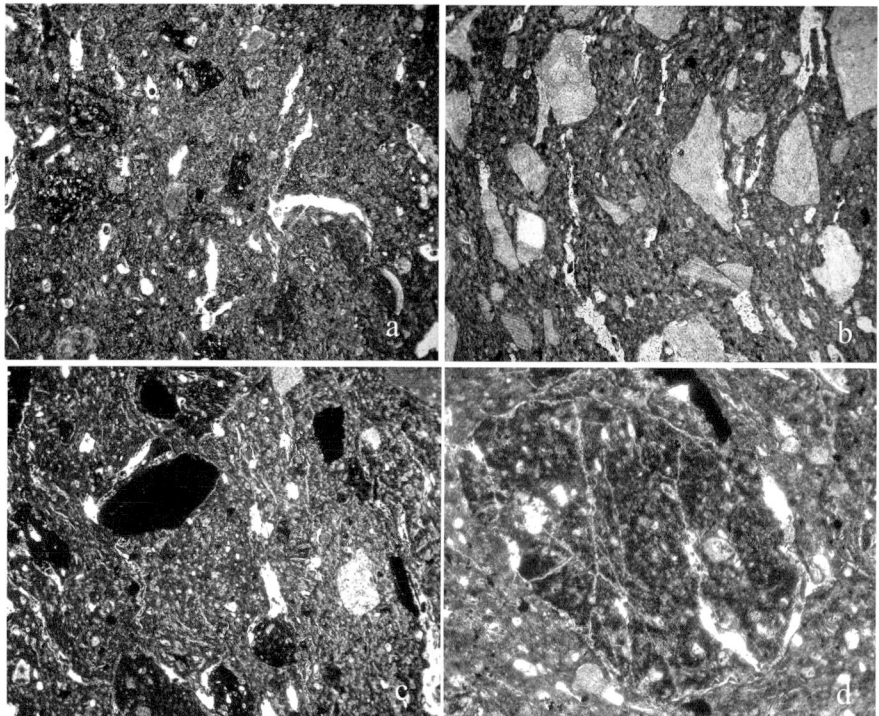

Figure 4. Photomicrographs of petrofabric groups of EB III ceramics from Tall al-'Umayri identified in this study. a) Sample UM 14.24 of PF 4a showing grog temper and micritic carbonates, b) Sample UM 10.13 of PF 4b showing calcite temper, c) Sample UM 6.1 of PF 5 showing grog temper, large opaques, and micrite, d) Sample UM 4.9 of PF 2d showing clay nodule inclusion. All images taken in plane polarised light. Image width = 3.8 mm, except d = 1.3 mm.

Sub-group 2a

The examples of this sub-group contain a mineral suite very similar to that of petrofabric group 1, although the low percentage of fine quartz occurring in the matrix of this group distinguish the two. Prominent inclusions include micrite and sparry calcite, grog, clay nodules, and chert. Other coarse inclusions include sedimentary rock, limestone, and sandstone (P419). Naturally occurring iron oxides identified as opaques are common, and trace accessory minerals in the form of plagioclase, amphibole, and clinopyroxene suggest natural mixing with alluvial sediments. Although sample UM 15.17 has one basalt inclusion, this may have been picked up during the processing of the clay. Further evidence of this is the lack of minerals associated with basic volcanic rocks within the sample (Figure 3A).

Sub-group 2b

Sample P453 is included in this calcite tempered sub-group as the calcite is described as sub-angular and ranges in size from fine to coarse sand in size (Nisar *et al.*, 2000) and is consistent with the other calcite-tempered samples analyzed in the "UM" series. Coarse inclusions include euhedral calcite showing twinning, sparry calcite, micrite, chert, and clay nodules (P453). Limited opaques, amphibole, and plagioclase are also present (Figure 3B).

Sub-group 2c

Only one sample belongs to this sub-group, which is tempered with sub-angular to angular fossilized mollusc shell. Micritic inclusions are also present along with opaques and trace amounts of coarse fossiliferous limestone composed of sparry calcite with silt-sized opaques, and allochems (Figure 3C).

Sub-group 2d

Sample UM 4.9 appears to be untempered as the inclusions have a unimodal distribution and occur at relatively low percentages. Inclusions include micrite, sparry calcite, micritic limestone, shell, opaques, clay nodules, fine quartz and trace plagioclase (Figure 3D).

Petrofabric group 3

This petrofabric group is also composed of calcareous clay (Table 2) and is determined by a relatively high percentage of silt to fine sand sized monocrystalline quartz. Three samples occur in the grog-tempered sub-group (3a), and one in the calcite-tempered sub-group (3B).

Sub-group 3a

The presence of grog temper is the distinguishing characteristic in this petrofabric group. The mineral suite closely resembles petrofabric group 1 with coarse grog, clay nodules, calcite, micrite, and opaques (Figure 3E). Although the accessory minerals identified in UM 7.19 may suggest this sample belongs to a separate petrofabric group, these minerals appear as silt to fine sand sized trace elements and appear in many of the "UM" series samples in all of the petrofabric groups identified at Tall al-'Umayri. This suggests that they occur naturally as alluvial sediments in the region. Their absence from the "P" series samples is due to the methodology used by Nisar *et al.* (2000).

Sub-group 3b

The dominant feature of this sub-group is the evenly distributed angular crystalline calcite showing clear cleavage indicating crushing and addition as temper. Also evident are micrite, coarse limestone, and opaques in the form of iron oxides. Trace elements of grog and amphibole also occur (Figure 3F).

Petrofabric group 4

This petrofabric group is identified by its calcareous clay and low percentage of fine sand to silt sized monocrystalline quartz (Table 2). Seven samples occur in the grog-tempered sub-group (4a) with one sample in the calcite-tempered sub-group (4b).

Sub-group 4a

As with petrofabric group 2a, this group is distinguished by coarse angular grog, rounded to sub-rounded micrite, clay nodules, and unimodal sparry calcite. Other coarse inclusions that occur in some samples include chert, limestone composed of sparry calcite, fine quartz, and opaques, with silt to fine sand sized quartz, plagioclase, opaques and accessory minerals such as amphibole and clinopyroxene (Figure 4A). Sample UM 2.28 has one inclusion of basalt showing trachytic texture with plagioclase laths and clinopyroxene, but should not be considered a separate petrofabric group due to its rare occurrence.

Sub-group 4b

Unimodal sub-angular to angular euhedral calcite showing twinning and clear cleavage is dominant throughout the fabric of this subgroup. Other mineral inclusions appear in the form of micrite, opaques, with trace elements of fine quartz, amphibole and muscovite (Figure 4B).

Petrofabric group 5

This petrofabric group is represented by only one sample and features a calcareous clay matrix with trace elements of monocrystalline quartz. It is tempered with grog and is clearly distinguished from petrofabric group 4 by the coarse-sized ferruginous ooliths occurring throughout the matrix.

Angular coarse grog is the dominant inclusion of this group, with coarse oolitic opaques (iron oxides) and prominent micrite also present. Unimodal sparry calcite occurs with trace elements of clay nodules and silt to fine sand sized quartz and clinopyroxene (Figure 4C).

Discussion

The clay matrices of the samples in petrofabric groups 1-4 are all calcareous. Some contain bioclasts and others not. All contain inclusions representative of the sedimentary rocks that match the description of the Wadi as-Sir Limestone and Amman Silicified Limestone formations that outcrop in the region around Tall al-'Umayri. Thus, we can suggest that these clays were locally procured as were the inclusions used as temper. Based on the lithology and mineralogy of petrofabric group 5, it would appear that this group should be identified with the Kurnub formation of the Lower Cretaceous period exposed to the west of the site (Figure 2). The use of this resource for pottery production is well documented (Goren, 1995, 1996; Greenberg and Porat, 1996), and suggests production at a site other than Tall al-'Umayri.

It appears that there is no correlation between the use of fossiliferous or non-fossiliferous clays and the production of the EB III vessels from Field Phase 4 at Tall al-'Umayri, as vessels of all functional groups are composed of both clay types (Tables 1 and 2). Both types of clays should be considered local, which suggests that a variety of different sources were used by local potters. The variability of clays in relation to vessel type might indicate that certain clay sources were exhausted through time, or that the presence or absence of fossils in the clay did not affect the manufacturing and function of the final product and therefore was not a concern of the potter. It also points to the diversity in production implying that various potters may have formed these vessels. However, when the percentage of naturally occurring silt to fine sand-sized quartz is compared between various vessel forms, a more consistent pattern becomes evident (Figure 5). The analysis conducted on the holemouth and flared rim jars indicates a preference for clays containing a low percentage of quartz. The five flared rim jars examined all show a percentage of naturally occurring quartz of 3% or less. Even more striking is that seven of the eight holemouth jars analysed have fine quartz that ranges between trace to 1%. Only one example (UM 8.24) exceeds that with 5%. The necked jars show a similar pattern to the holemouth and flared rim jars, with three of the four examples utilizing a clay source with less than 1% quartz. However, the sample size is slightly smaller and therefore less as conclusive. This is also true of a variety of other vessel types produced with clays containing limited amounts of quartz. Although these differences in percentage of fine quartz could be due to natural variation in the clay sources, the consistency, even with a relatively small sample size, implies that the potters collected less silty clays for the production of the holemouth and flared rim jars. The absence of quartz suggests that the potters who produced these two vessels types were particular in their choice of raw materials.

Chert appears in the fine fraction in many of the petrofabric groups and the variation in its abundance is likely to be the result of naturally occurring chert nodules in the source area (Benyon et al., 1986; Goren and Gilead, 1987). The only pattern that does stand out is the absence of chert in holemouth jars. Only two of the holemouth jars sampled contained chert: sample P453 has 1% sub-angular to sub-round chert, implying that it occurs naturally, whereas the chert in P458 occurs at 10%, is sub-angular and may be a tempering agent. Interestingly, this same holemouth jar was also tempered with grog, and therefore does not fit the pattern of shell and calcite tempering found in the majority of holemouth jars analyzed at Tall al-'Umayri (see below). Although the potter still utilized a clay source low in quartz as in the other holemouth jars, the grog and chert temper might indicate a different function for this particular vessel.

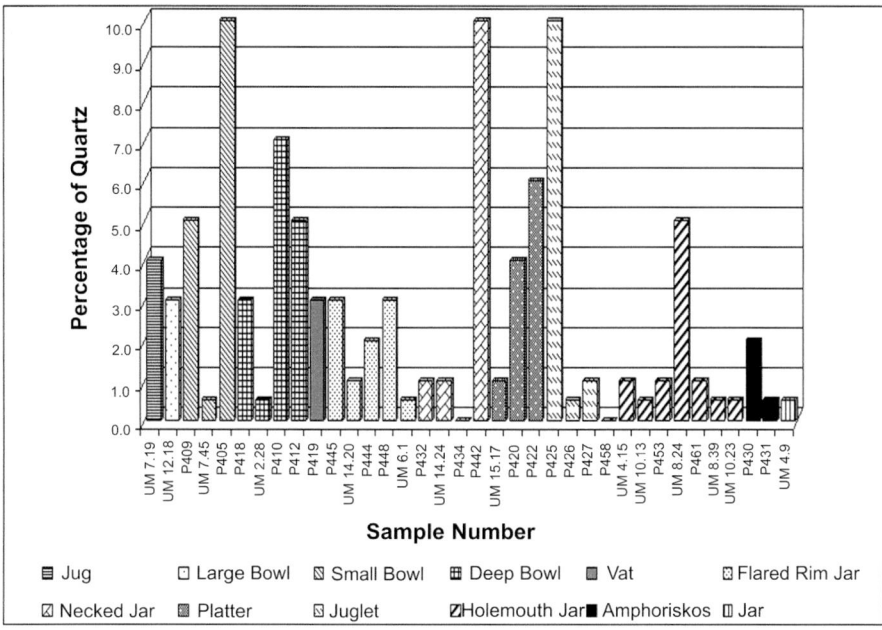

Figure 5. Histogram of relative abundance of quartz in ceramics from Tall al-'Umayri analyzed in this study.

Grog was the most common tempering agent used by the Field Phase 4 potters at Tall al-'Umayri, occurring in 81% of the vessels analyzed (Figure 6). There is no consistency in the percentage of grog used in relation to vessel type, nor in relation to the type of clay used (Tables 1 and 2). Although petrofabric group 2a has the highest number of grog-tempered vessels, the use of this tempering agent is evident in all four of the petrofabric groups identified as local. The dominance of grog as a tempering agent at Tall al-'Umayri suggests a sharing of knowledge related to the manufacturing of pottery among potters at the site. The vessel from petrofabric group 5, identified as coming from a production centre to the west of Tall al-'Umayri, was also grog-tempered, indicating that the use of grog as temper is a technical choice was well-established in the EB III potting industry as it is found at other sites in the region (Batiuk, 2005; London, 1988; Mazer *et al.*, 2000). The only vessel forms that are not consistently grog tempered are the holemouth jars, with only two samples (P458 and P461) containing grog.

The mineralogy of the ceramics used as grog temper was not uniform, suggesting potters crushed pottery from a variety of ceramic vessels. In some thin sections, up to five different sources of grog were found. This indicates that most potters did not discriminate in their use of grog, crushing any available pottery when needed, even non-local vessels. London *et al.* (1991, p. 436) initially suggested that grog temper might have been limited to the specialized production of storage jars. However, the present analysis indicates that a re-interpretation of this manufacturing technology is necessary. Grog temper is extremely common in the pottery recovered from Field Phase 4 at Tall al-'Umayri. It is therefore clear that grog temper was not used exclusively in specialized production, but was a widely used technique in pottery production conducted also at the household level.

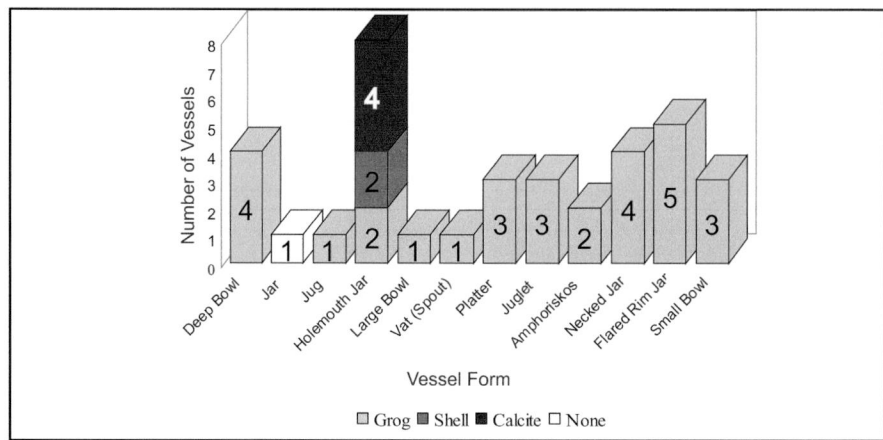

Figure 6. Histogram of types of temper used in vessel types of ceramics from Tall al-'Umayri analyzed in this study.

Similarly, clay nodules occur indiscriminately in petrofabric groups 1-4 (Figure 4D). These inclusions range from fine sand to granule in size and are usually rounded to sub-angular, showing either high or neutral optical density and sharp to clear boundaries (Whitbread, 1986). This type of inclusion was found in 21 of the 36 vessels analyzed (58%). Although some of the clay inclusions are difficult to discern from the matrix in PPL, in XP the distinction is clear. The matrices of the various clay nodules identified were not consistent. In some samples, only one type of clay nodule was evident, whereas in others, a number of different examples could be identified. Whitbread (1986) suggests caution in identifying clay nodules as temper as it is very difficult to determine if they are naturally occurring, caused by mixing clays, or added as a tempering agent. The appearance of such a variety of clay nodule types in the samples from Tall al-'Umayri suggests a technological application by the potters rather than a natural occurrence. This interpretation is reinforced by the noticeable pattern of holemouth jars lacking clay nodule inclusions (Figure 7). Of the eight-holemouth jars, only two (P461 and P453) have clay nodules added: P461 belongs to petrofabric group 2a (grog-tempered), and P453 belongs to petrofabric group 2b (calcite-tempered). Of these two, P461 does not fit the expected technique in the production of holemouth jars (see below). It would appear that the clay nodules are part of a technical tradition of clay mixing or tempering and is restricted (except for P453) to grog-tempered vessels.

Crushed calcite is the second most common aplastic inclusion, occurring in 14% of the samples studied, but is restricted to holemouth jars. The abundance of calcite ranges from 8-30% in the samples analysed, and the size fraction ranges from silt to granule. Angularity, poor sorting, and bimodal distribution of the calcite indicate that this inclusion was crushed and added as temper. The calcite is often in crystalline or cryptocrystalline form and typically shows breakage along cleavage. Other sub-rounded to angular, poorly sorted carbonates occur with the calcite in the form of micrite and micro-sparite suggesting the source of calcite is from calcite veins found in the limestone that is common to the area near Tall al-'Umayri (compare Porat, 1989b). While crystalline calcite occurs in samples belonging to other petrofabric groups (Tables 1 and 2), these inclusions tend to be sub-rounded to rounded. This suggests weathering, and the low percentage evident reflects a naturally occurring coarse inclusion (Rice, 1987, p. 410) rather than temper.

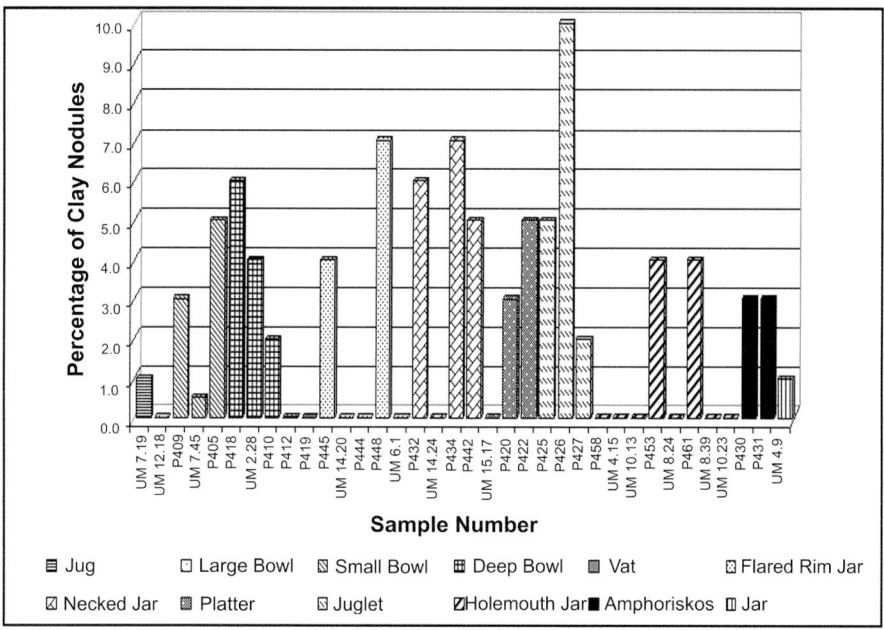

Figure 7. Histogram of relative abundance of clay nodules in ceramics from Tall al-'Umayri analyzed in this study.

Conclusions

The results of this petrographic analysis permits a more comprehensive view of the EB III potting industry at Tall al-'Umayri in Field Phase 4. Five petrofabric groups have been identified, with sub-groups based on various tempering agents added by the potter. The mineralogy of all but one sample, reflects the local geology, indicating that the majority of the vessels were produced in the vicinity of Tall al-'Umayri.

Several levels of ceramic production are evident at EBA Tall al-'Umayri. These include household production, specialization at the community level, and non-local production. Household production is apparent in the heterogeneous character of the majority of vessel types produced. They are composed of a wide variation of clay types, and the indiscriminate use of grog and clay nodules. This heterogeneity is characteristic of a dispersed potting industry. Specialized production is most evident in the production of the larger vessel forms, particularly the holemouth and flared rim jars. It is within these vessels that we find the consistent use of clay sources containing a low percentage of quartz. In the case of the holemouth jars, particular tempers such as calcite and fossilized shell were added, indicating the existence of a more specialized production process. The third and final level of production is represented by petrofabric group 5, a non-local composition that indicates the movement of either pottery or goods from outside the region.

The findings of this study support the conclusions of the metric analysis conducted by Harrison (in press) and expand our knowledge of the EB III ceramic industry in the northern part of the Madaba Plain Region. With a growing database and continued analysis, the diversity and regionalism of EBA pottery production in the Southern Levant is becoming more evident. In the future, a more comprehensive program of petrographic analysis would clarify fabric groups used in those vessel types under-represented in this study and the analysis of samples from both the earlier and later phases of the site could also help in recognizing possible changes in ceramic technology over the course of the EBA.

Acknowledgements

I wish to acknowledge Larry Herr, and the Madaba Plains Project for permitting access to the Tall al-'Umayri material, R. Thomas Schaub for allowing me to analyze the Tall al-'Umayri thin sections from his collection, and Abu-Jaber Nisar and Abu-Jaber al-Sa'ad for access to their unpublished report. I wish to thank Jill Hilditch and Heather Snow for their comments and feedback on this manuscript, and Tim Harrison for his helpful ideas and input, as well as his continued support and encouragement of the larger project.

References

Al-Hunjul, N. 1995. *The Geology of Madaba Area: Map Sheet (3153-II). Geological Mapping Series, Geology Bulletin No. 31*. Geology Directorate, Geological Mapping Division, Natural Resources Authority, Amman.

Batiuk, S. 2005. *Migration theory and the distribution of the early transcaucasian culture*. Unpublished doctoral dissertation, University of Toronto.

Bender, F. 1974. *Geology of Jordan. Contributions to the Regional Geology of the Earth. Supplementary Edition of Volume 7*. Gebrüder Borntraeger, Berlin.

Benyon, D.E., Donahue, J., Schaub, R.T., and Johnston, R.A., 1986. Tempering Types and Sources for Early Bronze Age Ceramics from Bab edh-Dhra' and Numeira, Jordan. *Journal of Field Archaeology*, 13: 297-305.

Bullard, R.G. 1972. Geological Study of the Hesban Area. *Andrews University Seminary Studies*, 10: 129-141.

Daviau, P.M. 1991. Field D: The Lower Southern Terrace. In: Geraty, L., Herr, L., LaBianca, Ø., and Younker R. (Eds.) *Madaba Plains Project 2: The 1987 Season at Tell el-'Umeiri and Vicinity and Subsequent Studies*, Andrews University Press, Berrien Springs: 87-155.

Dessel, J. P. 2009. *Lahav I. Pottery and Politics: The Halif Terrace Site 101 and Egypt in the Fourth Millennium B.C.E.* Eisenbrauns, Winona Lake.

Freestone, I.C. 1991. Extending Ceramic Petrology. In: Middleton, A. and Freestone I. (Eds.) *Recent Developments in Ceramic Petrology*, British Museum Occasional Papers No. 81, London: 399-410.

Goren, Y. 1995. Shrines and Ceramics in Chalcolithic Israel: The View through the Petrographic Microscope. *Archaeometry*, 37: 287-305.

Goren, Y. 1996. The Southern Levant in the Early Bronze IV: The Petrographic Perspective. *Bulletin of the American Schools of Oriental Research*, 303: 33-72.

Goren, Y. and Gilead, I. 1987. Petrographic analysis of Pottery from Shiqmim: a Preliminary Report. In: Levy, T.E. (Ed.) *Shiqmim I: Studies Concerning Chalcolithic Societies in the Northern Negev Desert, Israel (1982–1984)*. British Archaeological Reports, Oxford: 411-418.

Greenberg, R. and Porat, N. 1996. A Third Millennium Levantine Pottery Production Center: Typology, Petrography, and Provenance of the Metallic Ware of Northern Israel and Adjacent Regions. *Bulletin of the American Schools of Oriental Research*, 301: 5-24.

Harrison, T.P. 1995. *Life on the edge: human adaptation and resilience in the semi-arid highlands of Central Jordan during the early bronze age*. Unpublished doctoral dissertation, University of Chicago.

Harrison, T.P. 1997. Field D: The Lower Southern Terrace. In: Herr, L., Geraty, L., LaBianca, Ø., Younker, R., and Clark, D. (Eds.) *Madaba Plains Project 3: The 1989 Season at Tell el-'Umeiri and Vicinity and Subsequent Studies*. Andrews University Press, Berrien Springs: 99-175.

Harrison, T.P. 2000a. The Early Bronze III Ceramic Horizon for Highland Central Jordan. In: Philip, G. and Baird, D. (Eds.) *Ceramics and Change in the Early Bronze Age of the Southern Levant*. Levantine Archaeology 2. Sheffield Academic Press, Sheffield: 347-364.

Harrison, T.P. 2000b. Field D: The Lower Southern Terrace. In: Herr, L., Clark, D., Geraty, L., Younker, R., and LaBianca, Ø. S. (Eds.) *Madaba Plains Project 4: The 1992 Season at Tall al-'Umayri and Subsequent Studies*, Andrews University Press, Berrien Springs: 95-154.

Harrison, T.P. in press. Community Life, Household Production and the Ceramic Industry at EBA Tall al 'Umayri. In: Chesson, M. (Ed.) *The Archaeology of Early Bronze Age People*. Eisenbrauns, Winona Lake.

Harrison, T.P., and Savage, S.H. 2003. Settlement Heterogeneity and Multivariate Craft Production in the Early Bronze Age southern Levant. *Journal of Mediterranean Archaeology*, 16: 33-57.

Herr, L.G. 1997. Organization of Excavation and Summary of Results at Tall al'Umayri. In: Herr, L., Geraty, L., LaBianca, Ø., Younker, R., and Clark, D. (Eds.) *Madaba Plains Project 3: The 1989 Season at Tell el-'Umeiri and Vicinity and Subsequent Studies*. Andrews University Press, Berrien Springs: 7-20.

James H.E. Jr. 1976. Geological Study at Hesban. *Andrews University Seminary Studies*, 14: 165-169.

Lacelle, L. 1986. Bedrock Geology, Surficial Geology, and Soils. In: LaBianca Ø. and Lacelle, L. (Eds.) *Hesban 2. Environmental Foundations: Studies of Climatical, Geological, Hydrological, and Phytological Conditions in Hesban and Vicinity*. Andrew University Press, Berrien Springs: 23-58.

London, G. 1988. The Organization of the Early Bronze Age II and III Ceramics Industry at Tel Yarmouth: A Preliminary Report. In: de Miroschedji, P. (Ed.) *Yarmouth I*, Recherche sur les grandes civilizations. Mémoire 76. Éditions Recherche sur les Civilizations, Paris: 117-124.

London, G. 1991. Aspects of Early Bronze and Late Iron Age Ceramic Technology at Tell el-'Umeiri. In: Geraty, L., Herr, L., LaBianca, Ø., and Younker, R. (Eds.) *Madaba Plains Project 2: The 1987 Season at Tell el-'Umeiri and Vicinity and Subsequent Studies*, Andrews University Press, Berrien Springs: 383-419.

London, G. 1995. A Comparison of Bronze and Iron Age Pottery Production on Material from the Madaba Plains Region. *Studies in the History and Archaeology of Jordan*, 5: 603-606.

London, G., Plint, H., and Smith, J. 1991. Preliminary Petrographic Analysis of Pottery from Tell el-'Umeiri and Hinterland Sites, 1987. In: Geraty, L., Herr, L., LaBianca, Ø., and Younker R. (Eds.) *Madaba Plains Project 2: The 1987 Season at Tell el-'Umeiri and Vicinity and Subsequent Studies*, Andrews University Press, Berrien Springs: 429–439.

Mason, R.B. 2004. *Shine Like the Sun: Lustre-Painted and Associated Pottery from the Medieval Middle East*. Bibliotheca Iranica: Islamic Art and Architecture Series, 12. Mazda Press in association with the Royal Ontario Museum: Costa Mesa and Toronto.

Mazar, A., Ziv-Esudri, A., and Cohen-Weinberger, A. 2000. The Early Bronze Age II-III at Tel Beth Shean: Preliminary Observations. In: Philip, G. and Baird, D. (Eds.)

Ceramics and Change in the Early Bronze Age of the Southern Levant. Levantine Archaeology 2. Sheffield Academic Press, Sheffield: 255-278.

Mitchel, L. 1989. Field D: The Lower Southern Terrace. In: Geraty, L., Herr, L., LaBianca, Ø., and Younker R. (Eds.) *Madaba Plains Project 1: The 1984 Season at Tell el-'Umeiri and Vicinity and Subsequent Studies*, Andrews University Press, Berrien Springs: 282-295.

Nisar, A-J, and al-Sa'ad, A-J. 2000. *Preliminary report on the petrographic analyses of a collection of pottery sherds From Tell Umayri.* Unpublished report, Yarmouk University, Jordan.

Pettijohn, F.J., Potter, P.E., and Siever, R. 1987. *Sand and Sandstone.* 2nd ed. Springer-Verlag: New York and Berlin.

Philip, G. 2001. The Early Bronze I–III Ages. In: MacDonald, B., Adams, R., and. Bienkowski, P. (Eds.) *The Archaeology of Jordan*, Levantine Archaeology 1. Sheffield Press, Sheffield: 163-232.

Platt, E.E. and Herr, L.G. 2002. The Objects. In: Herr, L., Clark, D., Geraty, L., Younker R , LaBianca, Ø. (Eds.) *Madaba Plains Project 5: The 1994 Season at Tall al-'Umayri and Subsequent Studies*, Andrews University Press, Berrien Springs: 156-170.

Porat, N. 1989a. *Composition of pottery—application to the study of the interrelations between Canaan and Egypt during the 3rd millennium B.C.* Unpublished doctoral dissertation, Hebrew University of Jerusalem.

Porat, N. 1989b. Petrography of Pottery from Southern Israel and Sinai. In: de Miroschedji, P. (Ed.) *L'Urbanization de la Palestine a l'Age du Bronze Ancien*, BAR International Series No. 527. British Archaeological Reports, Oxford: 169-188.

Quinn, P.S. and Day, P.M. 2007. Calcareous microfossils in Bronze Age Aegean Ceramics: Illuminating Technology and Provenance. *Archaeometry*, 49: 775-793.

Rast, W.E. and Schaub, R.T. 2003. *Bab edh-Dhra': Excavations at the Town Site (1975-1981).* Reports of the Expedition to the Dead Sea Plains - REDSP 2. Eisenbrauns, Winona Lake.

Rice, P. 1987. *Pottery Analysis: A Sourcebook.* University of Chicago, Chicago.

Roux, V. 2003. Ceramic Standardization and Intensity of Production: Quantifying Degrees of Specialization. *American Antiquity*, 68: 768-82.

Schnurrenberger, D.W. 1991. Preliminary Comments on the Geology of the Tell el-'Umeiri Region. In: Geraty, L., Herr, L., LaBianca, Ø., and Younker R. (Eds.) *Madaba*

Plains Project 2: The 1987 Season at Tell el-'Umeiri and Vicinity and Subsequent Studies, Andrews University Press, Berrien Springs: 370-376.

Shawabekeh, K. 1998. *The Geology of Ma'in Area: Map Sheet (3153-III). Geological Mapping Series, Geology Bulletin No. 40.* Geology Directorate, Geological Mapping Division, Natural Resources Authority, Amman.

Terry, R.D. and Chillinger, G.V. 1955. Summary of 'Concerning some additional aids in studying sedimentary formations' by M. S. Shvetsov. *Journal of Sedimentary Petrology*, 25: 229-234.

Wentworth, C.K. 1922. A Scale of Grade and Class Terms for Clastic Sediments. *Journal of Geology*, 30: 377-92.

Whitbread, I.K. 1986. The Characterization of Argillaceous Inclusions in Ceramic Thin Sections. *Archaeometry*, 28: 79-88.

Whitbread, I.K. 1995. *Greek Transport Amphorae. A Petrological and Archaeological Study*. Fitch Laboratory Occasional Papers 4. British School at Athens: Athens.

COMPARISON OF VOLCANICLASTIC-TEMPERED INCA IMPERIAL CERAMICS FROM PARIA, BOLIVIA WITH POTENTIAL SOURCES

Veronika Szilágyi

Department of Nuclear Research, Institute of Isotopes,
Hungarian Academy of Sciences, Budapest, Hungary
(szilagyiv@iki.kfki.hu)

György Szakmány

Department of Petrology and Geochemistry, Institute of Geography and Earth Sciences, Eötvös Loránd University, Budapest, Hungary

Introduction

The Incas were one of the most developed cultures of medieval South America and for a century (1450-1535 AD) prior to the Spanish colonization, they had established a huge empire along the Pacific coastline extending inland to the Andean Cordilleran mountain range (Hyslop, 1984). Although there is evidence of well-organized population centers and an extended road system, there is limited archaeological knowledge about the southern territory of this civilization. Recent excavations in the Paria Basin of present day Bolivia (Figure 1) have shown that a settlement complex lying along the main road system leading from Cuzco to the south and located at the intersection of two important imperial roads, played a significant role in the Inca Empire (Gyarmati, 2006). This settlement complex included an Inca administrative centre called Paria and the surrounding inhabited areas.

During the field excavations, two different types of pottery were found, a classical Inca imperial type and a lower quality type thought to be either local Late Intermediate Period or local Inca (Gyarmati, 2006). Szilágyi *et al.* (2007) determined the lower quality type to be locally derived.

Inca imperial ceramics of the same high quality, with the same vessel shapes, decoration and surface treatment, can usually be found at every Inca settlement. D'Altroy and Bishop (1990) have queried how such standardized pottery manufacture could extend throughout the almost one million square kilometer extent of the Inca Empire. There appear to be little available evidence as to whether the Inca state organised and exported the products manufactured in certain pottery centres (Morris, 1978; D'Altroy, 1992), exported the knowledge needed for its manufacture, either by technique or artisans (Hayashida, 1998), or exported only the raw materials (clay, temper, fuel) for its manufacture. The present study evaluates further the origin of Inca imperial ceramics and addresses this uncertainty with respect to the finds from Paria. By investigating the provenance of the raw materials used to produce the Inca imperial ceramics found at this southern administrative center, the important question of state standardization of handicraft is readdressed with new scientific data.

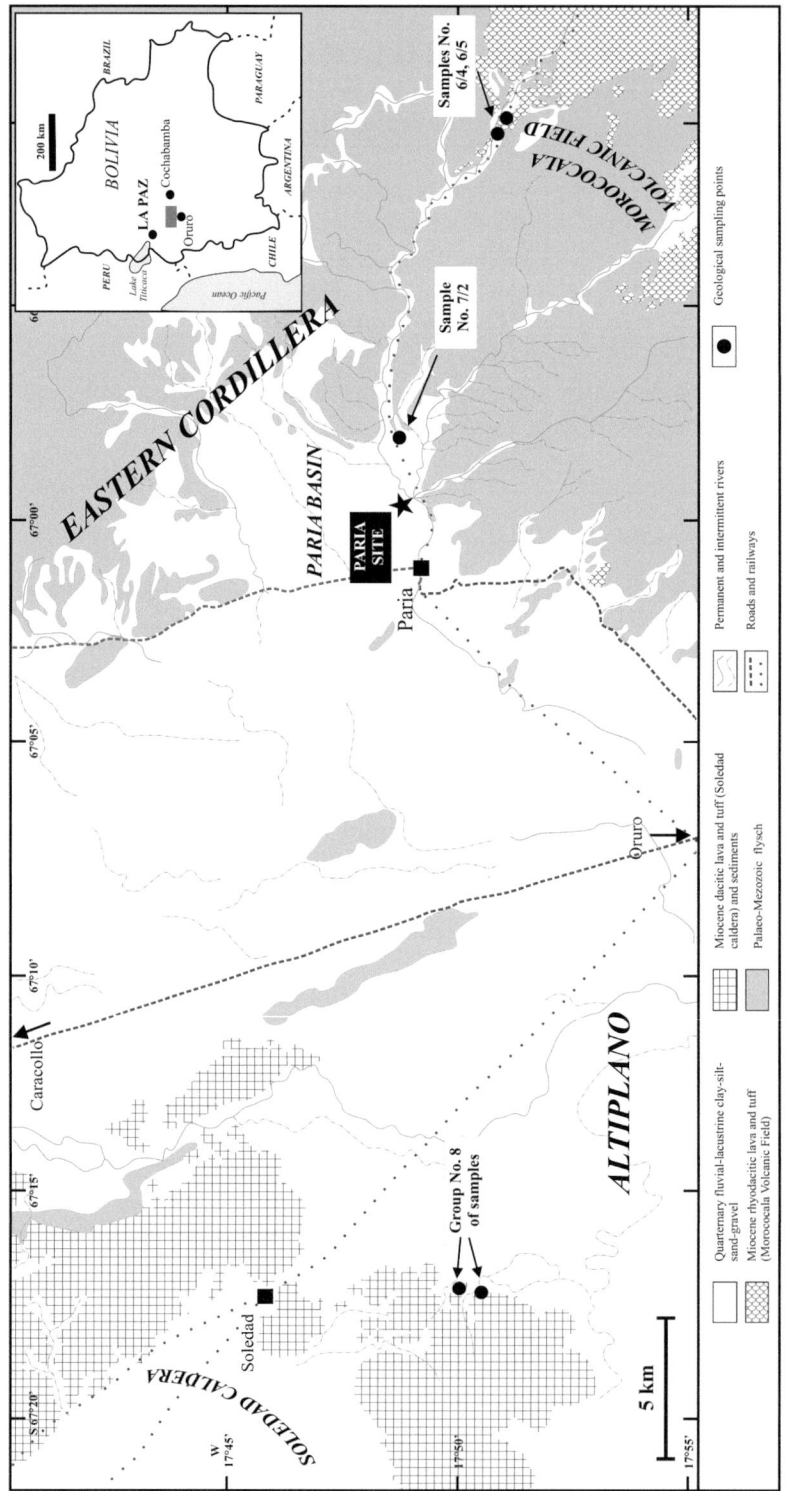

Figure 1. Geological map of the Paria Basin and its surroundings in Bolivia, with locations of ceramic and geological samples analysed in this study. Adapted from GEOBOL, 1992, 1994.

The present study is concerned with the qualitative and quantitative petrographic characterization of the Inca imperial ceramics of Paria. The results suggest that at Paria, Inca imperial ceramics were manufactured from local and nearby (<30 km) raw materials. Significant utilization of either raw materials or finished vessels exported from the heart of the empire can therefore be excluded in this case. The data from Paria support the idea of handicraft standardization in the Inca Empire resulting from the 'export' of the specific knowledge and skills required to manufacture Inca Imperial ceramics.

Materials and Methods

The excavation of Paria was conducted between 2004-2006 by Dr János Gyarmati of the Museum of Ethnography, Budapest, Hungary. Thousands of potsherds were collected during the systematic survey of the Paria Basin and the excavation of the capital site, Paria (Gyarmati, 2006). A subset of the entire assemblage from both phases of the fieldwork was selected on the base of style and function and subjected to archaeometric investigation. Half of the selected ceramic fragments were categorized as Inca imperial pottery, composed of bowls, plates, jars and large amphora-like containers, called *aribalos* (Figure 2). All of the Inca imperial ceramics were well fired, dense and hard, with a fine-grained texture. The surfaces were polished and smooth with a dominantly red slip. Surface decoration ranged from black-on-red, black and pale yellow/white-on-red to non-decorated styles.

Figure 2. Fragments of typical Inca imperial vessel forms. a) *Aribalo* (sample PA/6), b) Bowl of *puku* (sample I.10.4.), c) plate (sample PA/2). Scale bar = 1 cm.

Previous investigations by Szilágyi *et al.* (2005, 2007) indicated that the main aplastic raw materials added to the clay paste of Inca imperial pottery from Paria were volcaniclastic rock fragments. This observation indicated the need to sample the local and regional volcanic units. This paper deals only with the examination of the aplastic components of ceramics. The clay paste most probably derived from the local Quaternary formations, which do not contain volcanic related clasts. Only sandy sediments in the closest vicinity of the volcanic outcrops have significant volcanic related material content, but have little clay content and plasticity.

The site of Paria is situated on the western side of the Eastern Andean Cordillera and on the eastern margin of Central Altiplano (Figure 1). The basement rocks in this region are mainly sedimentary formations deposited during the Palaeozoic to Mesozoic. In the Palaeogene-Neogene, sedimentary strata related to a foreland continental basin were deposited. Miocene volcanism interrupted this depositional sequence and produced two regions of magmatic intrusions and volcanism: the Morococala Volcanic Field, which forms a great plateau at the edge of the Eastern Andean Cordillera, and the caldera structure of Soledad on the highland of the Altiplano (GEOBOL, 1992, 1994). To obtain representative samples of both volcanic territories, geological map based sampling was done.

The Soledad Caldera is a low-volume, non-resurgent 'ash-flow' caldera that started to form with the intrusion of small dacitic stocks (c. 15 Ma - million years), followed by small rhyolite domes uplifted along an outer ring fracture (c. 8.8 Ma). Caldera collapse (c. 5.4 Ma) resulted in the emplacement of variously welded and interbedded dacitic tuffs with an air-fall component (GEOBOL, 1992, 1994; Redwood, 1987). Because of the complex history of the Soledad caldera multiple sampling was necessary.

The formation of the Morococala Volcanic Field was initiated by quartz-latite porphirytic intrusions (c. 25-20 Ma) that were followed by extensive ignimbrite eruptions of variously welded rhyodacitic tuffs and lavas (c. 9-5 Ma) (GEOBOL, 1992, 1994; Morgan *et al.*, 1998; Barke *et al.*, 2007). Since the present-day surface of this volcanic field is dominated by the homogenous ignimbrite, the sampling resulted in a monotonous sample set.

To investigate the question of provenance, 93 Inca imperial ceramics and 17 geological samples (four from Morococala and 13 from Soledad) were analysed. To determine the similarities of the aplastic components, qualitative and quantitative examination of the samples was performed by petrography using a grid-based area counting technique. This method relies on the 'Delesse Relation', which holds that area proportions of the constituents in two-dimensional thin sections are equivalent to volumetric proportions of the constituents in three-dimensional bulk samples (Mouton, 2002; Baddeley and Vedel Jensen, 2005). In the area counting procedure, an ocular grid composed of 10 x 10 square units was used and a magnification at which the smallest aplastic inclusions represent half a square. In this work, the procedure was continued until 30 steps (~30 mm^2) were recorded in each sample. Due to the direct area measurement, this method is not loaded with counting error (Demirmen, 1971), and the restrictions of grain size (Gazzi, 1966; Dickinson, 1970) can be eliminated. From the data obtained by area counting, modal composition of the aplastic components was calculated to provide a model for geological provenance of the potential source rocks.

Sources of Volcaniclastic Temper in Inca Imperial Ceramics from Paria, Bolivia

	Inka Imperial Ceramic petrographic groups								
	I/A/a n=14	I/A/n n=16	I/B/a n=6	I/B/b n=20	I/B/c n=15	I/B/d n=4	I/C/a n=9
Pumice	+	+	+	+	+	+	-
Glass shard	curved	curved	curved	curved	curved	irregular	-
Volcanic lithic fr.	undef	undef	undef	undef	undef	undef	+ hyal
Other lithic fr.	volcanic+siltstone	volcanic+quartzite+ siltstone	-	siltstone	silt-sand stone	-	siltstone
Mineral fr.	Pl+Qtz+Bt+Am (Opx+Op)	Pl+Qtz+Bt (Am+Opx+Op)	Pl+Qtz (+Bt+Am) (Op+Ttn)	Pl+Qtz+Bt+Am (Opx+Op+Ttn)	Pl+Qtz+Bt (Am+Opx+Op+Tt n)	Pl+Qtz+Bt (Am+Opx+Op+Ttn)	Pl+/-Qtz +Bt +/-Am (Opx+Cpx+Op)
Dominant grain size (mm)	50-100	100-275	50-75	150-175	150-175	150-175	100-125
Discriminative parameters							
Feldspar	An30-45 polysynt twinned	An10-25 polysynt twinned	An25-45 polysynt	An25-45 polysynt	An20-30 polysynt	An20-35 polysynt	An<40 polysynt zoned
Amphibole	brown pleo. Hbl fine grained	-	brown pleo. Hbl fine grained	brown pleo. Hbl fine grained	-	-	brown pleo. Hbl/ Oxyam fine grained
Roundness	(S)A	(S)A	A	A	A	A	(S)A

	Petrographic groups continued		Source rocks		
	I/C/b n=4	I/D n=5	Morococala tuff n=1	Morococala volcanic rock n=1	Soledad tuff n=8	Soledad volcanic rock n=3
Pumice	-	+	+	-	+	-
Glass shard	-	-	unwelded glassy matrix	-	welded glassy matrix	-
Volcanic lithic fr.	+ vitro	+ undef	-	+ groundmass vitro	-	+ groundmass vitro/porph
Other lithic fr.	siltstone	-	-	-	-	-
Mineral fr.	Pl+/-Qtz +Bt Am (Op)	+/- Pl+Qtz+Bt+Am (Opx+Op)	Pl+Qtz+Bt (Kfs+Zm+Ap +Op)	Pl+Qtz+Bt (Kfs+Zm+Ap +Op)	Pl+Qtz+Bt +Am (Cpx+Op +Ttn+Zm)	Pl+Qtz+Bt+Am (Cpx+Op+Ttn +Zm)
Dominant grain size (mm)	100-125	50-100	75-1250	100-750	100-1000	100-900
Discriminative parameters						
Feldspar	An<40 polysynt zoned	An30-45 polysynt zoned	An15-20 polysynt zoned twinned zoned Kfs	An15-20 polysynt zoned twinned zoned Kfs	An35-40 fine/polysynt coarse/zoned twinned	An35-40 fine/polysynt coarse/zoned twinned
Amphibole	brown pleo. Hbl	brown pleo. Hbl	-	-	brown-green pleo. Hbl fine grained	brown-green pleo. Hbl fine grained
Roundness	SR	SR	A (fresh) (redeposit.) R	A (fresh) (redeposit.) R	A (fresh) (redeposit.) R	A (fresh) (redeposit.) R

Table 1. Petrographic data of Inca imperial ceramics from Paria and comparative geological samples analysed in this study. Am = amphibole, An = anortite, Ap = apatite, Bt = biotite, Cpx = clinopyroxene, Hbl = hornblende, Kfs = K-feldspar, Op = opaque mneral, Opx = orthopyroxene, Oxyam = oxyamphibole, Pl = plagioclase, Qtz = quartz, Ttn = sphene, Zrn = zircon, Pum = pumice, Gs = glass shard, Volc = volcanic rock fragment, Sed = sedimentary rock fragment, A = angular, R = rounded, S = sub-rounded, undef = undefined texture, hyal = hyalopilitic texture, vitro = vitrophiric texture, porph = porphyric microholocrystalline texture. Mineral names after Siivola and Schmid (2007).

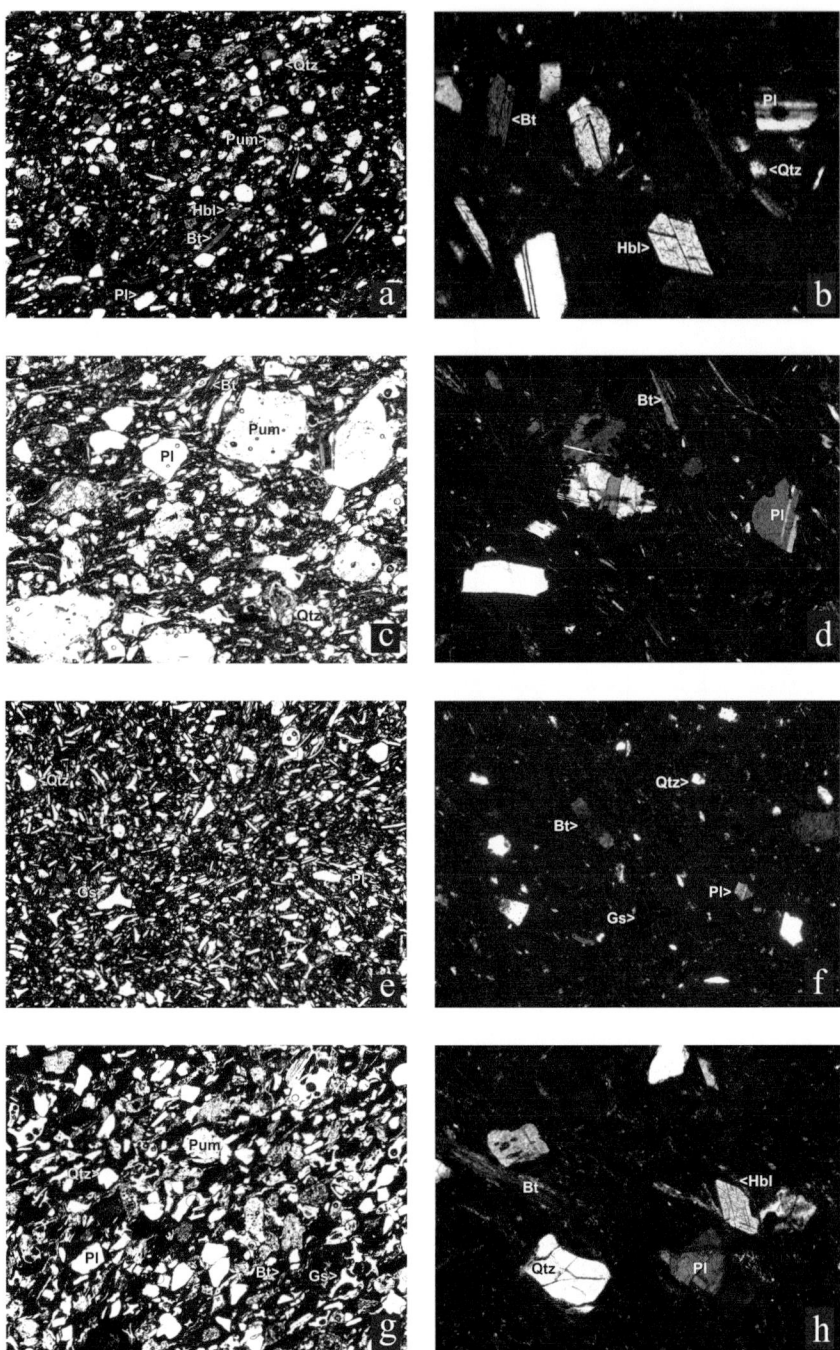

Figure 3. Thin section photomicrographs of Inca imperial ceramics from Paria, Bolivia. a,b) Petrographic subgroup I/A/a (sample 56.8) in PPL and XPL, c,d) Petrographic subgroup I/A/b (sample 11.23.) in PPL and XPL, e,f) Petrographic subgroup I/B/a (sample I.15.7) in PPL and XPL, g,h) Petrographic subgroup I/B/b (sample I.30.3) in PPL and XPL. Image width = 5.5 mm, except b,h = 0.85 mm and d,f = 2.3 mm. Pum = pumice. For other abbreviations see Table 1.

Sources of Volcaniclastic Temper in Inca Imperial Ceramics from Paria, Bolivia

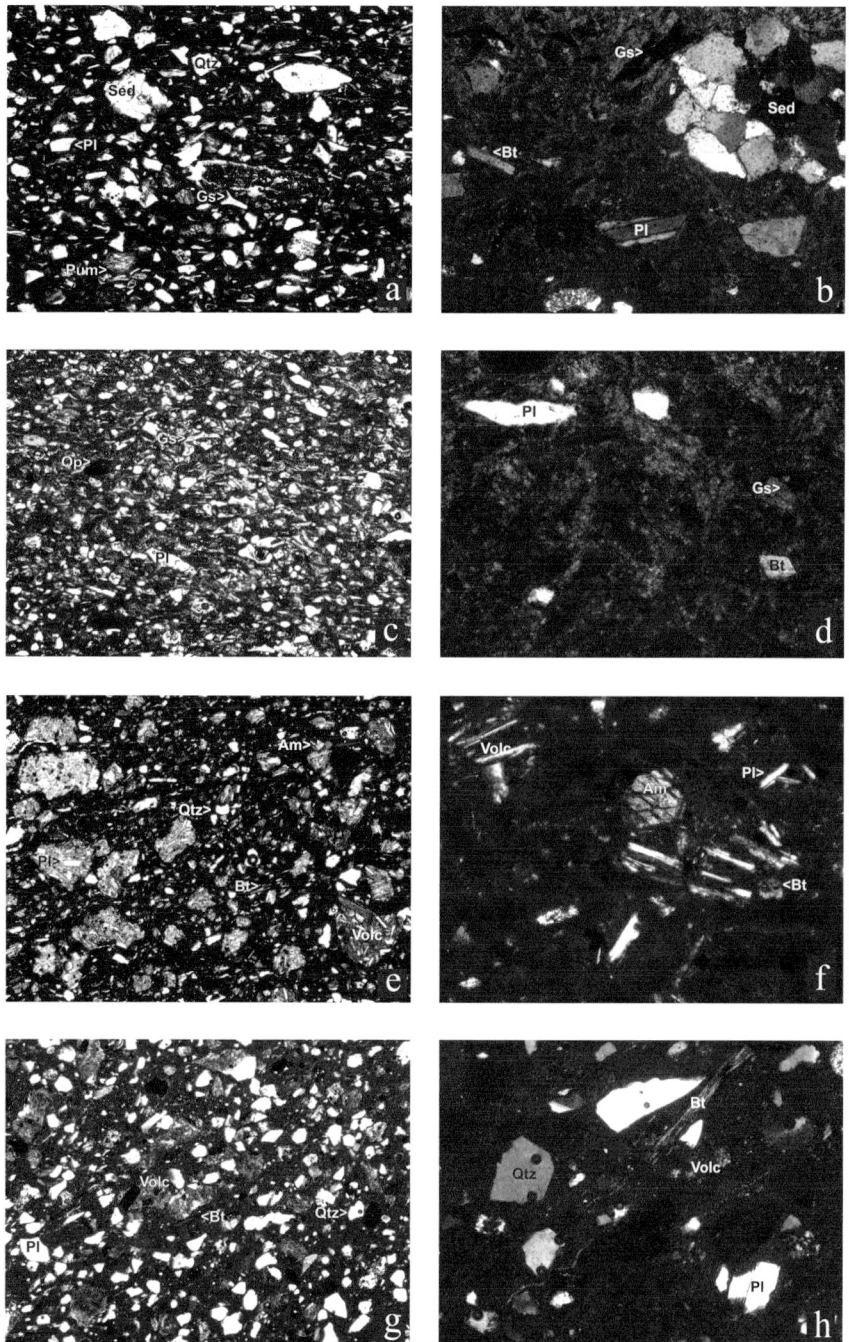

Figure 4. Thin section photomicrographs of Inca imperial ceramics from Paria, Bolivia. a,b) Petrographic subgroup I/B/c (sample I.10.3) in PPL and XPL, c,d) Petrographic subgroup I/B/d (sample 34.129) in PPL and XPL, e,f) Petrographic subgroup I/C/a (sample I.18.76) in PPL and XPL, g,h) Petrographic subgroup I/C/b (sample 18.24) in PPL and XPL. Image width = 5.5 mm, except b, h = 2.3 mm and d, f = 0.85 mm. Pum = pumice, Gs = glass shard, Sed = sedimentary rock fragment, Volc = volcanic rock fragment. For other abbreviations see Table 1.

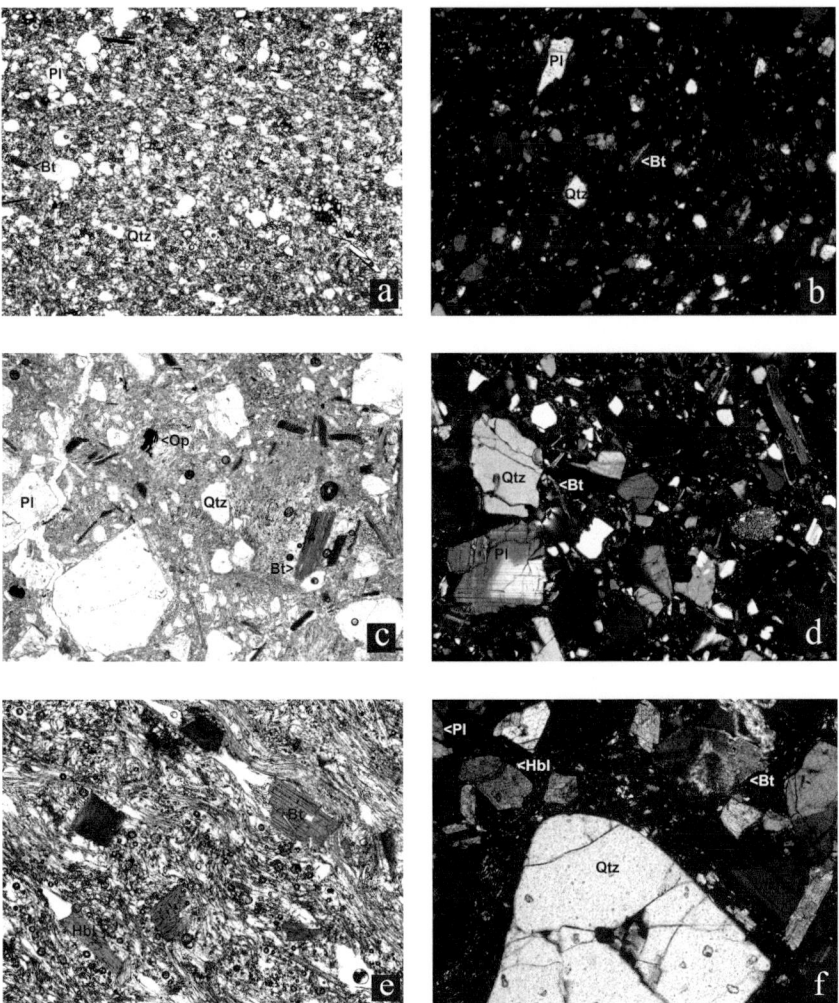

Figure 5. Thin section photomicrographs of Inca imperial ceramics from Paria, Bolivia and comparative volcaniclastic rocks. a,b) Petrographic group I/D (sample 42.9) in PPL and XPL. c) Sample 6/5 of rhyodacitic tuff from Morococala Volcanic Field in PPL, d) Sample 6/4 of rhyodacitic tuff from Morococala Volcanic Field in XP, e) Sample 8/2/A of dacitic tuff from Soledad Caldera in PPL, f) Sample 8/5/C A of dacitic tuff from Soledad Caldera in XPL. Image width = 5.5 mm, except b,f = 2.3 mm. For abbreviations see Table 1.

Results

Petrographic characterisation

The Inca imperial ceramics from Paria usually contain 20-35 modal percent aplastic inclusions that are primarily volcaniclastic in character. On the first level, identification of the petrographic groups was based on the average lithological composition of the samples (major rock types) regardless of their exact mineralogical composition. In this way the macroscopically observed physical properties preferred by the potters such as the utilization of porous tuff or compact volcanic rock fragments could most probably be comparable. Four petrographic groups were recognized: dominantly pumice (I/A), glass shard (I/B), volcanic rock (I/C) and mineral fragment (I/D) containing groups. Pumice fragments contain mineral crystals of feldspar (mainly plagioclase with An_{10-50}), quartz, coarse biotite with/without fine amphibole. The dominantly curved glass shards have mineral crystals of feldspar (An_{10-50}) and biotite. Volcanic lithic fragments are fresh or weathered, have a hyalopilitic (volcanic rock with groundmass containing fine-grained mineral crystals and glass) or vitrophyric texture (volcanic rock with glassy groundmass) and contain feldspar, biotite and rarely amphibole phenocrysts. The mineral fragments found in group I/D are quartz, feldspar, biotite and amphibole.

The inclusions of Group I/A ceramics are dominated by pumice fragments. These pumiceous tuff rock and mineral fragments are angular to sub-angular, generally unweathered and have grain sizes that range from 50-300 µm. These sherds have a hiatal fabric (discontinuous grain size distribution with several maxima), and a micaceous and fine-grained, silty paste that ranges in colour from red to grey and sometimes has a reduced core and oxidised rims. The pumiceous clasts appear to have been added as temper to this paste since they are unweathered and have fragmented shapes whose features cannot be explained by the use of a natural, weathered clay-pumice mixture. Group I/A can be divided into two subgroups (Table 1; Figure 3A-D; Figure 6A) based upon grain size, anortite and mafic mineral composition (biotite with or without amphibole).

Group I/B ceramics contain dominantly glass shard inclusions. These pyroclastic pummice fragments are very angular and unweathered, with bone-like curved shapes. They appear to have been added to the clay paste as temper and have fragmented shapes whose features cannot be explained by the use of a natural, weathered clay-pyroclastic mixture. The sherds have hiatal fabric and a micaceous and fine-grained, silty paste that ranges in colour from red to light grey. Group I/B can be divided into four subgroups (Table 1; Figure 3E-H; Figure 4A-D; Figure 6A) based upon their lithic fragments, mafic mineral content and grain size.

Group I/C ceramics are characterised by volcanic rock inclusions. These lithic fragments show a wider range in lithology than the volcaniclastic constituents of the former groups, from dacite to basaltic andesite. The sherds have a hiatal fabric and a micaceous and fine-grained, silty paste that is red-reddish in colour. The lithic clasts may have been added as temper to this paste since they form the largest inclusions (150-175 µm) in the grain size distribution. Group I/C can be split into two subgroups (Table 1; Figure 4E-H; Figure 6A) based upon the recrystallisation stage, degree of weathering, sphericity, roundness and the lithology of the volcanic rock fragments.

The inclusions of group I/D ceramics are almost exclusively mineral grains and lithic clasts are rare. These grains are moderately spherical, sub-angular to sub-rounded, unweathered mineral crystals. The sherds in group I/D have an almost continuous serial fabric and micaceous, fine-grained, silty clay paste that is red in colour (Table 1; Figure 5A,B; Figure 6A).

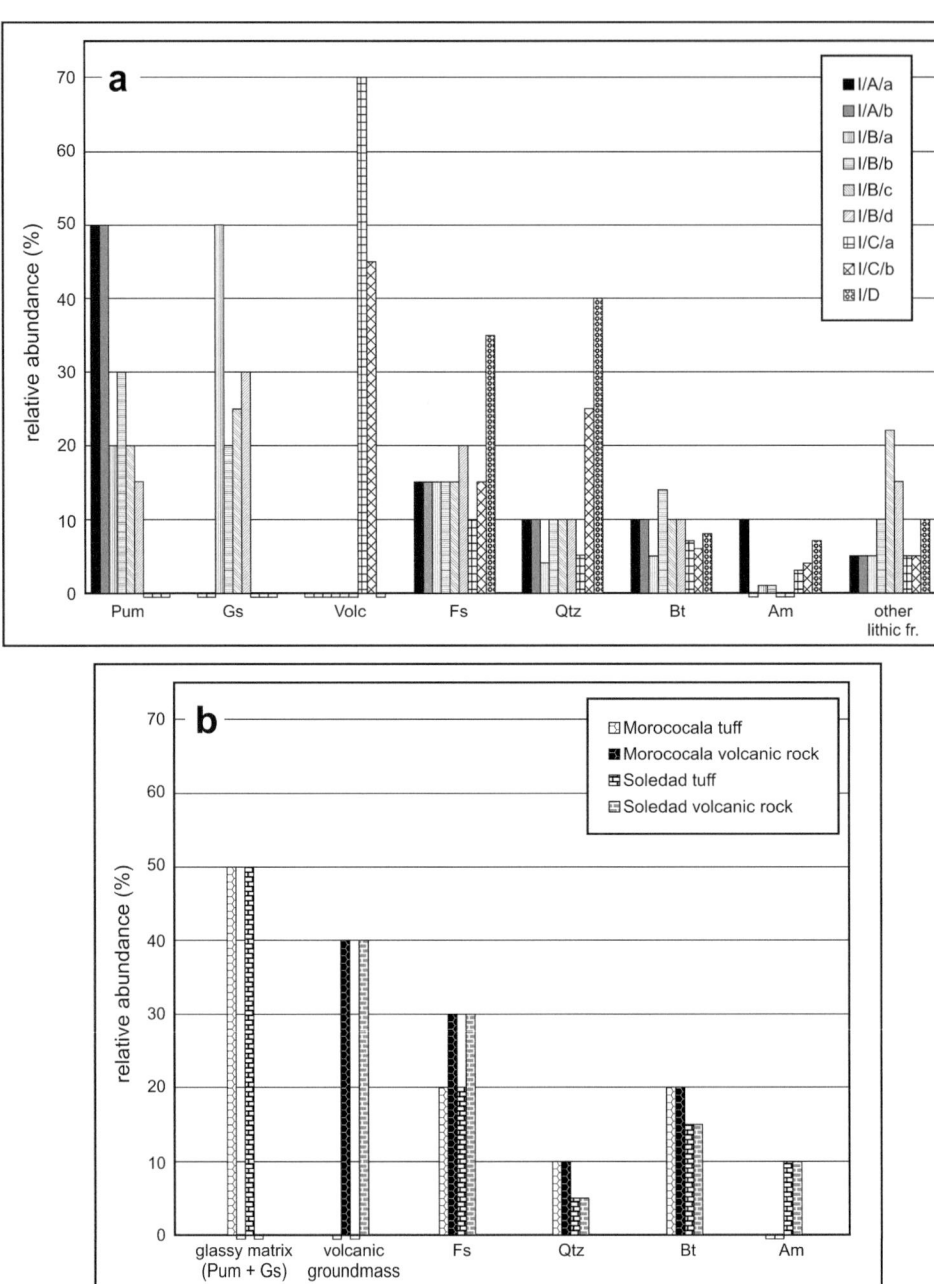

Figure 6. Quantitative modal analysis of a) Aplastic inclusions within petrographic subgroups of Inca imperial ceramics from Paria, Bolivia and, b) Potential raw material sources. For abbreviations see Table 1.

The Soledad dacitic tuffs and their alteration products are characterized by a pumiceous, unweathered, usually welded matrix, which has heterogeneous pore and tube size and high crystallite content (Table 1; Figure 5C,D; Figure 6B). The Soledad dacites and their weathered variations have vitroporphyric-porphyric microholocrystalline texture and two distinct grain sizes: phenocrysts and a partly or completely crystallized fine-grained groundmass (Table 1; Figure 6B).

The Morococala rhyodacitic tuffs and their alteration products consist of a pumiceous, unweathered, unwelded-weakly-welded matrix with heterogeneous pore and tube size (Table 1; Figure 5E,F; Figure 6B). The Morococala rhyodacites and their weathered variants are vitrophyric textured volcanic rocks (Table 1; Figure 6B).

Modal analysis

The material based classification of the ceramic assemblage resulted in several groups and subgroups which could be separated by their lithological composition on one level, and by their mineralogical composition on a second level. Using a grid-based area counting method, it was possible to compare quantitatively the average mineralogical composition of the aplastic inclusions in the sherds with that of the tuffs and volcanic rocks. The diagrams in Figure 6 summarize the results according to ceramic petrographic types and to volcanic territories. As is illustrated both visually (Figures 3-5) and with relative volume percentage values (Figure 6A,B), well-defined crystalline phase ratios are found for certain groups. However, it is a problematic task to match the pumice clast or glass shard content of pottery with the juvenile glassy matrix content of tuffs, or the volcanic lithic fragment content of the pottery with the more or less crystallized volcanic groundmass content of volcanic rocks. Another difficulty arises from the composite character of ceramics, since they also contain a proportion of other rock fragments, such as sedimentary rocks.

Taking these effects into consideration, it can be said that the glassy phase containing ceramic groups (I/A and I/B) and the comparative tuff samples have about 50% of this kind of matrix component. The only exception is subgroup I/B/a where this ratio increases to about 70%. Another situation is in the case of petrographic subgroup I/C of ceramics where the volcanic lithic fragment content is at minimum 45%, but it can reach 70%. It is somewhat inconsistent with the maximum 40% groundmass content of the volcanic rocks, but this can become accounted for by including the phenocryst content into the calculations.

The crystalline felsic (feldspar and quartz) and maphic (biotite and hornblende) constituent ratios are about 25%, and below 15% in almost all petrographic groups of ceramics, while these values are as low as 25% and 20% in the geological samples, respectively. The differences arise from the other inclusions content in the archaeological samples. The main distinguishing characteristic is the percentage of hornblende, which is absent in some ceramic subgroups (I/A/b, I/B/c-d), but varies from 1-10% in others (I/A/a, I/B/a-b, I/C/a-b, I/D). The same difference can be observed between the hornblende-bearing Soledad tuffs and volcanic rocks (10%) and the hornblende-free Morococala tuffs and volcanic rocks.

Discussion

Since it was the main goal of this study to survey the relationship between the aplastic inclusions of the Inka imperial ceramics from Paria and the two potential volcaniclastic-volcanic sources, both the physical appearance (grain size, optical behaviour) of the individual phases and the overall mineralogical composition are significant. The overview of the results of this analysis is summarised in Table 1 and Figure 6. Concerning the habit of the rock forming constituents, the main characters of the comparison were the relationship between the two mafic phases, biotite and hornblende, regarding both quantity and crystal size, the twinning, zoning and composition of plagioclase phenocrysts and the degree of welding of the glassy matrix of the tuffs, or the textural features of the volcanic rocks. The degree of welding can be easily compared to the ceramics in the case of the pumice containing groups. Another case is when we have to relate them to the bone-shaped glass shard content of the potteries. We have to take into consideration that welding modifies the primary pore content and shape of the glassy matrix, and in this way the conventional glass shards become less abundant.

Based on the above-mentioned features, it is possible that the raw materials of the ceramic subgroups I/A/a, I/B/a-b, I/D and perhaps I/C/a-b come from the Soledad volcanic area due to their hornblende content and the acidic-neutral composition of the plagioclase that they contain. The biotite and acidic plagioclase containing I/A/b and I/B/c-d ceramic types may have originated in the Morococala volcanic territory. To obtain further evidence on these provenance statements mineral chemical analyses will be necessary in the future. Another remark on the I/C subgroups is that the large heterogeneity of these groups of pottery may be due to some foreign originated pieces, especially in the I/C/a subgroup.

Based on the sphericity, roundness and weathering stage of the volcaniclastic-volcanic rock clasts, and especially the glassy/weakly crystalline matrix of these fragments, it is possible to reconstruct the technological choices involved in the production of the Inka imperial ceramics found at Paria. On the one hand, the unweathered, angular, volcaniclastic grains indicate the deliberate addition of these freshly quarried material and on the other, the less common weathered, sub-rounded, moderately spheric volcanic rock fragments are evidence of the use of raw material sources farther from the primary outcrops, such as sandy alluvial sediment. These later materials could also have been added as temper to the pottery. The fine grained fabric and the absence of lithic fragments in the ceramics belonging to group I/D might suggest that the potters chose to use a specific type of raw material for the production of fine pottery.

In summary, our petrographic investigations of the aplastic inclusions of the Inca imperial ceramics from Paria has permitted the identification of two probable raw material sources in the vicinity of the site. To the west is the biotite-hornblende bearing dacitic tuff and dacite outcrops of the Soledad Caldera and to the east is the biotite containing rhyodacitic tuff-rhyodacite outcrops of the Morococala Volcanic Field. The volcanic rock sources of the inclusions in the group I/C ceramic samples cannot as yet be determined.

Conclusions and Further Research

Petrographic comparison of Inca imperial ceramics from Paria and volcanic-volcaniclastic rock outcrops in the vicinity of the site suggest that suitable raw materials for the majority of the pottery would have been available within 20-30 km of the site. No significant quantity of long distance imported pottery was detected.

According to the petrographic characterisation and the modal analysis it is clear that almost all the Inca imperial ceramics contain volcaniclastic/volcanic material similar to either the pumiceous tuff and lava rocks of the Soledad caldera or the Morococala Volcanic Field. Inca Period potters from the Paria region had the possibility to use both *in situ* sandy talus occurring near the outcrops and the more altered sediments of rivers farther from the source area. This duality suggests a complex raw material supply system with several elements.

With regard to the organization of the highly standardized Inca imperial pottery, we can conclude that in this part of the Inca territory, pottery was produced using know-how 'exported' from Cuzco in the heart of the Empire applied to broadly local raw materials. This may have taken place by the transmission of technological knowledge, or more directly, by the movement of potters across Inca territory. It is important to emphasize that volcanic related material such as tuff and volcanic rock fragments were also used as components of the Inca imperial style ceramics found at other archaeological sites (e.g. Ixer and Lunt, 1999), and remain an important source of raw material for traditional pottery production (Arnold, 1972). It is therefore probable this volcanic material had a special and valuable meaning to the Incas.

Acknowledgements

This study was supported by grant T047048 from the Hungarian Scientific Research Fund (OTKA). The authors are grateful to archaeologist Dr. János Gyarmati for providing the ceramic samples his archaeological input. This paper has had the benefit of helpful grammatical suggestions by Jesse Weil.

References

Arnold, D.E. 1972. Mineralogical analyses of ceramic materials from Quinua, Department of Ayacucho, Peru. *Archaeometry*, 14: 93-102.

Baddeley, A. and Vedel Jensen, E.B. 2005. *Stereology for statisticians*. Monographs on Statistics and Applied Probability 103, Capman & Hall, CRC Press, Routledge, USA.

Barke, R., Lamb, S., and MacNiocaill, C. 2007. Cenozoic bending of the Bolivian Andes: New paleomagnetic and kinematic constraints. *Journal of Geophysical Research*, 112.

D'Altroy, T.N. 1992. *Provincial Power in the Inka Empire*. Smithsonian Institution Press, Washington, D.C.

D'Altroy, T.N. and Bishop, R.L. 1990. The Provincial Organization of Inka Ceramic Production. *American Antiquity*, 55: 120-138.

Demirmen, F. 1971. Counting error in petrographic point-counting analysis: a theoretical and experimental study. *Mathematical Geology*, 3: 15-41.

Dickinson, W.R. 1970. Interpreting detrital modes of graywacke and arkose. *Journal of Sedimentary Petrology*, 40: 695-707.

Gazzi, P. 1966. Le arenarie del flysch sopracretaceo dell'Appennino modenese; correlazioni con il flysch de Monghidoro. *Mineralogica e Petrografica Acta*, 12: 69-97.

GEOBOL 1992. Carta Geológica de Bolivia, Hoja Oruro. Publicación SGB Serie I-CGB-11 (Página 6140).

GEOBOL 1994. Carta Geológica de Bolivia, Hoja Bolivar. Publicación SGB Serie I-CGB-27 (Página 6240).

Gyarmati, J. 2006. PAP (Paria Archaeological Project) Report. Unpublished manuscript.

Hayashida, F. 1998. New insight into Inka pottery production. In: Shimada, I. (Ed.) *Andean Ceramics: Technology, Organization, and Approaches*. MASCA Research Papers, 15, University Museum of Archaeology and Anthropology, University of Pennsylvania, Philadelphia: 313-335.

Hyslop, J. 1984. *The Inka Road System*. Academic Press, Orlando.

Ixer, R.A. and Lunt, S. 1991. The Petrography of Certain Pre-Spanish Pottery from Peru. In: Middleton, A. and Freestone, I. (Eds.) *Recent Developments in Ceramic Petrology*, Occasional Paper 81, British Museum Publications, London: 137-164.

Morgan, G.B., London, D., and Luedke, R.G. 1998. Petrochemistry of Late Miocene Peraluminous Silicic Volcanic Rocks from the Morococala Field, Bolivia. *Journal of Petrology*, 39: 601-632.

Morris, C. 1978. The archaeological study of Andean exchange systems. In: Redman, C.L. (Ed.) *Social archaeology: beyond subsistence and dating*. Academic Press, New York: 315-328.

Mouton, P.R. 2002. *Principles and practices of unbiased stereology: an introduction for bioscientists*. Johns Hopkins University Press, Baltimore, USA.

Redwood, S.D. 1987. The Soledad Caldera, Bolivia: A Miocene caldera with associated epithermal Au-Ag-Cu-Pb-Zn mineralization. *Geological Society of America Bulletin*, 99: 395-404.

Siivola, J. and Schmid, R. 2007. *List of Mineral Abbreviations*. Recommendations by the IUGS Subcommission on the Systematics of Metamorphic Rocks. Web version available at: http://www.bgs.ac.uk/SCMR/docs/papers/paper_12.pdf

Szilágyi, V., Szakmány, Gy., and Gyarmati, J. 2005. Inka kori kerámiák petrográfiai vizsgálatának előzetes eredményei (Paria, Bolívia). *Archeometriai Műhely*, 2005/2: 42-47. (in Hungarian)

Szilágyi, V., Szakmány, Gy., Gyarmati, J., and Tóth M. 2007. Preliminary Comparative Archaeometric Results of Colonial and Inka Pottery in Paria (Oruro, Bolivia). In: Waksman, Y.S. (Ed.) *Archaeometric and Archaeological Approaches to Ceramics: Papers presented at EMAC'05, 8th European Meeting on Ancient Ceramics, Lyon 2005*. BAR International Series 1691: 195-199.

MULTI-VILLAGE SPECIALIZED CRAFT PRODUCTION & THE DISTRIBUTION OF HOKOHAM SEDENTARY PERIOD POTTERY, TUSCON, ARIZONA

James M. Heidke

Desert Archaeology Incorporated, Tucson, Arizona, USA
(jheidke@desert.com)

Introduction

A prehistoric Native American people, known to us archaeologically as the Hohokam, thrived throughout the Sonoran Desert region of Arizona from AD 50-1450. One distinctive form of public architecture found at many pre-Classic Hohokam villages is the ball court. These large, oval depressions surrounded by earthen embankments have been found at more than 160 sites in Arizona (Wilcox, 1991a). From the work of Emil Haury (1976) onward, many students of Hohokam archaeology have argued that ceremonies held at ball court villages would have facilitated the movement of goods between settlements (Abbott, 2001; Abbott *et al.*, 2007; Bayman, 2002; Heidke, 1996b; Heidke *et al.*, 2002; Wilcox, 1979). However, most archaeologists have found it easier to make that assertion, than to prove it. Here the role of ball courts is addressed through an analysis of Middle Rincon phase (c. AD 1000-1100) ceramics. The analysis involves three interrelated research questions: determining where the pottery was manufactured, describing spatial patterning in the provenance data, and reconstructing the organization of the economic system (Earle, 1982).

In this study, the provenance of Middle Rincon ceramics has been determined using a combination of binocular microscopic and petrographic analyses of sand temper (Heidke *et al.*, 2002). Ceramic provenance data recorded over the last two decades are employed. Afterward, regression analysis has been applied to describe spatial patterning in the temper data (Hodder and Orton, 1976; Shennan, 1990). Subsequently, a series of hypothesis tests were conducted in order to assess whether or not site type, site location, archaeological sample type, or sampling error may have caused or biased the regression analysis results. Finally, aspects of Middle Rincon settlement and ceremonial systems are considered in an attempt to document the way that pottery economics may have been embedded in broader social and political institutions. Although the data used in this study are case-specific, the general approach developed here, regression analysis followed by a series of hypothesis tests, should prove useful in evaluating pottery distribution patterns elsewhere.

Materials and Methods

Archaeologists generally make use of both direct and indirect evidence when reconstructing the organization of ceramic production (Costin, 1991). Direct evidence of pottery production includes the recovery of raw materials (clay, temper, pigments), forming and finishing tools (turntables, anvils, scrapers, polishers), facilities associated with production (clay storage and mixing basins, kilns, wind screens), and manufacturing debris (waster deposits) (Mills and Crown, 1995; Stark, 1985; Sullivan, 1988). Indirect evidence includes the provenance interpretation of ceramics, as well as morphological, and/or design data recorded from sherds that provide evidence of their place of origin (Costin, 1991). Successful provenance ascription requires detailed geological mapping and sampling of ceramic resources as well as technological analyses of ceramic pastes (Arnold, 2000; Costin, 2000; Lombard, 1987; Miksa and Heidke, 2001; Pool, 1992; Shepard, 1963).

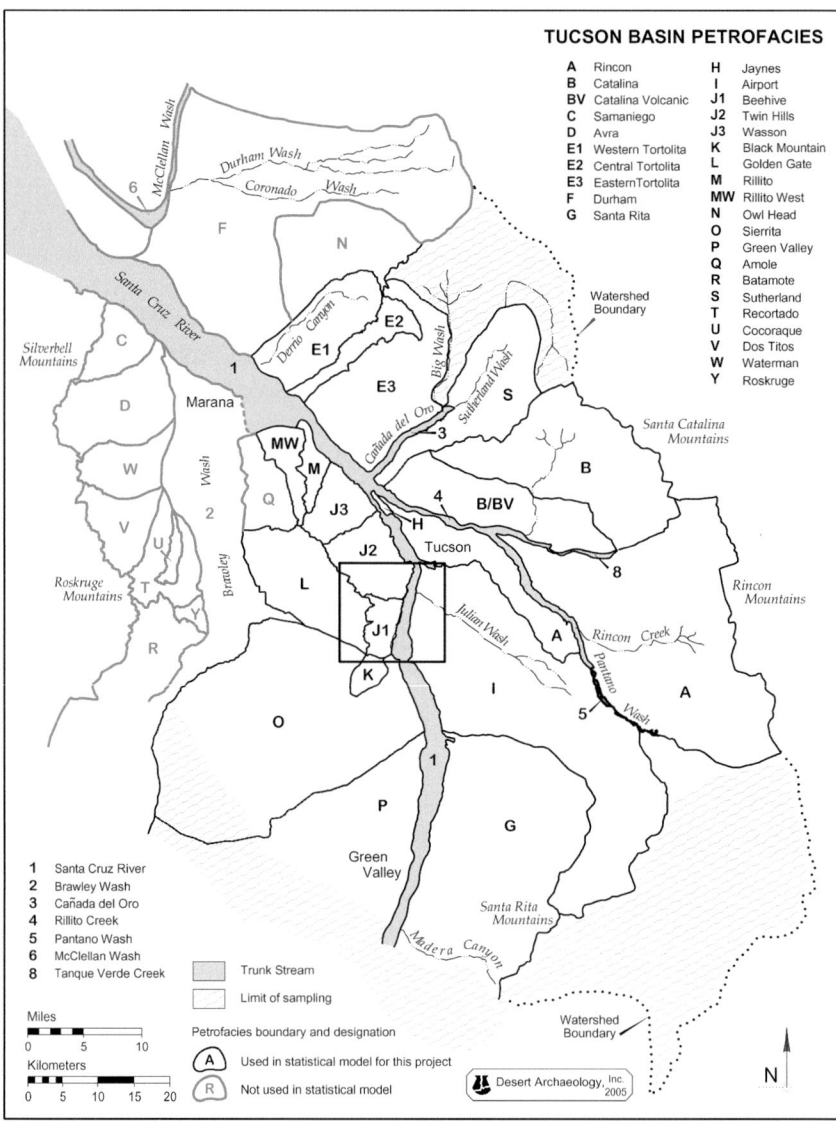

Figure 1. The current petrofacies map for the Tucson area showing letter designations and names. The box surrounding the Beehive Petrofacies (J1) indicates the area shown in detail in Figure 3.

The composition and provenance of sand temper observed in pot sherds was characterized with reference to a model of regionally available sand compositions, known as an actualistic petrofacies model (Heidke et al., 2002, p. 156-160, 165-166). Following a detailed review of the region's bedrock geology, sedimentary history, and geomorphology (Dickinson, 1991; Lipman, 1993) an intensive program of wash sand sampling was carried out which provides the evidence that many spatially discrete sand temper compositions were available to Hohokam potters living in the basin. The current petrofacies map for the greater Tucson Basin is shown in Figure 1. The Gazzi-Dickinson (Dickinson, 1970; Gazzi, 1966) point-counting method was used to count both the wash sand samples that comprise the Tucson Basin petrofacies model and the sand-tempered sherds recovered from archaeological sites. The petrofacies model was initiated by James Lombard in the mid-1980s (Lombard, 1987), with petrologists Elizabeth Miksa and Carlos Lavayen continuing to build upon his work to this day (Miksa et al., 2004).

Following the Gazzi-Dickinson approach, standard optical mineralogical techniques were used to identify and count monocrystalline mineral grains (Kerr, 1977; Phillips, 1971). Coarse-grained rocks were counted as the mineral phase to which they belong (quartz, hornblende, and so forth). Identification error was minimized through the use of stains that aid in the identification of different types of feldspars *and* help distinguish feldspars from quartz (Chayes, 1952). Sand maturity effects are minimized using this method. For example, a very immature granitic sand collected near the upper reaches of a drainage might contain large numbers of sand-sized granite grains and very few free minerals while further downstream the sand would be more mature, consisting primarily of free minerals and few granite fragments. The Gazzi-Dickinson technique facilitates comparison of the two sands' compositions because all sand-sized mineral grains in the immature sand are counted as minerals, not as rocks. Fine-grained rock fragments were identified using the joint criteria of overall mineral composition and internal texture and fabric (Dickinson, 1985) and placed into a taxonomic system pioneered by Dickinson (1970), with improvements to the nomenclature and procedure made by others (Graham et al., 1976; Ingersoll, 1983; Ingersoll and Suczek, 1979; Lombard, 1987).

Miksa et al. (2004) list the parameters used to point count the rock and mineral types encountered during analyses of Tucson Basin sands and sherds. All grains 0.0625 mm-2.0 mm were counted, with an average of 400 points counted per thin-section. Following the procedure described in Heidke and Miksa (2000), discriminant analysis has been used to demonstrate that the petrofacies compositions are statistically distinguishable from one another and to determine the set of functions that best discriminates among them (Miksa et al., 2004). In total the point-counted composition data recorded from 249 wash sand samples are included in the current Tucson Basin model, and the provenance of 214 of these samples (86%) can be predicted correctly by the discriminant model.

To date, the sand temper composition of nearly 8,300 Middle Rincon phase sherds recovered from 38 sites has been characterized with the aid of a binocular microscope and the petrofacies model. Some 159 sherds have also been point-counted in order to verify their binocular microscopic assessment following the approach advocated by Shepard (1942, p. 140, 227). Generally speaking, a large proportion of the pottery recovered from all investigated sites has contained sand temper of the Beehive Petrofacies (Figure 2). The Beehive Petrofacies can be identified easily in pottery by the presence of sand-sized, multi-colored, irregular, sub-angular felsite fragments with attached biotite in association with potassium and plagioclase feldspar grains and quartz, all derived from the tuff of Beehive Peak (Lipman, 1993). Sand temper provenance data, summarized in Heidke (1996a, b, 1999, 2000a, b; Heidke *et al.*, 2002), has suggested that potters residing at the West Branch site were the area's primary producers of three Middle Rincon wares, Middle Rincon Red-on-brown, Rincon Red, and Rincon Polychrome, and that at least five additional Middle Rincon phase villages produced pottery that was not distributed as widely throughout the basin (Heidke, 1999, figure 3.5, p. 58).

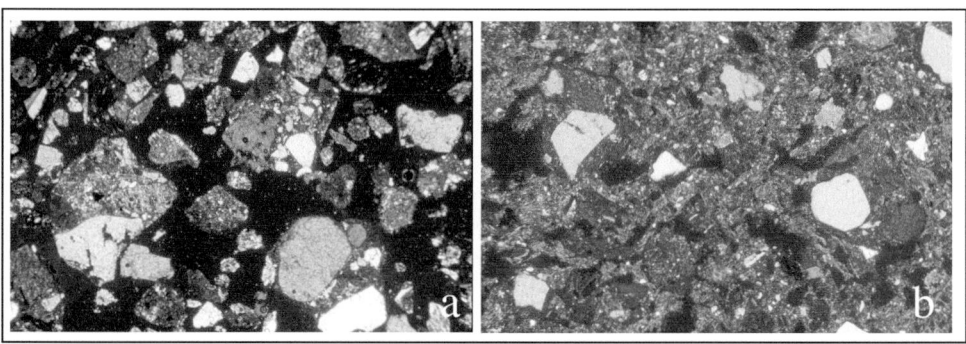

Figure 2. Thin section photomicrographs of Beehive Petrofacies (J1) sand sample and sand-tempered ceramics. a) Epoxy mount of sand from the Beehive Petrofacies, Tucson Mountains, Arizona. This sand is rich in felsic volcanic rock fragments with quartz, feldspar, and biotite phenocyrsts, b) Ceramic sample tempered with Beehive Petrofacies sand fragments in a heterogeneous micaceous clay. Added sand is approximately 50% of the paste by volume, with voids at 5-7%. Images taken in crossed polars. Image width = 9.0 mm.

This view has been modified as direct and indirect evidence of production recovered from the Julian Wash and Valencia sites (including the hamlet designated BB:13:74) has indicated that potters residing at these villages also tempered their pottery with Beehive Petrofacies sand (Heidke, 2003a, b, 2009a). Unfired pots, sand-tempered clays, processed red and white pigments, and supplemental micaceous schist/gneiss tempering agents have been recovered from many Middle Rincon phase features at these sites. Furthermore, approximately one-third of the ground stone tools recovered from each of them appear to have been used in pottery production, whereas pottery production tools typically represent only about five percent of the ground stone assemblages recovered from other Middle Rincon sites (Table 1). It therefore seems likely that potters residing at the West Branch, Julian Wash, and Valencia sites made the short trip (<3.0 km) to the West Branch of the Santa Cruz River in order to gather clay from deposits exposed along its bank and added Beehive Petrofacies sand to it while they were there (Figure 3).

Potters residing in all three of these villages were regional specialists in the production of red-on-brown, red-slipped, and polychrome vessels, although the emphasis placed on making each ware seems to have varied from village to village as well as within different portions of the West Branch site. Following Costin's (1991) typology for the organization of specialized craft production, both the direct and indirect archaeological evidence for pottery manufacture at these sites is consistent with 'community specialization,' which is defined by independent context, nucleated concentration, household constitution, and part-time intensity (Costin and Hagstrum, 1995). That is to say, pottery production took place in individual households and these households in the aggregate created what is known anthropologically as community specialization (Hagstrum, 1989).

Types of ground stone tools and raw materials	West Branch	Valencia	Julian Wash	Seven other Middle Rincon sites
Pottery production tools				
Polishing stone without pigment staining	147	5	34	46
Polishing stone with pigment staining	6	1	0	1
Anvil with/without pigment staining	5	0	2	0
Other pigment-stained tools	22	5	8	6
Other associated tools	11	0	0	0
Total ground stone tools recovered (all types)	593	33	111	1,091
Percentage of ground stone tools devoted to pottery production	32.2	33.3	39.6	4.9
Raw materials and unfired pottery				
Schist/gneiss temper	Present	Present	Present	Present
Unfired, sand-tempered clay	Absent	Absent	Present	Absent
Processed hematite (red slip or paint)	Present	Present	Present	Present
White pigment (processed caliche?)	Present	Absent	Absent	Absent
Unfired pottery	Absent	Absent	Present	Absent

Table 1. Summary of direct evidence for ceramic production at ten Middle Rincon sites. All primary data citations are reported in Heidke and Ryan (2005).

The anthropological term community specialization need not imply that all households residing at these three settlements contain a craft specialist. Indeed, a review of ethnographic, ethnohistoric, and ethnoarchaeological literature indicates that the percentage of households involved in pottery production can vary widely between different specialist communities. Information regarding the percentage of households involved in pottery production in 17 communities characterized as pottery-making specialists suggests that we should expect to find archaeological evidence of pottery production in 33-100% of all households present in a community that specialized in pottery manufacture (Heidke, 2009b, table 15.7). Furthermore, there is reason to believe that the annual production rate of all potters residing in a village may be highly variable (Deal, 1998; Nash, 1961). For example, Stark (1993, p. 297) reports that the least active potters working in the Kalinga community of Dalupa, Northern Luzon, Philippines in 1988 produced just three pots that year while the most active potter there made 368 pots. Nor does craft specialization at the community level require that specialist households abandon other economic activities. Indeed, it is likely that farming and the fulfillment of other important domestic tasks (Gosselain, 1998; Sillar, 2000) were the primary economic concerns of most West Branch, Valencia, and Julian Wash site households on a daily basis.

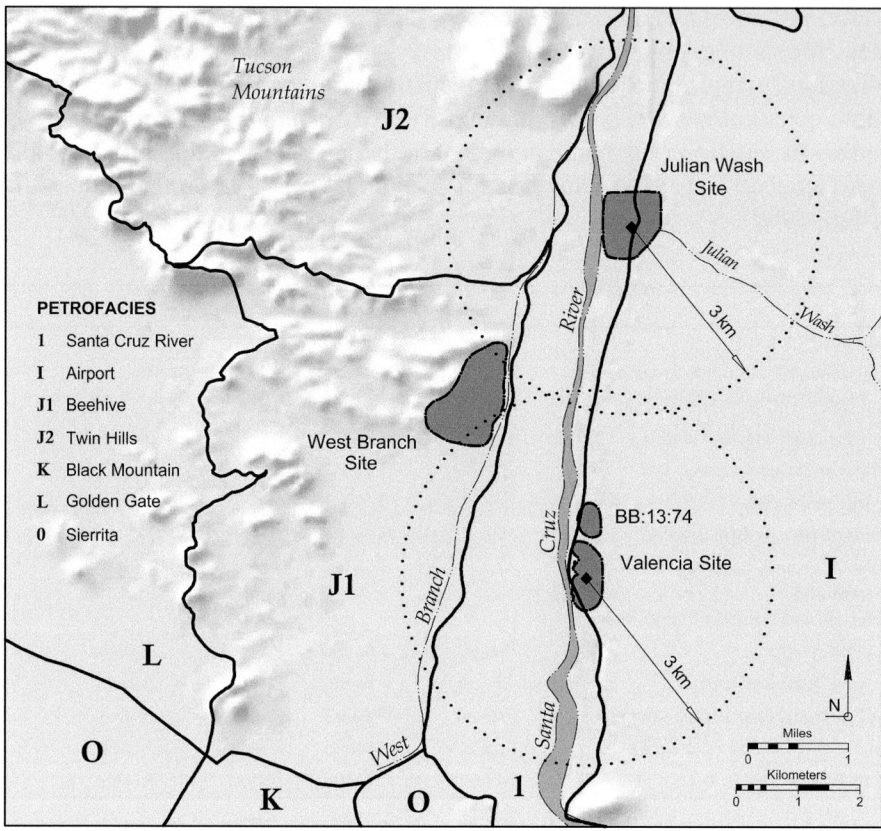

Figure 3. Location of the West Branch, Valencia, and Julian Wash sites analyzed in this study in relation to the Beehive Petrofacies.

Results

Distribution is the act that links producers and consumers (Pool, 1992). The simplest possible relationship between a producer and consumer occurs when they are one and the same person (Pool, 1992, p. 283). If producer and consumer are the same person and that person's lifeway involved a seasonal round, then some, or all, of the 'nonlocal' material recovered from an archaeological site may reflect that individual's yearly pattern of mobility (Earle, 1994). More complicated relationships occur when producer and consumer are not the same person (Renfrew, 1977). Pottery made at the West Branch, Valencia, and Julian Wash sites must have been distributed throughout the basin by foot once it was manufactured. Consumers may have traveled to potters, or potters to consumers, although, as I have argued elsewhere (Heidke, 1996b, 1999, 2000b; Heidke *et al.*, 2002), many of the pots may have been exchanged at 'markets' or 'fairs' (Haury, 1976, p. 78; Wilcox, 1979, p. 111) that took place in conjunction with ceremonies that occurred at villages with one or more ball courts.

One type of quantitative information that can help us understand the nature of ceramic distribution is the distance between sites inferred to be manufacturers and those inferred to be consumers. Here regression analysis is used to describe and aid in the interpretation of spatial patterning in the provenance data (Earle, 1982; Hodder and Orton, 1976). Regression analysis procedures are straightforward. First, the frequency of wares tempered with Beehive Petrofacies sand were calculated, using a sample of sites in the study area. Second, as a measure of transport cost, the straight-line distance from each site to the Beehive Petrofacies source was measured. Third, data from all sites were plotted on graphs where frequency is represented on the y axis and distance on the x axis. Fourth, the relationship between distance and frequency is established by the regression line that best predicts the point scatter.

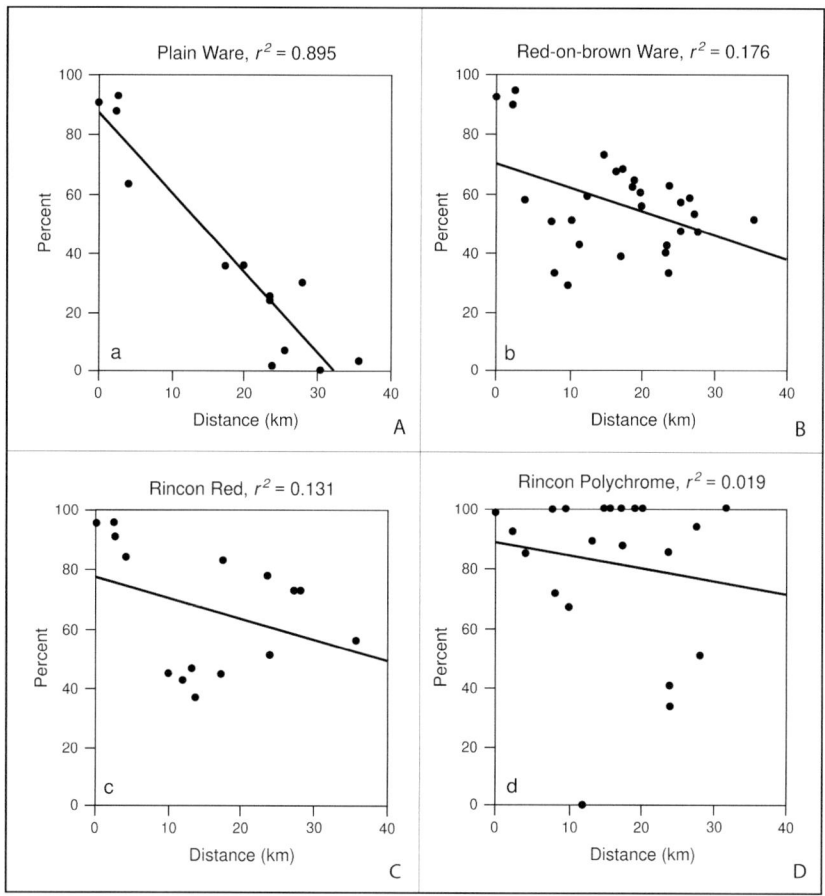

Figure 4. Distance-decay functions of Middle Rincon phase plain ware. a), b) red-on-brown, c) red ware, and d) polychrome pottery tempered with Beehive Petrofacies sands. All distance measurements between sites located to the east of the Tucson Mountains are geodesic, or straight line; distance measurements between sites located on both sides of the Tucson Mountains used gaps and passes to cross the mountains but are otherwise geodesic.

Scattergrams of four ware-based data batches and the regression lines that fit each set of data points best is displayed in Figure 4. It is clear that the plain ware pattern is quite different from those of the other three wares. The plain ware pottery shows a normal distance-decay function with the amount of pottery tempered with Beehive Petrofacies sand decreasing markedly the farther away from that source an archaeological site is located. Nearly 90% of the variation is related to distance from source (correlation coefficient $r^2 = 0.895$). In contrast, the amount of Middle Rincon Red-on-brown pottery tempered with Beehive Petrofacies sand is similar regardless of the recovery site's distance from that source with only 18% of the variation accounted for by the regression model ($r^2 = 0.176$). The distribution of Rincon Red pottery shows a similar pattern, although in that case only 13% of the variation is related to distance ($r^2 = 0.131$). The distribution of Rincon Polychrome shows an even weaker relationship with less than two percent of the variation related to distance-decay ($r^2 = 0.019$).

Discussion and Conclusions

Because most of the red-on-brown, red ware, and polychrome variability is not related to distance from source, other explanations must be searched for. However, before looking for social or political explanations to this economic pattern, it is logical to first assess whether or not any other differences present in the data sets may have caused the results. Site type, location of a site within the basin, archaeological sample type, and archaeological sampling error represent real differences in the ware-based data batches used in the regression analyses. Up to four hypothesis tests were conducted on each ware in order to determine whether or not one, or more, of these differences might have biased the results. Prior to conducting the tests, each ware-based data set was partitioned into two qualitative groups: those cases that had more Beehive Petrofacies sand-tempered pottery than expected given their distance from the source (i.e., the regression analyses' positive residuals) and those that had less (the negative residuals).

Archaeologists working in the Tucson Basin often utilize a hierarchically-ordered, five stage site typology. Ranked from least to most complex it includes resource procurement sites, field houses, farmsteads, hamlets, and villages (Elson, 1986). The first hypothesis test examined the relationship between settlement type and residual value. The null hypothesis (H_0) for this test was that the distribution of site types across the two residual categories is the same and the alternate hypothesis (H_1) was that there is a difference. Utilizing a significance level of 0.05, a chi-square test of the relationship between settlement type and residual value category indicated that no significant difference exists in the red-on-brown data ($\chi^2 = 2.036$; *degrees of freedom* = 3; *probability* = 0.565).

The second hypothesis test examined the relationship between settlement location and residual value. Tucson Basin archaeologists recognize that the resource potential of, and rainfall patterns in, eastern and western portions of the basin are fundamentally different, and have suggested that this difference played an important role in the nature of prehistoric settlement in the two subareas (Elson, 1986). The null hypothesis (H_0) for this test was that the distribution of settlement locations across the two residual categories is the same and the alternate hypothesis (H_1) was that there is a difference. A Fisher's exact test of the relationship between settlement location and residual value category indicated that there is no significant difference in the red-on-brown data (*probability* = 0.678).

The provenance data were collected from two different types of archaeological samples, referred to here as typological and contextual samples. The third hypothesis test examined the relationship between sample type and residual value. *Typological* samples only contain sherds typed as Middle Rincon Red-on-brown, and usually were drawn from multicomponent site or surface collections. *Contextual* samples were drawn from excavated deposits assigned to the Middle Rincon phase. Typically, these deposits also include sherds classified as 'Early or Middle Rincon red-on-brown', 'Middle or Late Rincon red-on-brown', and so on as well as sherds typed as Middle Rincon Red-on-brown. The third hypothesis test, therefore, examined the relationship between sample type and residual value. The null hypothesis (H_0) for this test was that the distribution of sample types across the two residual categories is the same and the alternate hypothesis (H_1) was that there is a difference. A Fisher's exact test of the relationship between archaeological sample type and residual value category indicated that there is no significant difference (*probability* = 1.000).

Archaeologists generally recognize that small samples may be prone to sampling error and may provide less reliable information on which to base conclusions (Robertson, 1999). For this reason a minimum sample size threshold of 30, 50, or 100 artifacts is often chosen (Hodder and Orton, 1976, p. 105; Robertson, 1999, p. 139). The minimum sample size used in the regression analysis was 11 red-on-brown sherds per site and the maximum was more than 1,250. The fourth hypothesis test compared the population distribution of the standard error for the estimate of a proportion (Robertson, 1999) grouped by residual value category. The null hypothesis (H_0) for this test was that the population distribution of the standard error for the estimate of a proportion is the same in the two categories and the alternate hypothesis (H_1) was that there is a difference. A Mann-Whitney U test of the data showed no significant difference between the two population distributions (*probability* = 0.695).

The results of these four hypothesis tests indicate that real differences in the data are unlikely to have caused, or biased, the regression model's result as the null hypothesis was accepted in every one of the red-on-brown ware tests (Table 2). With these concerns removed, an explanation as to why some settlements had more Beehive Petrofacies sand-tempered red-on-brown pottery than expected, while others had less, must be searched for elsewhere. Not surprisingly, the West Branch and Valencia sites yielded the highest positive residuals (Table 3). They are followed closely by the Hodges Ruin and two hamlets located near it, Sunset Mesa and Rillito Fan. These three sites are located anywhere from 15 km to 17.5 km from the Beehive Petrofacies and are unlikely to have been ceramic producers. In simulation studies Hodder and Orton (1976, p. 146-153) demonstrated that a 'secondary peak' in abundance, such as the Hodges Ruin's data point, is characteristic of places where redistribution occurred. The importance of the Hodges Ruin in the Tucson Basin's local system is attested to by the fact that its ball court shares the same orientation as Ball Court 1 at Snaketown, an orientation shared with only two other Hohokam ball courts (Wilcox, 1991b), and it is also of comparable length (Doelle and Wallace, 1986, figure 6.1, p. 93). The Hodges Ruin's location near the confluence of the Santa Cruz River, Rillito Creek, and the Cañada del Oro would have facilitated the distribution of pottery (produced at West Branch, Julian Wash, and Valencia) into the eastern and northeastern portions of the basin, following the natural corridors formed by Rillito Creek and Cañada del Oro Wash.

Test Statistic	Ware			
	Plain Ware	Red-on-brown Ware	Red Ware	Polychrome
Sample size (Number of individual site samples)	13	28	15	22
Regression				
r^2	0.895	0.176	0.131	0.019
Site type by residual[a]				
Chi-square	5.985	2.036	3.611	0.055
Degrees of freedom	3	3	2	2
Probability	0.114	0.565	0.164	0.973
Site location by residual[b]				
Fisher's Exact Test probability	1.000	0.678	1.000	1.000
Sample type by residual				
Fisher's Exact Test probability	N/A[c]	1.000	0.011	0.350
Comparisons of the population distribution of the standard error for the estimate of a proportion grouped by residual				
Probability (parametric t-test)	0.119	0.654	0.593	N/A[d]
Probability (nonparametric Mann-Whitney U test)	0.242	0.695	0.814	N/A[d]

Table 2. Summary of the amount of variation related to distance from temper source and potential sources of variation in the distribution of Middle Rincon pottery wares. [a]In this test the red-on-brown n = 26, redware n = 13, and polychrome n = 19 because site type was not always specified in the site reports. [b]In this test the plain ware n = 9, red ware n = 9, red-on-brown n = 24, and polychrome n=16 because some sites are located in the southern, northern, and northwestern portions of the basin (rather than eastern or western). [c]Test is not applicable; all plain ware samples are contextual. [d]Test is not applicable; all polychrome samples contain fewer than 100 sherds.

Sites with the lowest negative residuals appear to reflect localized ceramic production traditions in other parts of the basin. Previously, I have hypothesized that at least five settlements—the St. Mary's Ruin, Punta de Agua, the Hardy site, Honey Bee Village, and a site located somewhere in the western Avra Valley (Cocoraque Petrofacies)—contained some potters whose annual production rate exceeded the needs of their own household (Heidke, 1999). Three of these settlements are represented in the red-on-brown data. All three are villages, and each one yielded very low negative residuals in the regression analysis (including the two lowest values). In addition, other studies have shown the average percentage of locally-produced red-on-brown pottery to be about 50% at these sites (Heidke, 1999, 2000a, 2003b). Therefore, one plausible interpretation of the regression model's negative residuals is that they often denote other villages that were actively involved in pottery manufacture or sites closely allied with these villages. For example, the seasonal farmstead known as the Cienega site is located near the Hardy site (Elson, 1986), and both of these settlements yielded negative residual scores.

Identical suites of hypothesis tests were run on the Rincon Polychrome and Rincon Red data, and they produced results similar to those of the red-on-brown pottery. No significant differences were obtained in any of the polychrome tests (Table 2). However, a Fisher's exact test of the relationship between archaeological sample type and residual category of Rincon Red ware indicated a significant difference (*probability* = 0.011). Review of the cases showed that most typological samples yielded negative residuals, whereas most contextual samples yielded positive residuals. One of the six negative cases represents the Punta de Agua site, while four of the remaining five negative cases are sites located close to that village. So, as in the red-on-brown data, the negative residual scores denote another pottery producing village and allied settlements—it just so happens that, in this case, all of these samples are typological.

Common Name	Producer or Consumer?	Site Type	Residual	Percent Beehive Pf.	Distance to Beehive Pf. (km)
Punta de Agua	Consumer	Village	-33.291	28.08	10.0
St. Mary's Ruin	Consumer	Village	-30.831	32.04	8.1
Santa Cruz Bend	Consumer	Hamlet	-18.452	41.67	11.6
Tanque Verde Wash	Consumer	Hamlet	-18.146	32.30	23.9
Hardy	Consumer	Village	-17.799	37.84	17.3
Presidio	Consumer	Village	-13.185	50.00	7.7
AZ AA:16:408 (ASM)	Consumer	Farmstead	-11.600	39.13	23.5
Stone Pipe	Consumer	Unknown	-10.983	50.00	10.5
Cienega	Consumer	Farmstead	-9.210	41.47	23.6
Julian Wash	Producer	Village	-8.637	57.42	4.0
AZ AA:15:120 (ASM)	Consumer	Farmstead	-2.373	46.67	25.7
Duval Mine Road	Consumer	Farmstead	-1.246	45.98	28.0
EK Ranch	Consumer	Village	-0.501	58.82	12.6
Seneca Terrace II	Consumer	Farmstead	1.111	54.54	20.1
Bosque	Consumer	Village	4.788	52.33	27.6
Vista del Rio	Consumer	Hamlet	6.190	59.78	19.9
Lonetree	Consumer	Field house	7.140	56.25	25.6
AZ AA:12:20 (ASM)	Consumer	Unknown	7.574	61.90	19.0
AZ AA:12:785 (ASM)	Consumer	Farmstead	9.286	50.40	35.8
Observatory	Consumer	Farmstead	9.415	63.64	19.1
Whispering Wings	Consumer	Farmstead	9.728	57.89	26.8
Rillito Fan	Consumer	Hamlet	10.582	66.67	16.7
Toland	Consumer	Farmstead	11.536	61.90	24.0
Sunset Mesa	Consumer	Hamlet	11.730	67.29	17.4
Hodges	Consumer	Village	15.204	72.73	14.9
AZ BB:13:74 (ASM)	Producer	Hamlet	21.906	89.34	2.3
West Branch	Producer	Village	22.879	92.12	0.0
Valencia	Producer	Village	27.185	94.37	2.6

Table 3. Red-on-brown ware residuals ranked from lowest negative value to highest positive value.

In summary, regression analyses of the percentage of Beehive Petrofacies sand-tempered wares recovered from sites located throughout the basin has shown that most of the variation in the plain ware data can be attributed to distance from production source, whereas most of the Middle Rincon Red-on-brown, Rincon Red, and Rincon Polychrome variation can not. The findings of multiple hypothesis tests showed that real differences in the data sets are unlikely to have caused, or biased, the regression models' results; accordingly, social and/or political explanations seem highly likely. One of these explanations is that a 'secondary peak' in the abundance data denotes a settlement that facilitated the movement of Beehive Petrofacies sand-tempered red-on-brown, red ware, and polychrome pottery within the basin. Another explanation is that low negative residuals often denote the presence of other, smaller-scale ceramic production and distribution systems.

Why did households at the West Branch, Valencia, and Julian Wash sites become specialists in pottery production? Community-based specialization occurs when particular communities become recognized for the production or procurement of particular goods, be they wild or cultivated plant products, raw materials, or crafts (Stark, 1993). Since these goods are exchanged between communities within a regional economic system, community-based specialization in small-scale societies involves economic integration at the regional level (Chávez, 1992; Costin, 1991; Hendry, 1957; Nash, 1961; Ogundiran, 2001; Sillar, 2000; Specht, 1972; Stark, 1993).

The clay resource that was locally available to the West Branch, Valencia, and Julian Wash site potters is reported to be amongst the best in the area (Reid and Whittlesey, 1997). Potters began using it by Agua Caliente times, c. AD 50-500, to make simple storage vessels (Heidke *et al.*, 1998), and, likely, continued to learn the clay's properties during the subsequent Tortolita phase (AD 500-700) when the production of both cooking and group serving vessels first began in earnest (Heidke, 2003c). By the time the first ball courts were built in the late Pioneer or early Colonial period (c. AD 700-850), their descendants would have mastered those properties. Once the ball court system was fully established, manufacturing surplus pottery for exchange would have reduced the risk of farming along the Santa Cruz River. Farming is always a risky way to make a livelihood (Hunt, 1962). Exchanging pottery would have provided a means to diversify and/or supplement a household's harvest in good years as well as bad (Chávez, 1992; Hunt, 1962). The basin-wide distribution of Beehive Petrofacies sand-tempered pottery ended by the beginning of the Classic period, circa AD 1150, after which most pottery was locally produced (Heidke, 1996c, 2009b; Heidke *et al.*, 1994; Wallace, 1957). The Tucson Basin Hohokam's ball court system had already been abandoned for at least 50 years by that time, and the last remnants of intra-regional economic, social, and political integration, which the ball game's ceremonies and market places had once enabled, finally ceased.

References

Abbott, D.R. 2001. Conclusions for the GARP ceramic analysis. In: Abbott, D.R. (Ed.) *The Grewe Archaeological Research Project: Vol. 2. Material Culture: Part I.*

Ceramic Studies. Anthropological Papers No. 99-1. Northland Research, Inc., Flagstaff and Tempe, Arizona: 263-272.

Abbott, D.R., Smith, A.M., and Galllaga, E. 2007. Ballcourts and ceramics: The case for Hohokam marketplaces in the Arizona desert. *American Antiquity* 72:461-484.

Arnold, D.E. 2000. Does the standardization of ceramic pastes really mean specialization? *Journal of Archaeological Method and Theory* 7:333-375.

Bayman, J.M. 2002. Hohokam craft economies and the materialization of power. *Journal of Archaeological Method and Theory* 9:69-95.

Chávez, K.L.M. 1992. The organization of production and distribution of traditional pottery in south highland Peru. In: Bey, G.J., III and Pool, C.A. (Eds.) *Ceramic production and distribution: An integrated approach.* Westview Press, Boulder: 49-92.

Chayes, F. 1952. Notes on the staining of potash feldspar with sodium cobalt nitrite on thin sections. *American Mineralogist* 14:290-292.

Costin, C.L. 1991. Craft specialization: Issues in defining, documenting, and explaining the organization of production. In: Schiffer, M.B. (Ed.) Archaeological method and theory, Vol. 3. University of Arizona Press, Tucson: 1-56.

Costin, C.L. 2000. The use of ethnoarchaeology for the archaeological study of ceramic production. *Journal of Archaeological Method and Theory* 7:377-403.

Costin, C., and Hagstrum, M.B. 1995. Standardization, labor investment, skill, and the organization of ceramic production in late prehistoric Highland Peru. *American Antiquity* 60:619-639.

Deal, M. 1998. *Pottery Ethnoarchaeology in the Central Maya Highlands.* Foundations of Archaeological Inquiry Series. University of Utah Press, Salt Lake City.

Dickinson, W.R. 1970. Interpreting detrital modes of Graywacke and Arkose. *Journal of Sedimentary Petrology* 40:695-707.

Dickinson, W.R. 1985. Interpreting provenance relations from detrital modes of sandstones. In: Zuffa, G.G. (Ed.) *Provenance of Arenites.* NATO Advanced Science Institute Series C, No. 148. D. Reidel, Boston: 333-361.

Dickinson, W. R. 1991. *Tectonic setting of faulted Tertiary strata associated with the Catalina Core Complex in southern Arizona.* Special Paper No. 264. Geological Society of America, Boulder.

Doelle, W.H., and Wallace, H.D. 1986. *Hohokam settlement patterns in the San Xavier Project Area, southern Tucson Basin.* Technical Report No. 84-6. Institute for

American Research, Tucson.

Earle, T.K. 1982. Prehistoric economics and the archaeology of exchange. In: Ericson, J.E. and Earle, T.K. (Eds.) *Context for prehistoric exchange*. Academic Press, New York: 1-12.

Earle, T.K. 1994. Positioning exchange in the evolution of human society. In: Baugh, T.G. and Ericson, J.E. (Eds.) *Prehistoric exchange systems in North America*. Plenum Press, New York: 419-437.

Elson, M.D. 1986. Previous research. In: Elson, M.D. (Ed.) *Archaeological investigations at the Tanque Verde Wash Site, a Middle Rincon settlement in the eastern Tucson Basin*. Anthropological Papers No. 7. Institute for American Research, Tucson: 11-42.

Gazzi, P. 1966. Le arenarie del flysch sopracretaceo dell'appennino modenese: Correlazioni con il flysch di monghidoro. *Mineralogica e Petrografica Acta* 12:69-97.

Gosselain, O.P. 1998. Social and technical identity in a clay crystal ball. In: Stark, M. T. (Ed.) *The Archaeology of Social Boundaries*. Smithsonian Institution Press, Washington, D.C.: 78-106.

Graham, S.A., Ingersoll, R.V. and Dickinson W.R. 1976. Common provenance for lithic grains in carboniferous sandstones from Ouachita Mountains and Black Warrior Basin. *Journal of Sedimentary Petrology* 46:620-632.

Hagstrum, M.B. 1989. *Technological continuity and change: Ceramic ethnoarchaeology in the Peruvian Andes*. Doctoral dissertation, Department of Anthropology, University of California, Los Angeles. University Microfilms International, Ann Arbor, Michigan.

Haury, E.W. 1976. *The Hohokam: Desert Farmers & Craftsmen. Excavations at Snaketown, 1964-1965*. University of Arizona Press, Tucson.

Heidke, J.M. 1996a. Ceramic artifacts from the Cook Avenue Locus. In: Dart, A. and Swartz, D.L. (Eds.) *Archaeological Data Recovery Project at the Cook Avenue Locus of the West Branch Site, AZ AA:16:3 (ASM)*. Technical Report No. 96-8. Center for Desert Archaeology, Tucson: 53-76.

Heidke, J.M. 1996b. Production and distribution of Rincon phase pottery: Evidence from the Julian Wash site. In: Mabry, J.B. (Ed.) *A Rincon phase occupation at Julian Wash, AZ BB:13:17 (ASM)*. Technical Report No. 96-7. Center for Desert Archaeology, Tucson: 47-71.

Heidke, J.M. 1996c. Qualitative temper characterization of potsherds from the Gibbon Springs site. In: Slaughter, M. C. and Roberts, H. (Eds.) *Excavation of the Gibbon*

Springs site: A Classic period village in the northeastern Tucson Basin. Archaeological Report No. 94-87. SWCA, Inc., Tucson: 259-266.

Heidke, J.M. 1999. Ceramic consumption at AZ BB:13:535 (ASM). In: Mabry, J. B., Lindeman, M.W. and Wöcherl, H. (Eds.) *Prehistoric uses of a developing floodplain: Archaeological investigations on the east bank of the Santa Cruz River at A-Mountain.* Technical Report No. 98-10. Desert Archaeology, Inc., Tucson: 46-60.

Heidke, J.M. 2000a. Ceramic data. In: Lindeman, M. W. (Ed.) *Excavations at Sunset Mesa Ruin.* Technical Report No. 2000-02. Desert Archaeology, Inc., Tucson: 249-260.

Heidke, J.M. 2000b. Middle Rincon phase ceramic artifacts from Sunset Mesa. In: Lindeman, M.W. (Ed.) *Excavations at Sunset Mesa Ruin.* Technical Report No. 2000-02. Desert Archaeology, Inc., Tucson: 69-118.

Heidke, J.M. 2003a. Middle Rincon phase ceramics from AZ BB:13:74 (ASM). In: Lindeman, M.W. (Ed.) *Excavations at AZ BB:13:74 (ASM): An examination of three Middle Rincon phase loci.* Technical Report No. 2000-01. Desert Archaeology, Inc., Tucson: 35-62.

Heidke, J.M. 2003b. *Preliminary results of Arizona State Museum Loan Agreement DT-2003-30: A study to assess the likelihood of Sedentary period ceramic production at the Punta de Agua site complex.* Petrographic Report No. 2004-1. Desert Archaeology, Inc., Tucson.

Heidke, J.M. 2003c. Tortolita phase ceramics. In: Wallace, H.D. (Ed.) *Roots of sedentisim: Archaeological excavations at Valencia Vieja, a founding village in the Tucson Basin of southern Arizona.* Anthropological Papers No. 29. Center for Desert Archaeology, Tucson: 145-191.

Heidke, J.M. 2009a. Prehistoric pottery containers from the Julian Wash site, AZ BB:13:17 (ASM). In: Wallace, H.D. (Ed.) *Craft specialization in the southern Tucson Basin: Archaeological excavations at the Julian Wash site, AZ BB:13:17 (ASM): Part 1. Introduction, excavation results, and artifact investigations.* Anthropological Papers No. 40. Center for Desert Archaeology, Tucson: Draft.

Heidke, J.M. 2009b. Sedentary period ceramic production and distribution: New evidence from the Julian Wash site, AZ BB:13:17 (ASM). In: Wallace, H.D. (Ed.) *Craft specialization in the southern Tucson Basin: Archaeological excavations at the Julian Wash Site, AZ BB:13:17 (ASM): Part 2. Synthetic studies.* Anthropological Papers No. 40. Center for Desert Archaeology, Tucson: Draft.

Heidke, J.M., Goetze, C.E. and Dart, A. 1994. Schuk Toak project ceramics: Chronology, formation processes, and prehistory of the Avra Valley. In: Dart, A. (Ed.) *Archaeological studies of the Avra Valley, Arizona: Excavations in the Schuk Toak*

District: Vol. 2. Scientific studies and. Anthropological Papers No. 16. Center for Desert Archaeology, Tucson: 11-76.

Heidke, J.M. and Miksa, E.J. 2000. Correspondence and discriminant analyses of sand and sand temper compositions, Tonto Basin, Arizona. *Archaeometry* 42:273-299.

Heidke, J.M., Miksa, E.J. and Wallace, H.D. 2002. A petrographic approach to sand-tempered pottery provenance studies: Examples from two Hohokam local systems. In: Glowacki, D.M. and Neff, H. (Eds.) *Ceramic production and circulation in the greater Southwest: Source determination by INAA and complementary mineralogical.* Monograph No. 44. Cotsen Institute of Archaeology, University of California, Los Angeles: 152-178.

Heidke, J.M., Miksa, E.J. and Wiley, M.K. 1998. Ceramic artifacts. In: Mabry, J. B. (Ed.) *Archaeological investigations of early village sites in the middle Santa Cruz Valley: Analyses and synthesis, Part II.* Anthropological Papers No. 19. Center for Desert Archaeology, Tucson: 471-544.

Heidke, J.M. and Ryan, S. L. 2005. Sedentary period pottery from the West Branch site, AZ AA:16:3 (ASM): Provenance and function. In: Swartz, D. L. (Ed.) *Results of Phase 2 data recovery at the southern margin of the West Branch site, AZ AA:16:3 (ASM), Pima County, Arizona.* Technical Report No. 2005-01. Desert Archaeology, Inc., Tucson: 43-79.

Hendry, J.C. 1957. *Atzompa: A pottery producing village of southern Mexico.* Doctoral dissertation, Cornell University, Ithaca. University Microfilms International, Ann Arbor, Michigan.

Hodder, I. and Orton, C. 1976. *Spatial analysis in archaeology.* New Studies in Archaeology No. 1. Cambridge University Press, Cambridge, England, and New York.

Hunt, M.E. 1962. *The dynamics of the domestic group in two Tzeltal villages: A contrastive comparison.* Unpublished doctoral dissertation, Department of Anthropology, University of Chicago, Chicago.

Ingersoll, R.V. 1983. Petrofacies and provenance of late Mesozoic fore-arc basin, northern and central California. *The American Association of Petroleum Geologists Bulletin* 67(7):1125-1142.

Ingersoll, R.V. and Suczek, C.A. 1979. Petrology and provenance of Neogene sand from Nicobar and Bengal fans, DSDP Sites 211 and 218. *Journal of Sedimentary Petrology* 49:1217-1228.

Kerr, P.F. 1977. *Optical mineralogy.* 2nd ed. McGraw-Hill, New York.

Lipman, P.W. 1993. *Geologic map of the Tucson Mountains Caldera, southern Arizona*. Map No. I-2205. Miscellaneous Investigations Series, Geological Survey, U.S. Department of the Interior, Washington, D.C.

Lombard, J.P. 1987. Provenance of sand temper in Hohokam ceramics, Arizona. *Geoarchaeology* 2:91-119.

Miksa, E.J. and Heidke, J.M. 2001. It all comes out in the wash: Actualistic petrofacies modeling of temper provenance, Tonto Basin, Arizona, USA. *Geoarchaeology* 16:177-222.

Miksa, E.J., Heidke, J.M., Lavayen, C. and Lombard, J.P. 2004. *Ceramic petrography laboratory research results: Tucson Basin Petrofacies model summary*. On-line report available at: http:www.desert.com/petroweb/petrology.php?proj=TUC

Mills, B.J. and Crown, P.L. (Eds.) 1995. *Ceramic production in the American Southwest*. University of Arizona Press, Tucson.

Nash, M. 1961. The social context of economic choice in a small society. *Man* 61:186-191.

Ogundiran, A.O. 2001. Ceramic spheres and regional networks in the Yoruba-Edo region, Nigeria, 13th-19th Centuries A.C. *Journal of Field Archaeology* 28:27-43.

Phillips, W.R. 1971. *Mineral optics*. W. H. Freeman, San Francisco.

Pool, C.A. 1992. Integrating ceramic production and distribution. In: Bey, G. J., III and Pool, C. A. (Eds.) *Ceramic production and distribution: An integrated approach*. Westview Press, Boulder: 275-313.

Reid, J.J., and Whittlesey, S.M. 1997. *The archaeology of ancient Arizona*. University of Arizona Press, Tucson.

Renfrew, C. 1977. Production and exchange in early state societies: The evidence of pottery. In: Peacock, D. P. S. (Ed.) *Pottery and early commerce: Characterization and trade in Roman and later ceramics*. Academic Press, London: 1-20.

Robertson, I.G. 1999. Spatial and multivariate analysis, random sampling error, and analytical noise: Empirical Bayesian methods at Teotihuacan, Mexico. *American Antiquity* 64:137-152.

Shennan, S. 1990. *Quantifying Archaeology*. Academic Press, New York.

Shepard, A.O. 1942. Rio Grande Glaze Paint Ware: a study illustrating the place of ceramic technological analyses in archaeological research. In: *Contributions to American Anthropology and History No. 39*. Carnegie Institution, Washington, D.C.: 129-262.

Shepard, A.O. 1963. *Beginnings of ceramic industrialization: An example from the Oaxaca Valley.* Notes from a Ceramic Laboratory No. 2. Carnegie Institution, Washington, D.C.

Sillar, B. 2000. *Shaping culture: Making pots and constructing households. An ethnoarchaeological study of pottery production, trade, and use in the Andes.* BAR International Series No. 883. Hadrian Books, Ltd., Oxford, England.

Specht, J. 1972. The pottery industry of Buka Island, Territory of Papua, New Guinea. *Archaeology and Physical Anthropology of Oceania* 7:125-144.

Stark, B.L. 1985. Archaeological identification of pottery production locations: Ethnoarchaeological and archaeological data in Mesoamerica. In: Nelson, B.A. (Ed.) *Decoding Prehistoric Ceramics.* Southern Illinois University Press, Carbondale: 158-223.

Stark, M.T. 1993. *Pottery economics: A Kalinga ethnoarchaeological study.* Doctoral dissertation, Department of Anthropology, University of Arizona, Tucson. University Microfilms International, Ann Arbor, Michigan.

Sullivan, A.P., III 1988. Prehistoric Southwestern ceramic manufacture: The limitations of current evidence. *American Antiquity* 53:23-35.

Wallace, R.M. 1957. Petrographic analysis of pottery from University Indian Ruin. In: Hayden, J.D. (Ed.) *Excavations, 1940, at the University Indian Ruin, Tucson, Arizona.* Technical Series No. 5. Southwestern Monuments Association, Globe, Arizona: 209-219.

Wilcox, D.R. 1979. The Hohokam regional system. In: Rice, G.E., Wilcox, D., Rafferty, K. and Schoenwetter, J. (Eds.) *An Archaeological test of sites in the Gila Butte-Santan Region, south-central Arizona.* Anthropological Research Papers No. 18. Arizona State University, Tempe: 77-116.

Wilcox, D.R. 1991a. Hohokam religion: An archaeologist's perspective. In: Noble, D. G. (ed.) *The Hohokam, ancient people of the desert.* School of American Research Press, Santa Fe: 47-59.

Wilcox, D.R. 1991b. The Mesoamerican ballgame in the American Southwest. In: Scarborough, V.L. and Wilcox, D.R. (Eds.) *The Mesoamerican Ballgame.* University of Arizona Press, Tucson: 101-125.

A PRELIMINARY EVALUATION OF THE VERDE CONFEDERACY MODEL: TESTING EXPECTATIONS OF POTTERY EXCHANGE IN THE CENTRAL ARIZONA HIGHLANDS

Sophia E. Kelly

School of Human Evolution and Social Change, Arizona State University, Tempe, Arizona. USA
(Sophia.Kelly@asu.edu)

David R. Abbott

School of Human Evolution and Social Change, Arizona State University, Tempe, Arizona, USA

Gordon Moore

Department of Chemistry & Biochemistry, Arizona State University, Tempe, Arizona, USA

Christopher Watkins

School of Human Evolution and Social Change, Arizona State University, Tempe, Arizona, USA

Caitlin Wichlacz

Department of Anthropology, Washington State University, Pullman, Washington, USA

Introduction

Regional demographic movements during the mid 13th-14th centuries signaled corresponding changes to social and economic networks throughout the American Southwest. In the high mesa country of central Arizona large, masonry pueblos were constructed around AD 1250–1300 overlooking the vertical walls of Perry Mesa, in the Bloody Basin, and along the middle Verde River valley (Figure 1). As these settlement clusters coalesced, a 45 km expanse of empty land opened between the upland pueblos and the densely packed Hohokam population centers in the Phoenix Basin.

A current theory termed the Verde Confederacy Model contends that this gap in settlement reflected mounting social tensions between populations in the irrigated river valleys of the Phoenix Basin and people living in the rugged high country to the north (Wilcox *et al.*, 2001). The model posits that defensive alliances formed between northern settlements to deter attacks from Hohokam invaders. A well-documented web of line-of-sight relationships across the Verde Confederacy territory enabled settlements to rapidly communicate for defensive purposes and to support social and economic networks among villages in the confederacy (Wilcox, 2005; Wilcox *et al.*, 2001).

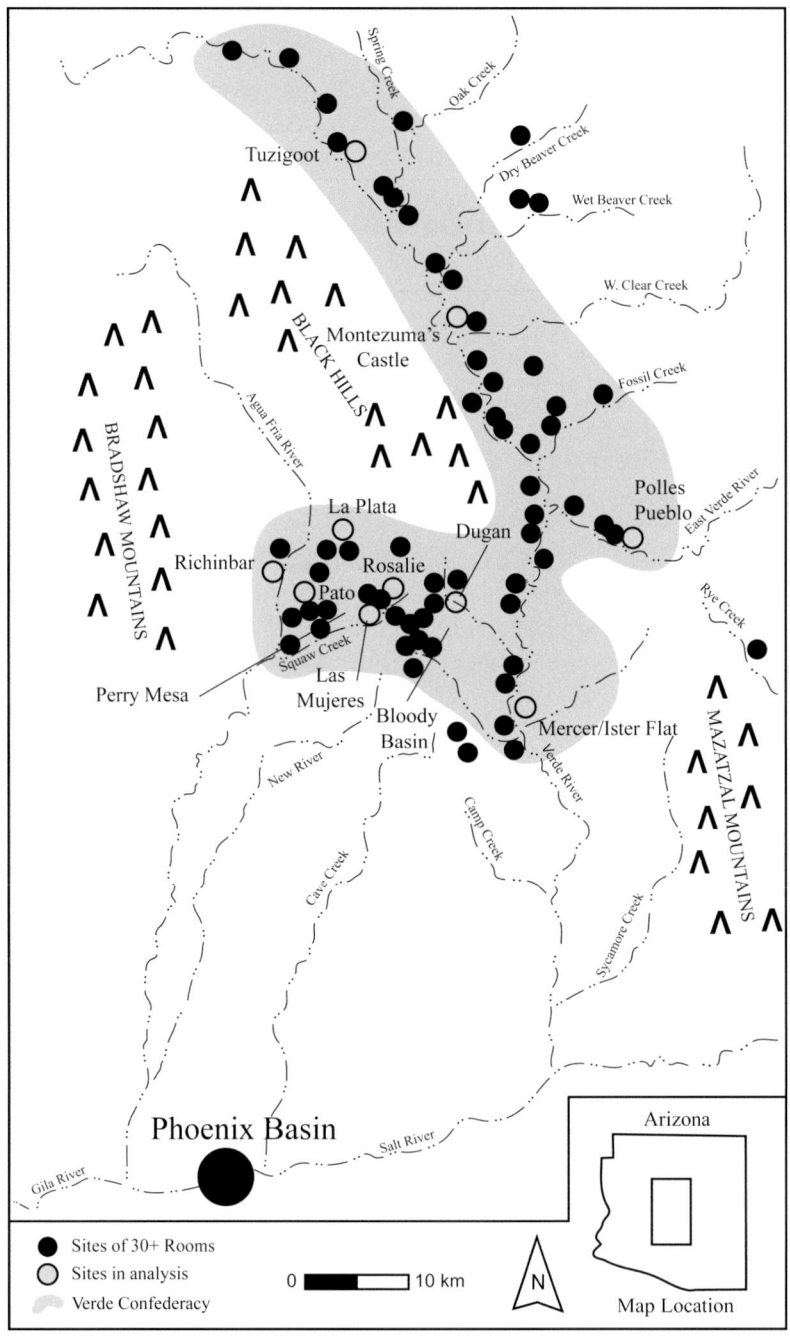

Figure 1. The proposed Verde Confederacy (after Wilcox, 2005; Wilcox *et al.*, 2001).

This study evaluates two expectations of the Verde Confederacy model using ceramic provenance and exchange data from Perry Mesa plainware pottery assemblages. First, the Verde Confederacy model suggests that Perry Mesa inhabitants may have been dependent on other Verde Confederacy communities to the east for basic necessities, such as craft goods and foodstuffs (Wilcox *et al.*, 2001; Wilcox and Holmlund, 2007). This dependence was supposedly rooted in the role that Perry Mesa settlements played in protecting the western flank of the alliance during the AD 1300s. The defensive role of Perry Mesa in the alliance would presumably have out-weighed concerns over the low potential of the mesa-top environment to sustain the resident population and the lack of natural water sources on top of the mesa. The model anticipates that Perry Mesa settlements were reliant on non-local producers to provision them with various necessities, perhaps including clay bowls and jars.

Second, the Verde Confederacy Model predicts that members of the confederacy were closely allied for defensive purposes (Wilcox *et al.*, 2001). As such, settlements throughout the alliance would have been united within a single interaction sphere, marked by persistent connectivity between people across the region. In the American Southwest, intrusive pottery composed of temper and clay that originated elsewhere is a commonly cited archaeological signature for exchange and interaction among settlements (e.g. Abbott, 2000; Mills and Crown, 1995).

The Verde Confederacy Model

Initially proposed by David Wilcox and his colleagues in a seminal paper (Wilcox *et al.*, 2001), the Verde Confederacy Model uses settlement pattern data to evaluate relationships between Perry Mesa settlements and other villages in the Bloody Basin and along the middle Verde River during the 14th century. The model suggests that rapid political aggregation of approximately 10,000 to 13,000 people in this region was a response to regional factionalism and warfare. The resulting network of over 135 sites was presumably linked through line-of-sight relationships, wherein an attack on one part of the alliance could receive prompt response from other settlements (Wilcox 2005, p. 26; Wilcox *et al.*, 2001, table 7.4, p. 164).

The Verde Confederacy Model suggests that threats from Hohokam populations in the Phoenix Basin prompted the formation of defensive alliances in the central Arizona highlands (Wilcox *et al.*, 2001). Around 1250 AD, a settlement gap 45 km wide separated upland villages from large Hohokam population centers in the irrigated river valleys of the Phoenix Basin (Figure 1) (Wilcox *et al.*, 2001; Wilcox, 2005; Wilcox and Holmlund, 2007, appendix B, table 11, p. 838-850). Wilcox and his colleagues argue that this spatial break in occupation created a buffer zone, which separated upland populations from would-be aggressors. Vastly outnumbered by the Phoenix Basin Hohokam to the south, settlements in the rugged Arizona highlands were at a tactical disadvantage. Hohokam population centers had twice the number of residents as settlements in the Verde Confederacy, and had up to seven times as many residents as Perry Mesa. Wilcox *et al.* (2001, p. 154-155) suggest that armies of up to 1,000 Hohokam warriors could have been assembled to siege upland settlements to the north.

The motivation for Hohokam assaults on Perry Mesa and other settlements in the Verde Confederacy may have been retaliation for raiding (Wilcox et al., 2001), or to capture people from northern settlements to help maintain the extensive irrigation networks along the lower Salt River (Wilcox and Holmlund, 2007, p. 39-40).

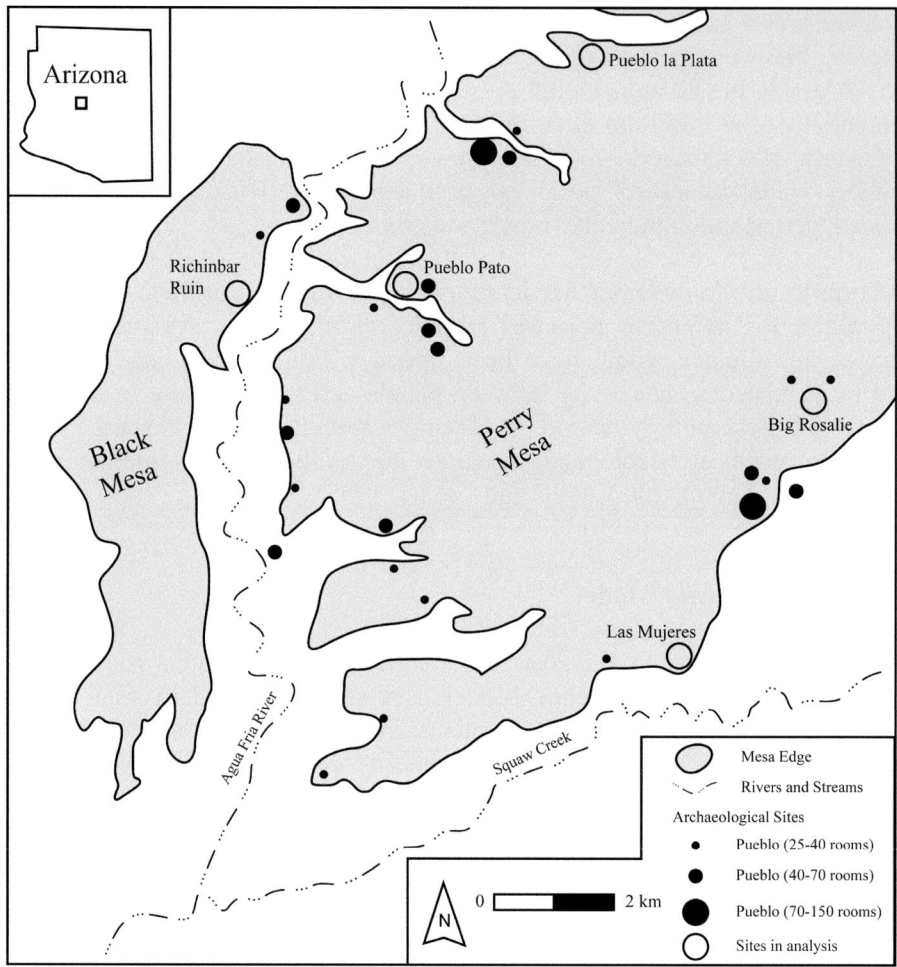

Figure 2. Perry Mesa with the location of its main settlements including those analyzed in this study.

In the early 1300's, the Verde Confederacy may have launched a coordinated strategic response to mounting threats from the Phoenix Basin by constructing large, masonry pueblos on Perry Mesa (Figure 2). The possibility that these pueblos were constructed rapidly as well as their limited access to water on Perry Mesa may suggest that the sites were positioned to protect the western flank of the Verde Confederacy (Wilcox et al., 2001, p. 155, 167-168). Several features of Perry Mesa sites have been interpreted as defensive, such as the high masonry walls, limited number of exterior entryways, walls that extend over mesa edges, and loop holes that look down onto access routes (Wilcox et al., 2001; Wilcox and Holmlund, 2007). The positioning of sites and lookouts on the mesa top would have allowed residents to spot attackers from a distance. Large pueblos located adjacent to the steep mesa walls and limited numbers of access routes would have also made the mesa virtually inaccessible to unwanted visitors. In addition, one of the largest sites on Perry Mesa, Las Mujeres, may have served as a command center that could have communicated with other villages on Perry Mesa through a series of line-of-sight connections (Wilcox *et al.*, 2001).

Evaluating the Model

If a vast alliance between Perry Mesa and other upland communities did exist, as proposed by the Verde Confederacy Model, we would expect that Perry Mesa settlements participated in a network of social and economic ties with settlements to the east, in Bloody Basin, and along the middle Verde River (Wilcox and Holmlund, 2007, p. 107). Frequent communication and coordinated political action between Perry Mesa settlements would have been necessary to establish a comprehensive regional defensive system (after Upham *et al.*, 1994). Finally, the model implies economic interdependence between settlements, and posits that villages depended on others to supply them with essential goods and resources, as well as for protection (Wilcox and Holmlund, 2007, p. 21). For instance, the Verde Confederacy Model suggests that Perry Mesa played a particularly important role in establishing an "impenetrable" western perimeter for the alliance, but could have had limited access to essential resources for agricultural and craft production. Perry Mesa settlements would have therefore relied on other villages in the Verde Confederacy to supply them with basic necessities in return for their role in protecting the alliance.

In this study, we address the proposition that the Perry Mesa residents relied on populations elsewhere in the alliance for basic necessities, which may have included plainware pottery. An integrated network of communities within the confederacy would have made reliance on non-local plainware pottery feasible, and, perhaps, essential to residents on Perry Mesa. If Perry Mesa settlements did depend on non-local pottery producers to supply their ceramic inventories, we would expect the plainware pottery to be made with tempers that are not locally available on Perry Mesa. Alternatively, if Perry Mesa settlements produced the majority of their plainware locally, the pottery would be made with readily available tempering materials. In particular, we would expect local production to be marked by temper dominated by the basaltic sands eroded from the c.100 m-thick Late Miocene sub-alkaline basalt flows that cap the entire mesa (Leighty, 1997, p. 121-129).

In the second stage of our analysis, we address the proposition that central Arizona settlements communicated regularly and cooperated within a coordinated defensive alliance. We presume that interconnectivity between settlements across the region was often associated with the exchange of plainware between members of the confederacy. Plainwares are the dominant ceramic type recorded at Perry Mesa sites, as well as at all other sites in the proposed Verde Confederacy. The exchange of these ubiquitous bowls and jars would have likely been associated with close and regular interactions between people across the region. If this expectation of the Verde Confederacy Model holds true, then we can expect to find plainware pots from sites across the entire region at each site within the confederacy. On the other hand, the existence of pottery assemblages with only locally produced pottery, or evidence of spatial patterns of exchange that suggest social boundaries within the confederacy, would not support the expectations of the model.

Sites and Study Materials

To evaluate these two expectations of the Verde Confederacy Model, we use qualitative petrographic analysis to characterize the mineralogy of sand tempers from five sites on Perry Mesa and nearby Black Mesa, one site in Bloody Basin, and five sites located along the middle Verde River (Figure 1). Settlements in the analysis from Perry Mesa include Big Rosalie and Las Mujeres on the eastern end of Perry Mesa and Pueblo Pato and Pueblo la Plata on the western end of the mesa (Figure 2). Richinbar is located directly west from Pueblo Pato across the Agua Fria on Black Mesa. All of these sites were constructed and occupied from AD 1250 to AD 1450 (Ahlstrom and Roberts, 1995, p. 83; North and Foster, 2002, p. 19). However, the occupation span of some pueblos, such as Las Mujeres, may have been shorter than other sites (Schafssma et al., 2006). All sites tested from Perry Mesa are of masonry construction and include over 50 rooms. The largest of these sites is Pueblo Pato with approximately 200 rooms and Las Mujeres with approximately 150 rooms (Wilcox et al., 2001, appendix 7.1, p. 173-176). Big Rosalie includes over 80 rooms (Wilcox et al., 2001, appendix 7.1, p. 173-176), and Pueblo la Plata has over 69 rooms (Mapes, 2005; Schollmeyer, 2004). The site of Richinbar is the smallest pueblo tested in this analysis and consists of approximately 30 to 40 rooms (North, 2002; Schollmeyer, 2005).

Sites included in the analysis from other regions of the proposed Verde Confederacy were also occupied between the mid to late 1200s until the mid 1400s. These pueblos were all constructed using masonry and include more than 50 rooms. Dugan Pueblo in the Bloody Basin consists of approximately 50 rooms. Along the southern Verde River, the nearby sites of Ister Flats and Mercer consist of more than 40 rooms, and between 225-300 rooms respectively. The Mercer settlement also includes a platform mound located within a two-story roomblock and five plaza areas. Polles Pueblo, located on Polles Mesa to the north of the East Verde River, is composed of more than 184 rooms and an enclosed plaza (Wilcox et al., 2001, Appendix 7.2B). Montezuma's Castle is located in the Verde Valley to the north, and consists of approximately 65 rooms spread between several room blocks. Finally, Tuzigoot, adjacent to the Verde River to the north, includes more than 93 rooms (Wilcox et al., 2001, appendix 7.2A, p. 176-180).

Plainware pottery constitutes the primary ceramic vessels that were produced and used by the inhabitants on Perry Mesa and surrounding regions. Plainware pottery is unslipped and does not have painted decorations. Although plainware vessels appear in a variety of sizes and forms, medium-sized bowls and jars constitute the major vessel forms (Wood, 1987). The ubiquity of and use-wear on plainware vessels suggests that these containers were an important part of daily household cooking, storage, and serving activities in central Arizona during prehistory. Imported decorated pottery such as White Mountain Red ware, Jeddito and Awatovi Black-on-yellow, Gila and Tonto polychromes, and Tusayan Black-on-white from Puebloan and Salado groups in northeastern Arizona appears at sites in the Verde Confederacy in variable frequencies (North and Foster, 2002, p. 11; Stone, 2000, p. 208; Wilcox and Holmlund, 2007, p. 105).

Methodology

This analysis utilizes a two-pronged approach that incorporates both ceramic petrography and chemical data from electron microprobe analysis. Chemical and petrographic techniques have been combined successfully in several recent archaeological studies in the New World (e.g. Glowacki and Neff, 2002) and the Old World (e.g. Baxter *et al.*, 2008; Day *et al.*, 1999; Montana *et al.*, 2003). In particular, several recent analyses have used petrographic and electron microprobe analysis to analyze ceramic temper and other inclusions in pottery paste (e.g. Abbott, 2000; Abbott and Schaller, 1994; Dorais and Shriner, 2002; Faber *et al.*, 2008; Neff *et al.*, 2006; Shriner and Dorais, 1999). Ceramic petrography and chemical data are often considered as useful complements, or sometimes as interchangeable modes of inquiry. In this study, ceramic petrography is used to identify the mineral and lithic constituents of pottery temper, and the electron microprobe provides data that distinguish the geologic origin of some temper grains.

Petrographic analysis

Ceramic petrography has been successfully applied to studies of ceramic provenance and exchange studies in the Phoenix (e.g. Heidke, 2004; Miksa *et al.*, 2004; Schaller, 1994), Tucson (e.g. Miksa, 2003), and Tonto Basins (e.g. Heidke and Miksa, 2000; Miksa and Heidke, 2001) of Arizona due to remarkable geologic diversity in these areas. These analyses have relied on the detailed characterization of sand compositions across the landscape to model the production and exchange of plainware pottery. Sand temper can be a sensitive indicator of production locale. Ethnographic data has shown that potters typically collect sand temper within a small radius (c. 3 km) of where they produce their wares (Arnold, 1985).

Although petrographic analyses of raw materials and ceramics from Perry Mesa and other settlement areas in the proposed Verde Confederacy are in their infancy, preliminary characterizations indicate that sufficient geologic diversity exists to permit relatively high-resolution pottery provenance determinations (Castro-Reino, 2004; see also Wilcox and Holmlund, 2007, p. 76-79) (Figure 3). Moreover, preliminary analyses of the temper composition of plainware pottery on Perry Mesa suggest sufficient compositional variation exists to identify multiple production sources for plainware pottery (Wichlacz, 2006; Wood, 1987).

The geology of Perry Mesa is dominated by basaltic lava flows that cap the mesa (Figure 3). Thus, one signal for local plainware production might be the use of the basaltic sands derived from these flows. In addition to the basalt that covers Perry Mesa, outcrops of granite and schist are also exposed in the steep river canyons that surround the mesa (Lindgren, 1926; Jaggar and Palache, 1905; Wilson *et al.*, 1958). Even though a significant effort would have been required to procure the granite-based or schist-rich materials, they may have represented an alternative temper source for Perry Mesa potters.

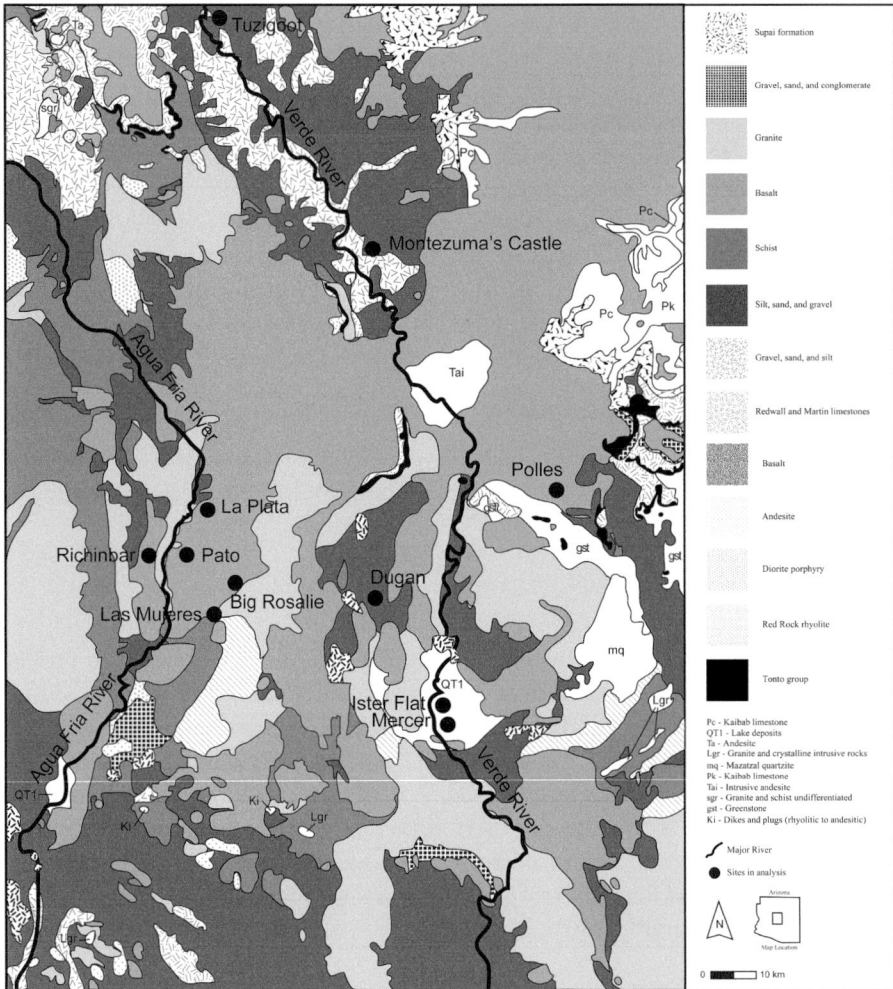

Figure 3. Simplified bedrock geology of Verde Confederacy region with sites analyzed in this study (after Wilson *et al.*, 1957, 1958, 1959).

The geology of the Bloody Basin and the Verde River differs from that of Perry Mesa, and we expect that local pottery production in these regions would be signaled by geologically distinct temper varieties (Figure 3). Sites located in the Bloody Basin such as Dugan Pueblo rest on large quaternary silt, sand, and gravel deposits that have eroded from basalt-capped mesas that once surrounded the basin (Rhys-Evans, 2007). Local pottery production may be signaled by temper dominated by eroded basalt or by a mixture of eroded basalt and small parts weathered granite that has washed down stream from Precambrian granite exposures to the north.

Sites in the southern portion of the Verde Confederacy, such as Mercer and Ister Flats, are located along the Verde River in an area characterized by Tertiary stream and lake sediments (Figure 3). These sediments are well stratified and range between fine-grained silt and coarse sand and gravel grains that were eroded from Tertiary and Quaternary basalt lava flows from nearby mesas (Pearthree, 1993; Royce and Wadell, 1970; Wilson *et al.*, 1957). Local pottery production at these sites may be distinguished by temper with a mixture of eroded granitic and basaltic grains. Local sand temper might also include a mixture of other mineralic and lithic grains that washed down the Verde River from formations to the north.

Further up the Verde River, the terrain becomes more mountainous and is characterized by basalt-capped mesas separated by gorges cut through Precambrian granite and metamorphic rocks (Pearthree, 1993; Royce and Wadell, 1970) (Figure 3). Polles Pueblo is located on one of these basalt-capped mesas to the north of a major tributary to the Verde River, the East Verde River. Local plainware production at Polles Pueblo may be distinguished by temper primarily composed of basalt grains. In addition, local potters could have also used sand temper eroded from metamorphic rocks exposed just below the rim of the mesa.

Finally, the northern reaches of the Verde Confederacy include Montezuma's Castle and Tuzigoot along the northern Verde River valley. This area is separated from other areas of the Arizona Central Highlands by the Verde Fault, which runs approximately 35 kilometers northwest-southeast (Figure 3). The geology surrounding Montezuma's Castle and Tuzigoot consists of Miocene-Pliocene limestones and siltstones deposited by the Verde River (Nations *et al.*, 1981; Pearthree, 1993; Royce and Wadell, 1970). Although the local geology is dominated by limestone, crushed limestone or sand containing limestone fragments were not favored as pottery temper. Calcium carbonate, which is a major constituent of limestone, expands and induces spalling when pottery is fired to temperatures exceeding 870°C (Rice, 1987, p. 98). Local pottery production in the northern Verde Valley may be marked by temper composed of small, heavily weathered sand grains that were washed into the area by the Verde River and tributary streams. Locally available temper may be more restricted and potters may have relied on stream and river beds for sand suitable for temper.

In this study, petrographic analysis is used to identify temper reference groups for 185 plainware pottery samples from pueblos in the Verde Confederacy. Temper reference groups represent provenance-related varieties that are tentatively linked with specific production sources. Particular reference groups are then matched to specific sites by virtue of their dominance within that site's assemblage, as well as their correlation to the distinctive geologic setting of each settlement area. Finally, the movement of pottery between sites is identified by the presence of non-local plainware reference groups.

Microprobe analysis

The electron microprobe was used to examine ten sherds from sites on Perry Mesa to determine if the pyroxene grains in their temper are derived from the basalt flows that cap the mesa, or if they are from the plutonic units that are located along the mesa edges. Preliminary analysis suggests that basalt grains themselves are not present in the pottery, but mafic minerals are present and sometimes abundant.

Sherds containing abundant mafic minerals were commonly encountered during the petrographic analysis. Some mafic minerals, such as pyroxenes, appear in both plutonic granitic and basaltic rocks. Fortunately, chemical point-characterization can distinguish between pyroxenes from these two rock units by their magnesium-iron ratios. Pyroxenes in granites have low Mg/Fe ratios, whereas pyroxenes in basalts have high Mg/Fe ratios (McBirney, 2006). We selected a sample of seven specimens from those sherds with relatively abundant mafic grains for microprobe analysis. We then analyzed between four and eight mafic grains in each sample.

A Joel JXA-8600 electron microprobe with an automated energy-dispersive analysis system was used to perform the analysis. Each potsherd was cut to extract a thick slice of its cross section that was then mounted on a circular glass slide. The thick section was then ground, polished, and coated with 150-250 angstroms of carbon.

All samples were analyzed using 15-Kv filament voltage and a 10-nA beam current, and the beam was defocused to a ~10 micron diameter. The x-ray detector was mounted at a take-off angle of 40°. Matrix effects were corrected with a ZAF algorithm, and the equipment was calibrated with a Kakanui hornblende standard. The detector live-counting time was 50 seconds. The percentages of eight chemical elements (Na, Mg, Al, Si, Ca, K, Ti, and Fe) were determined.

Results

Petrographic analysis

Petrographic analysis of 185 thin sections of plainware pottery from sites on Perry Mesa, Bloody Basin, and along the Verde River revealed nine distinct sand temper varieties (Table 1; Figure 4). Five of these sand varieties were granitic and were defined by their mineral constituents, grain size, and extent of weathering. In addition to the granite sand temper groups, sherd temper, sand consisting of a mixture of schist and granite, and sand consisting of a mixture of schist and phyllite temper were identified.

Samples from Perry Mesa sites were divided in composition between sites on the west side of the mesa and those on the east side of the mesa (Table 2). The temper compositions of samples from the western Perry Mesa sites of Pueblo la Plata, Richinbar and Pato Pueblo were dominated by granitic sand characterized by large, relatively unaltered quartz and plagioclase feldspar grains (Granite I), and a granitic sand defined by heavily weathered plagioclase feldspar grains and no potassium feldspar or mafic minerals (Granite II). Pottery samples from Pueblo Pato were particularly associated with the Granite I reference group. The eastern Perry Mesa sites of Las Mujeres and Big Rosalie, on the other hand, were dominated by a mixture of schist and arkosic sands (Schist & Granite).

Temper Reference Group	Main Characteristics
Granite I	Characterized by large, relatively unaltered quartz and plagioclase feldspar grains. Very little potassium feldspar. Large pieces of biotite are present as well as large, altered mafic grains.
Granite II	No potassium feldspar and only trace amounts of mafic minerals. Plagioclase fledspar is heavily weathered.
Granite III	Large plagioclase feldspar and quartz grains dominate. Smaller and less abundant potassium feldspar grains present. Pyroxene grains appear in trace amounts. Temper is composed of both lithic and mineralic grains.
Granite IV	Large monomineralic grains of plagioclase feldspar with distinctive wavy alteration. No potassium feldspar and very few mafic minerals. Most temper grains are monomineralic.
Granite V	Large potassium feldspars represent half or more of the temper composition. Wavy alteration of plagioclase feldspar grains.
Schist and Granite	Smaller grain size than other reference groups. Charcaterized by schist mixed with heavily weathered arkosic sands. Almost no phyllitic textures present.
Schist and Phyllite	Large pieces of schist and phyllite dominate the temper. Relatively few individual mineral grains, although large unaltered quartz grains are present.
Sherd Temper	Characterized by sherd temper as well as a variety of other lithic grains, including basalt.
Volcanics	Temper dominated by a mixture of porphyritic basalt with vitrophyric texture and fine-grained, felty volcanics. Various stages of alteration present. Composition varies and the group can likely be subdivided.

Table 1. Main characteristics of plainware reference groups.

Region	Site	Granite I	Granite II	Granite III	Granite IV	Granite V	Schist & Granite	Schist & Phyllite	Sherd	Volcanic	Total
E. Perry Mesa	Big Rosalie	-	-	-	-	-	15	-	-	-	15
	Las Mujeres	-	-	-	-	-	5	-	-	-	5
W. Perry Mesa	La Plata	10	6	-	-	-	8	4	-	-	28
	Pato	18	1	-	-	-	2	-	-	-	21
	Richinbar	6	3	-	-	-	1	-	-	-	10
Bloody Basin	Dugan	-	-	-	3	1	-	20	-	-	24
Verde River	Ister	-	-	-	-	14	-	-	-	-	14
	Mercer	-	-	-	-	7	-	-	-	-	7
	Montezuma's Castle	3	-	16	-	-	-	-	-	-	19
	Polles	-	-	-	-	-	-	-	2	18	20
	Tuzigoot	-	-	2	-	1	-	-	16	3	22
Grand Total		37	10	18	3	23	31	24	18	21	185

Table 2. Reference groups for Verde Confederacy plainware sherds.

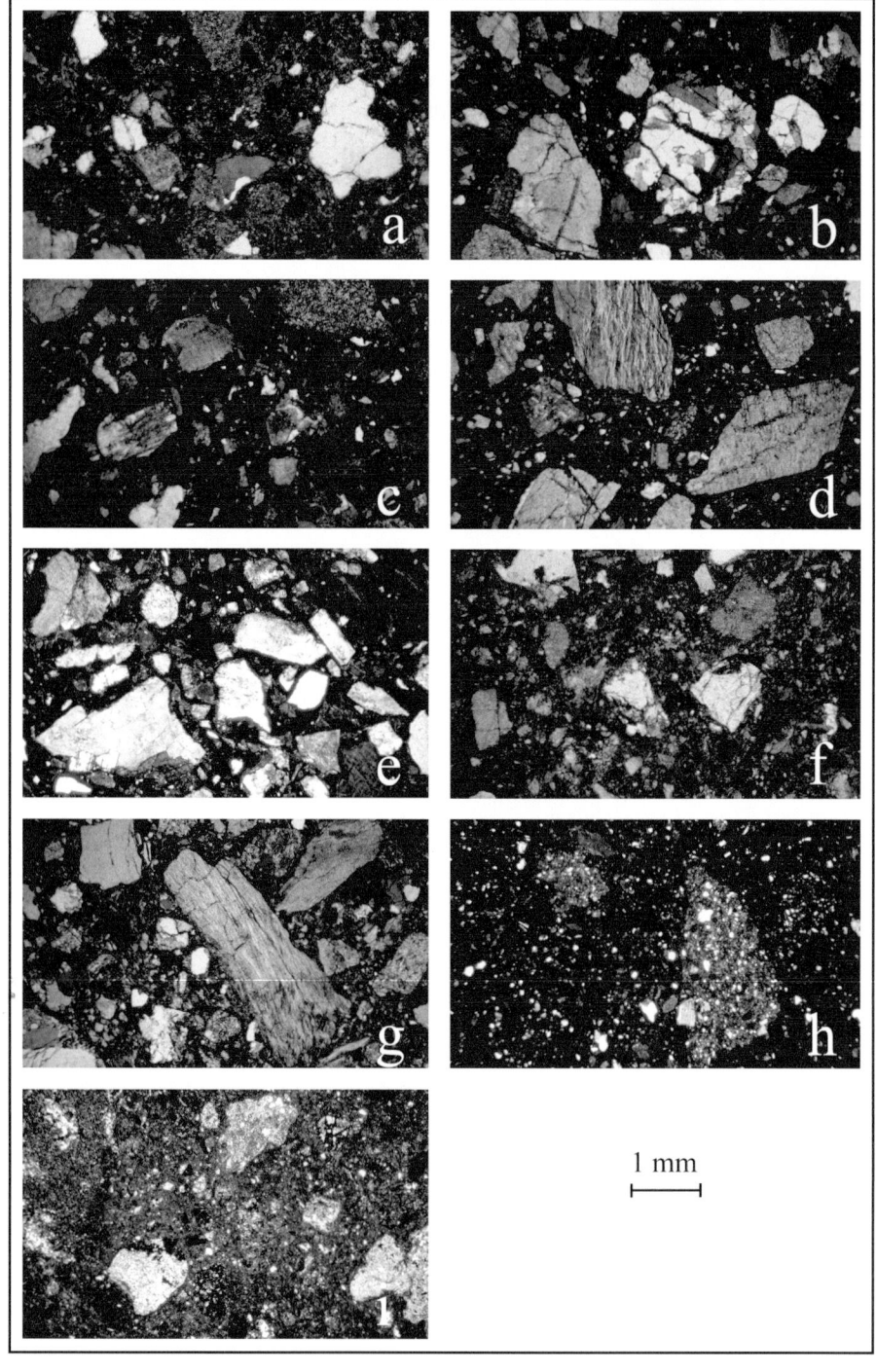

Figure 4. Micrographs of plainware temper reference groups from Perry Mesa, Bloody Basin, and along the Verde River. a). Granite I, b). Granite II, c). Granite III, d). Granite IV, e). Granite V, f). Granite & Schist, g). Schist & Phyllite, h). Sherd, i). Volcanics. All images taken with crossed polars. Potassium and plagioclase feldspars stained.

Each of the Bloody Basin and middle Verde assemblages were dominated by a different temper reference group (Table 2). Samples from Dugan in the Bloody Basin were predominantly composed of the same schist and phyllite temper group identified in some samples from Pueblo La Plata. In addition, a few samples contained distinct granitic sand, which primarily consisted of large monomineralic grains of plagioclase feldspar and little to no potassium feldspar or mafic minerals (Granite IV). Samples from Montezuma's Castle were mostly tempered with granitic sand distinguished by the presence of large plagioclase feldspar and quartz grains (Granite III). Samples from Tuzigoot were primarily tempered with fragments of sherds or 'grog' that were mixed with a variety of other lithic grains including basalt. Samples from Polles Pueblo were dominated by felsic volcanic temper. Finally, samples from the nearby sites of Ister Flats and Mercer along the Verde River were primarily tempered with granitic sand characterized by large grains of potassium feldspar (Granite V).

Microprobe analysis

Microprobe analysis of eight samples from Pueblo la Plata on Perry Mesa and two samples from Richinbar on Black Mesa established that they did not have chemical signatures consistent with a basalt source. Five of the samples from Pueblo la Plata were petrographically linked with the Granite I temper reference group and three were linked with the Granite II reference group. Both samples from Richinbar were ascribed to the Granite I reference group. Microprobe analysis confirmed that all of the pyroxene grains in these samples had low Mg/Fe ratios, which are consistent with derivates from granitic sources.

Discussion

Plainware production on Perry Mesa

The first expectation of our analysis was that the majority of Perry Mesa plainwares were not produced locally. Plainware pottery produced at settlements elsewhere in the Verde Confederacy would be signaled by the presence of non-local temper. Optical petrography and electron microprobe analysis established that Perry Mesa plainware pottery was tempered with materials such as granite and schist that were unavailable on the mesa top. Petrographic analysis did not detect basalt textures in temper in any Perry Mesa ceramics. A small group of seven samples contained one or two inclusions of basalt, but not in the quantities expected if the sherds had been tempered with the basaltic sands derived from the basalt flows that cap Perry Mesa. In addition, microprobe analysis of the mafic minerals in Perry Mesa plainware temper suggested that they do not derive from basalt.

Although none of the Perry Mesa plainwares sampled in this analysis contained the locally available basaltic sands, it is possible that most plainware pots used at the Perry Mesa sites were produced locally using alternative raw materials. The dominant temper groups used to produce Perry Mesa plainwares, Granite I and Schist & Granite, could have derived from the schist and granite bedrock underlying the basalt lava flows. This bedrock is now exposed in the steep canyons that surround Perry Mesa (Figure 3). Even though we cannot link these temper groups to specific geologic units, we can use the distribution of bedrock units in the Perry Mesa vicinity to develop hypotheses about where the sands may have been obtained (Figure 3). Sand and rock samples collected along the side of Perry Mesa beneath Pueblo Pato closely match the composition of the Granite I temper group. In addition, extensive schist deposits interbedded with granite plutons are present to the south of Perry Mesa, and could have provided the mixture of granite and schist sands that characterize the Granite & Schist temper reference group. These hypotheses will be tested in future analyses, which will include systematic sand sampling in the drainages surrounding Perry Mesa. If sands collected from these locales closely match the Granite I and the Schist & Granite temper groups, we can be reasonably sure that these materials are associated with the local production of plainwares on Perry Mesa.

Our results thus far leave open the possibility that Perry Mesa communities may have produced most of their own plainware pottery. The ceramic data thus far fails to support the first expectation of the Verde Confederacy Model, which posits that Perry Mesa communities may have been heavily reliant on other settlements in the Verde Confederacy to supply them with basic necessities. Petrographic and microprobe analyses established that plainware sherds were not tempered with the basalt sand that covers Perry Mesa. However, tempering materials in Perry Mesa pottery were consistent with nearby granite and schist outcrops below the mesa top, and were possibly derived from these sources.

Plainware exchange

In our second stage of analysis, we evaluate if settlements in the Verde Confederacy regularly interacted with each other. Sites from the eastern and western sides of Perry Mesa as well as sites located in Bloody Basin and along the middle Verde River were dominated by different temper reference groups, which are defined by mineralogically distinct sand temper. Regionally distinct temper compositions suggest that potters at each site probably used specific local tempering materials (Table 2; Figure 4). Given the distance between the sites and the geologic diversity in the region, this result is not surprising (Castro-Reino, 2004; Wilcox and Holmlund, 2004, p. 76-79).

The diversity of sand tempers used by potters across the region enabled us to begin charting the movement of pottery across the landscape by identifying the presence of plainware pottery produced using a temper type associated with other sites or groups of sites (Figure 5). We tentatively consider the dominant temper group at each site to represent local production of pottery at that site. This assumption is based on a close association between the dominant temper types at each site and the raw materials available within five kilometers of each settlement (Figure 3). In addition, each site or group of sites located in close proximity to each other was marked by a distinct temper variety. Therefore, at this juncture, we consider the dominant temper groups at each site to represent local production at that site. Non-local pottery is identified by the presence of a pot with temper linked with another site.

Testing Expectations of Pottery Exchange in the Central Arizona Highlands

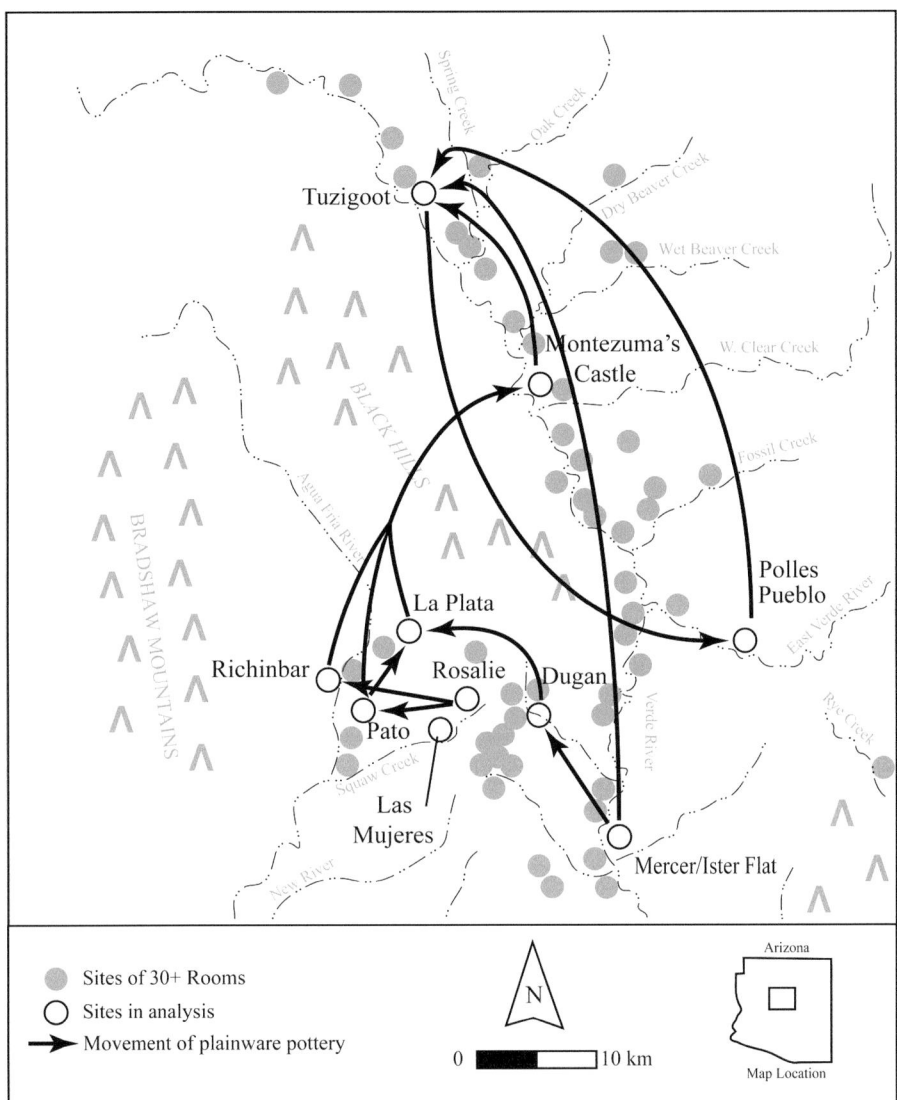

Figure 5. Plainware pottery exchange within the Verde Confederacy interpreted by petrographic and microprobe analysis in this study.

The preliminary results of plainware pottery exchange among pueblos within the proposed Verde Confederacy suggest that there was relatively frequent interaction between settlements in adjacent regions. Approximately fifteen percent of pottery sampled in this analysis (n = 27) may have been moved between sites associated with different temper reference groups (Table 2). The strongest pattern of interaction was identified between eastern Perry Mesa sites (Big Rosalie and Las Mujeres) and western Perry Mesa sites (Pueblo Pato, Pueblo la Plata, and Richinbar). In addition, the data suggest relatively strong patterns of interaction among sites up and down the Verde River (Mercer, Ister Flats, Polles Pueblo, Montezuma's Castle, and Tuzigoot). However, there is less substantial evidence for interaction between Perry Mesa sites and sites in other regions of central Arizona. Only three samples of pottery that may have been produced on Perry Mesa appeared at sites elsewhere in the Verde Confederacy; only four samples may represent pottery that was brought from other Verde Confederacy sites to Perry Mesa.

The exchange patterns identified in this analysis are not necessarily consistent with the original formulation of the Verde Confederacy Model, which posits that there was regular and continuous interaction among all pueblos in the confederacy. In particular, the model contends that the alliance was constructed around strong ties between settlements on Perry Mesa and settlements along the Verde River to the east (Wilcox *et al.*, 2001; Wilcox and Holmlund, 2007). As of yet, ceramic evidence for persistent connectivity between Perry Mesa settlements and other regions of the Verde Confederacy is tentative.

Conclusions

The Verde Confederacy Model presents an intriguing characterization of the social and economic relationships between upland settlements in central Arizona during the 14th century. Although little material evidence addresses this proposition thus far, the Verde Confederacy Model has provided a direction for current and future research (see Wilcox and Holmlund, 2007). This analysis uses pottery provenance and exchange data to address the two primary expectations for a coordinated alliance among central Arizona settlements: 1) persistent connectivity, and 2) coordinated social and economic action (after Upham *et al.*, 1994). The results of our analysis do not readily support these two expectations of the Verde Confederacy Model. First, Perry Mesa potters may have produced the majority of their own plainware vessels rather than relying on sites in Bloody Basin and along the Verde River to supply them with ceramics. In addition, the movement of pottery between sites within the proposed Confederacy may not correspond to the continuous web of social and economic relationships between sites in the region predicted by the model.

The results of this study provide a fruitful foundation to evaluate the social and economic relationships between settlements on Perry Mesa and across central Arizona. In addition, this analysis joins a growing body of scholarship that explores the use of petrographic and chemical sourcing techniques. Although ceramic petrography and chemical analysis provided by the electron microprobe have been used effectively on their own in many archaeological studies, this case study illustrates a situation where both of these methods were necessary to address the social questions posed. We anticipate that our forthcoming work will continue to delineate how, and in what circumstances these methods can be used effectively to address pottery provenance and exchange.

Future Research

Our results provide a foundation to direct future research on the relationships between sites in the rugged Arizona highlands. In subsequent analyses, we will continue to rely on a combination of petrographic and microprobe analysis to explore the relationships between sites in the Verde Confederacy. The nine delineated reference groups will be evaluated by chemical analysis of the clay fraction using the electron microprobe. The integrity of our reference groups would be supported by an association between clay compositional groups and petrographically defined temper varieties.

In future analyses, microprobe analysis of the clay fraction will assess the provenance ascription of the reference groups established through ceramic petrography. We expect that the ceramics from a particular production source should have similar clay chemistry as well as temper mineralogy. For example, three granite-tempered sherds categorized as Granite I were among the samples analyzed from Montezuma's Castle (see Table 2). Granite I specimens have been tentatively associated with production on Perry Mesa and the presence of Perry Mesa pottery at sites along the middle Verde River would imply closer connections between the two areas than we have surmised (Figure 4). Granite I sherds are characterized by granitic sand consisting largely of unaltered quartz and feldspar grains that set them apart from all of the other temper groups in our analysis. Nevertheless, it is easy to imagine that similar granitic material may have been exploited by a production group not on Perry Mesa, but somewhere closer to Montezuma's Castle. If so, the clay chemistry of the three Granite I sherds at Montezuma's Castle, as measured by the microprobe, should be distinctive from that of the clay used to fabricate the Granite I sherds recovered from the Perry Mesa pueblos.

Finally, systematic sand sampling and characterization strategies are necessary to document the composition and distribution of sand types in the region. These samples should provide resolution to the varieties of sand tempers available to Perry Mesa potters, as well as a check on the reference groups delineated in this analysis. Provenance studies of the additional samples will enable our future research to establish production locales with greater precision.

Acknowledgements

Funding for this project was provided by a grant from the National Science Foundation: "Alliance and Landscape: Perry Mesa, Arizona in the Fourteenth Century" (BCS-0613201). Additional funding was provided by the Bureau of Land Management for archaeological research in the Agua Fria National Monument. The authors would like to thank Assistant Manager for the Hassayampa Field Office, Connie Stone, and Agua Fria National Monument Manager, D. Remington Hawes. We also greatly appreciate the assistance of J. Scott Wood for work on the Tonto National Forest. Kristin Higgins, Melissa Kruse, Brian Medchill, Will Russell, Colleen Strawhacker, Jason Vega, Jason Walker, JoAnn Wallace, David Wilcox, and J. Scott Wood all graciously spent time assisting with ceramic collections at several sites on Perry Mesa and on the lands of the Tonto National Forest. Joshua Watts designed and executed a sampling scheme for collecting sand from the Perry Mesa area. Kim Beckwith and John Schroeder of the National Park Service provided access to the collections at the Western Archaeological and Conservation Center from Tuzigoot and Montezuma Castle National Monuments. Scott Thompson provided invaluable comments on the final manuscript. Finally, we are grateful to the Ak-Chin Indian Community, Gila River Indian Community (special thanks to Barnaby Lewis and J. Andrew Darling), the Hopi Tribe (special thanks to the Cultural Resources Advisory Task Team), the Yavapai Prescott Tribe, the Salt River Pima-Maricopa Indian Community, and the Tohono O'odam Nation for their willingness to discuss their perspectives with us on the ancient occupation of Perry Mesa.

References

Abbott, D.R. 2000. *Ceramics and Community Organization among the Hohokam.* University of Arizona Press, Tucson.

Abbott, D.R. and Schaller, D.M. 1994. Ceramics among the Hohokam: Modeling Social Organization and Exchange. In: Scott, D.A. and Meyers, P. (Eds.) *Archaeometry of Pre-Columbian Sites and Artifacts*. Getty Conservation Institute, Los Angeles: 85-109.

Ahlstrom, R.V.N. and Roberts, H. 1995. *Prehistory of Perry Mesa: The Short-Lived Settlement of a Mesa-Canyon Complex in Central Arizona, ca. A.D. 1200-1450.* Arizona Archaeologist No. 28, Arizona Archaeological Society, Phoenix.

Arnold, D.E. 1985. *Ceramic Theory and Cultural Process*. Cambridge University Press, Cambridge.

Baxter, M.J., Beardah C.C., Papageorgiou, I., Cau, M.A., Day, P.M. and Kilikoglou, V. 2008. On Statistical Approaches to the Study of Ceramic Artefacts using Geochemical and Petrographic Data. *Archaeometry*, 50: 142-157.

Castro-Reino, S.F. 2004. *Predicted Petrofacies Map of Perry Mesa and the Adjacent Agua Fria Drainage Basin with Inferred Sand Compositions.* Unpublished map. Desert Archaeology, Tucson.

Day, P.M., Kiriatzi, E., Tsolakidou, A. and Kilikoglou, V. 1999. Group Therapy: A Comparison Between Analyses by NAA and Thin Section Petrography of Early Bronze Age Pottery from Central and East Crete. *Journal of Archaeological Science*, 26: 1025-1036.

Dorais, M.J. and Shriner, C.M. 2002. An Electron Microprobe Study of P645/T390: Evidence for an Early Helladic III Lerna-Aegina Connection. *Geoarchaeology*, 17: 755–778.

Faber, E., Knight, D., Carney, J. and Marsden, P. 2008. Microanalysis of Later Prehistoric Granodiorite-tempered Pottery from the East Midlands. In: Quinn, P.S. (Ed.) *Petrography of Archaeological Materials, Department of Archaeology, University of Sheffield, Abstracts and Programme*: 31.

Glowacki, D.M. and Neff, H. (Eds.) 2002. *Ceramic Production and Circulation in the Greater Southwest: Source Determination by INAA and Complementary Mineralogical Investigations*. Costen Institute of Archaeology, UCLA, Los Angeles.

Heidke, J.M. 2004. Temper Characterization. In: Harry, K.G. and Whittlesey, S.M. (Eds.) *Pots, Potters, and Models - Archaeological Investigations at the SRI Locus of the West Branch Site, Tucson, Arizona: Vol. 1. Feature Descriptions, Material Culture, and Specialized Analyses* (CD-ROM). Technical Series No. 80. Statistical Research, Inc., Tucson.

Heidke, J.M., and Miksa, E.J. 2000. Correspondence and Discriminant Analyses of Sand and Sand Temper Compositions, Tonto Basin, Arizona. *Archaeometry*, 42: 273-299.

Jaggar, T.A. Jr. and Palache, C. 1905. Description of Bradshaw Mountains Quadrangle. In: *Geologic Atlas of the United States: Bradshaw Mountains Folio*. Department of the Interior, U.S. Geological Survey, USGS Folio 126.

Leighty, R.S. 1997. *Neogene Tectonism and Magmatism Across the Basin and Range-Colorado Plateau Boundary, Central Arizona*. Unpublished doctoral dissertation, Arizona State University, Tempe.

Lindgren, W. 1926. *Ore Deposits of the Jerome and Bradshaw Mountains Quadrangles, Arizona*. Department of the Interior, U.S. Geological Survey, Bulletin 782.

Mapes, S.D. 2005. *The Walls Still Stand: Reconstructing Population at Pueblo La Plata*. Unpublished senior honors thesis, School of Human Evolution and Social Change, Arizona State University, Tempe.

McBirney, A. R. 2006. *Igneous Petrology*, 3rd Edition. Jones & Bartlett Publishers, Boston.

Miksa, E.J. 2003. Petrographic Analyses of Tucson Basin Pottery. In: Wallace, H.D. and Lindeman, M.W. (Eds.) *Archaeological Excavations at Valencia Vieja: Appendices and Supplemental Data*. Technical Report No. 2001-11. Desert Archaeology, Inc., Tucson: 53-77.

Miksa, E.J., and Heidke, J.M. 2001. It All Comes Out in the Wash: Actualistic Petrofacies Modeling of Temper Provenance, Tonto Basin, Arizona. *Geoarchaeology*, 16: 177-222.

Miksa, E.J., Castro-Reino, S. and Lavayen, C. 2004. A Combined Petrofacies Model for the Middle Gila and Phoenix Basins, with Application to Pottery from the Sky Harbor Site. In: Henderson, T. K. (Ed.) *Hohokam Farming on the Salt River Floodplain: Refining Models and Analytical Methods*. Anthropological Papers No. 43. Center for Desert Archaeology, Tucson. Anthropological Papers No. 10. Pueblo Grande Museum, City of Phoenix Parks, Recreation and Library Department, Phoenix: 7-44.

Mills, B.L. and Crown, P.L. 1995. *Ceramic Production in the American Southwest*. University of Arizona Press, Tucson.

Montana, G., Mommsen, H., Iliopoulos, I., Schwedt, A. and Denaro, M. 2003. The Petrography and Chemistry of Thin-walled Ware from a Hellenistic Roman Site at Segesta (Sicily). *Archaeometry*, 45: 375–389.

Nations, J.D., Hevly, D.W., Blinn, D.W., and Landye, J. J. 1981. Paleontology, Paleoecology, and Depositional History of the Miocene-Pliocene Verde Formation, Yauapai County, Arizona. *Arizona Geological Society Digest*, 13: 133-149.

Neff, H., Blomster, J., Glascock, M.D., Bishop, R.L., Blackman, M.J., Coe, M.D., Cowgill, G.L., Diehl, R.A., Houston, S., Joyce, A.A., Lipo, C.P., Stark, B.L., and Winter, M. 2006. Methodological Issues in the Provenance Investigation of Early Formative Mesoamerican Ceramics. *Latin American Antiquity*, 17: 54-76.

North, C.D. 2002. *Farmers of Central Arizona's Mesa-Canyon Complex: Archaeology Within and Adjacent to the Agua Fria National Monument*. SWCA Cultural Resource Report No. 02-339. SWCA, Inc. Environmental Consultants, Phoenix.

North, C.D. and Foster, M.S. 2002. The Agua Fria National Monument/Bradshaw Planning Area Survey: Introduction and Background. In: North, C.D. (Ed.) *Farmers of Central Arizona's Mesa-Canyon Complex: Archaeology Within and Adjacent to the Agua Fria National Monument*. Report prepared for the Bureau of Land Management, SWCA Cultural Resource Report No. 02-339, Phoenix.

Pearthree, P.A. 1993. *Geological and Geomorphic Setting of the Verde River from Sullivan Lake to Horseshoe Reservoir*. Arizona Geological Survey Open-File Report 93-4, 1:24,000.

Rhys-Evans, G. 2007. *Geology of the Bloody Basin: Central Arizona's Transition Zone*. CR-07-B, 1:24,000.

Rice, P. 1987. *Pottery Analysis: A Sourcebook*. University Of Chicago Press, Chicago.

Royce, C.F. and Wadell, J.S. 1970. Geology of the Verde Valley, Yavapai County, Arizona. In: Smith, C.T. (Ed.) *Guidebook to the Four Corners, Colorado Plateau, and Central Rocky Mountain Region*. National Association of Geology Teachers, Southwest Section, Cedar City, Utah: 35-39.

Schafssma, H., Kruse, M. and Johnson, K. 2006. The Palimpsest Landscape: Ancient Land Use Intensities Revealed by Modern Soils and Plant Communities. *Presented at the Society for American Archaeology Annual Meeting.* San Juan, Puerto Rico.

Schaller, D.M. 1994. Geographic Sources of Phoenix Basin Hohokam Plainware Based on Petrographic Analysis. In Abbott, D.R. (Ed.) *Pueblo Grande Project: Vol. 3. Ceramics and the Production and Exchange of Pottery in the Central Phoenix Basin.* Publications in Archaeology No. 20. Soil Systems, Inc., Phoenix: 17-90.

Schollmeyer, K.G. 2004. *Spring 2004 Architecture Studies at Pueblo La Plata.* Report submitted to the Bureau of Land Management and National Park Service, Phoenix.

Schollmeyer, K.G. 2005. Architecture Studies at Richinbar Ruin, Spring 2005. In Spielmann, K.A. (Ed.) *Report of the Spring 2005 Field Season, Legacies on the Landscape: Archaeological and Ecological Research at Agua Fria National Monument.* Report submitted to the Bureau of Land Management and National Park Service: 59-64.

Shriner, C. and Dorais, M.J. 1999. A Comparative Electron Microprobe Study of Lerna III and IV Ceramics and Local Clay-rich Sediments. *Archaeometry,* 41: 25-49.

Stone, C.L. 2000. The Perry Mesa Tradition in Central Arizona: Scientific Studies and Management Concerns. In: Motsinger, T.N., Mitchell, D.R., and McKie, J.M. (Eds.) *Archaeology in West-Central Arizona: Proceedings of the 1996 Arizona Archaeological Council Prescott Conference.* Sharlott Hall Museum, Prescott: 205-214.

Upham, S., Crown, P.L. and Plog, S. 1994. *Alliance Formation and Cultural Identity in the American Southwest.* In: Gumerman, G. (Ed.) Themes in Southwest Prehistory. School of American Research Press, Santa Fe: 183-210.

Wichlacz, C. 2006. *A Compositional Analysis of Plain Ware Pottery from Pueblo La Plata and Richinbar Ruin, Agua Fria National Monument, Arizona.* Unpublished senior honors thesis, School of Human Evolution and Social Change, Arizona State University, Tempe.

Wilcox, D.R. 2005. Big Issues, New Syntheses. *Plateau* 2(1): 8-19.

Wilcox, D.R., Robertson, G.J., and Wood, J.S. 2001. Organized for War: The Perry Mesa Settlement System and Its Central-Arizona Neighbors. In Rice, G.E. and LeBlanc, S.A. (Eds.) *Deadly Landscapes: Case Studies in Prehistoric Southwestern Warfare.* University of Utah Press, Salt Lake City: 141-194.

Wilcox, D.R. and Holmlund, J. 2007. *The Archaeology of Perry Mesa and its World.* Bilby Research Center Occasional Papers No. 3., Northern Arizona University, Flagstaff.

Wilson, E.D., Moore, R.T., and Pierce, H.W. 1957. *Geologic Map of Maricopa County, Arizona*. Prepared for the Arizona Bureau of Mines, 1:375,000. University of Arizona, Tucson.

Wilson, E.D., Moore, R.T., and Pierce, H.W. 1958. *Geologic Map of Yavapai County, Arizona*. Prepared for the Arizona Bureau of Mines, 1:375,000. University of Arizona, Tucson.

Wilson, E.D., Moore, R.T., and Pierce, H.W. 1959. *Geologic Map of Gila County, Arizona*. Prepared for the Arizona Bureau of Mines, 1:375,000. University of Arizona, Tucson.

Wood, J.S. 1987. *Checklist of Pottery Types for the Tonto National Forest: An Introduction to the Archaeological Ceramics of Central Arizona*. The Arizona Archaeologist No. 21, Tonto National Forest Cultural Resources Inventory Report No. 87-01, Phoenix.

CERAMIC PETROGRAPHY & THE RECONSTRUCTION OF HUNTER-GATHERER CRAFT TECHNOLOGY IN LATE PREHISTORIC SOUTHERN CALIFORNIA

Patrick Quinn

Department of Archaeology, University of Sheffield, UK
(patrick.quinn@sheffield.ac.uk)

Margie Burton

San Diego Archaeological Center, Escondido, California, USA

Introduction

Plain, undecorated ceramic sherds are a common component of archaeological assemblages from villages, temporary camps and resource processing sites across southernmost California (Figure 1). Ceramic technology arrived in this area during the last 1000-1300 years (Laylander, 1992; Campbell, 1999, p. 119; Griset 1996) and perhaps as recently as 1450-1500 AD in western San Diego County, where its appearance is used as a chronological marker for the Late Prehistoric period. Based on archaeological and ethnohistoric evidence, indigenous societies of the San Diego area practiced a mobile hunter-gatherer lifestyle with seasonal movements across environmental zones to exploit a range of plant, animal and geological resources. The manufacture and use of ceramics by these groups thus provides another example, within a growing corpus of recently studied cases (e.g. Sassaman, 1993, 2000; Eerkens *et al.*, 2002; Eerkens, 2003; Skibo and Schiffer, 2008; Thompson *et al.*, 2008) of pottery technology among hunter-gatherers.

Despite their abundance, southern California ceramics have, to date, contributed only limited information towards reconstruction of the region's prehistory. Important ethnographic studies document 'traditional' southern California pottery making as it existed in the 20th century (e.g. Rogers, 1936; Wilken, 1982), however very little attention has been given to the nature of prehistoric ceramic technology, with the entire occurrence of pottery often thought of as a single paddle-and-anvil 'tradition' (Griset, 1996, p. 9) (Figure 1A). In addition, a number of technological, depositional and post-depositional factors have worked against the definition of clear-cut and useful typological categories. First, Late Prehistoric ceramic assemblages in this area are dominated by plain, undecorated brown and buff coloured sherds (Figure 1B,C) representing a restricted range of simple, round-bottomed vessels, such as jars or 'ollas' and bowls (Figure 1D). The many large, non-standardised, globular forms with restricted necks result in archaeological accumulations with high ratios of body sherds to rim sherds, a high degree of within vessel-type variability and a general absence of decoration, making typological classification of sherd assemblages difficult. Lack of well-preserved site stratigraphy and the apparent long use-life of many vessels tend to further obscure any chronological patterning in forms, manufacturing methods, or decoration. Numerous attempts at classifying southern California ceramics based upon characteristics such as form, rim shape and the nature of their paste in hand specimen (Schroeder, 1958; May, 1978; Van Camp, 1979; Waters, 1982; Laylander, 1997) have only added to the debate regarding the replicability of proposed typologies and their cultural, chronological and technological significance.

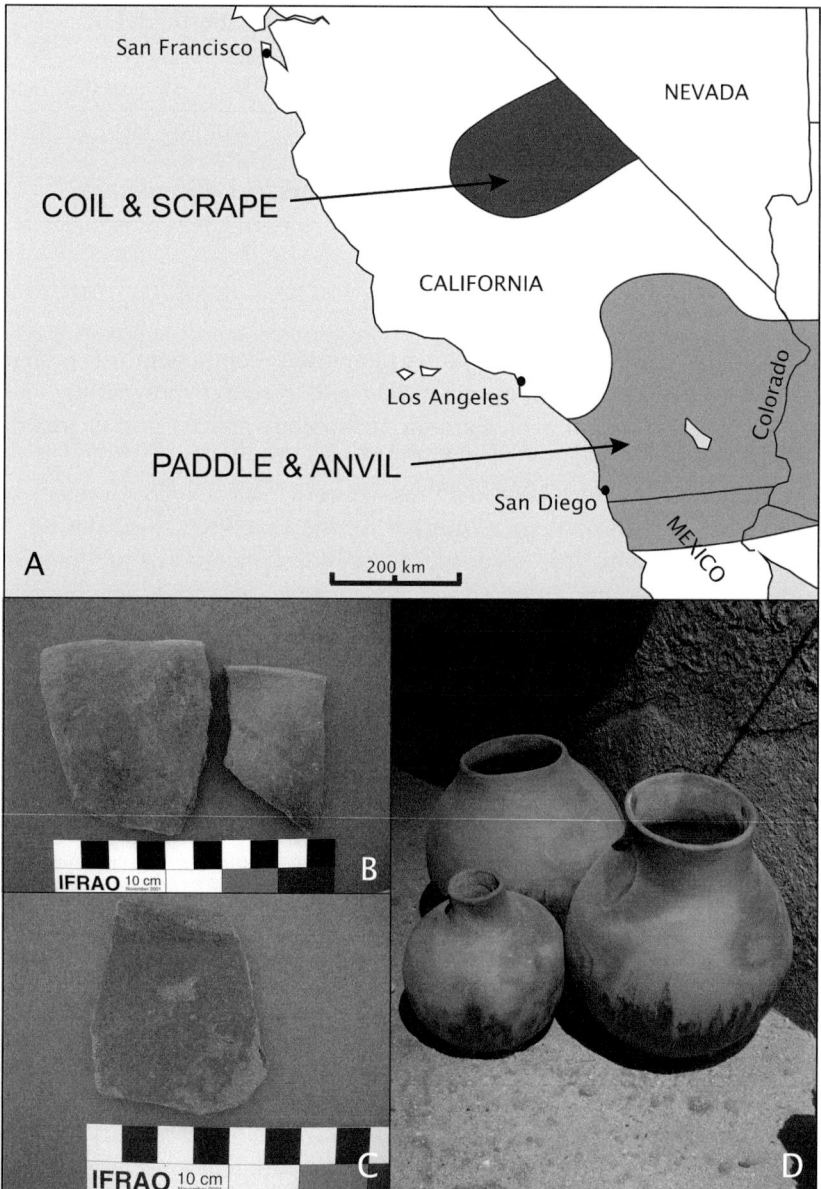

Figure 1. Late Prehistoric plainware ceramics and their distribution in southern California. a) The approximate geographic distribution of the 'paddle and anvil' ceramic tradition (modified from Griset, 1996, figure 2, p. 9), b,c) Plainware sherds from sites analysed in this study, d) Typical plainware ceramic vessels including holemouth jar and 'olla', in museum display at Anza-Borrego Desert State Park Headquaters, Borrego Springs, California.

More recently, a small number of scientific analyses (Pymale-Schneeberger, 1993; Hildebrand *et al.*, 2002; Gallucci, 2004) have highlighted the use of compositional data to detect archaeologically meaningful patterns in the region's ceramics, such as large-scale shifts in population density (Arnold *et al.*, 2004, p. 47) and social group movements during the Late Prehistoric Period (Hildebrand *et al.*, 2002). Documenting the composition of southern California ceramics is a first step towards understanding these artefacts. However, previously described compositional groupings remain broad and generalised and, more crucially, associated raw materials, technology and cultural affiliations remain poorly understood.

Given the high geological heterogeneity and historic ethno-linguistic diversity of the southern California region, it is likely that a more thorough and refined analysis of the ubiquitous Late Prehistoric ceramics might reveal much archaeological meaningful variation about how they were made and the people who made them. Inspired by this, the present study has applied thin section ceramic petrography to reconstruct in detail the craft of pottery manufacture within a restricted area of southern California during the Late Prehistoric period. By interpreting the specific technological choices and steps taken by potters in the past, it has been possible to decipher valuable information about their knowledge and exploitation of the natural environment, their technological skills and their appreciation of the performance characteristics of raw materials. Documenting some of the different recipes or styles of southern California pottery making for the first time has also provided valuable data with which to begin to examine the cultural identities and traditions of the hunter-gatherer groups that made ceramics, used them and eventually left them behind.

Archaeological Sites and Samples

The present study focuses on the inland region of eastern San Diego County. Much of this area is now contained within the boundaries of the Anza-Borrego Desert State Park (Figure 2). This vast, relatively undeveloped landscape is rich in archaeological sites, but remains only superficially investigated. During the Late Prehistoric period, groups ancestral to the present-day Kumeyaay, Luiseño and Cahuilla Indian tribes appear to have established seasonal winter camps in this area (Schaefer, 1994a). Known trails run westwards into the mountains and eastwards into the low-lying desert plain of the Salton Basin (Cline, 1979, p. 17, 18, 21, 1984, p. 13, 16-17), which in Late Prehistoric times was intermittently occupied by the freshwater Lake Cahuilla (Waters, 1983). For this study, ceramics were selected from seven sites situated on the western margin of the Colorado Desert and close to the eastern base of the Peninsular Range mountains. The studied sites lie within the traditional territories of the Shoshonean-speaking Cahuilla Indians in the north (sites CA-SDI-343 and CA-SDI-2336) and Yuman-speaking Tipai linguistic group of Kumeyaay Indians in the south (sites CA-SDI-955, CA-SDI-956, CA-SDI-963, CA-SDI-10571 and CA-SDI-10573) (Figure 2).

Figure 2. Late Prehistoric sites analysed in this study. a) Location of the seven sites in San Diego County with environmental/landscape zones, the Anza-Borrego Desert State Park and traditional ethno-linguistic boundaries, b) Collins Valley and Coyote Canyon with the locations of the northern sites analysed, c) Indian Valley and Bow Willow Canyon with the location of the southern sites analysed.

With the exception of CA-SDI-343, which was part of the large Cahuilla village site of 'Los Coyotes', all sites appear to have been temporary winter camps associated with food collecting and processing activities. Bedrock milling features occur at most sites (Figure 3) and both CA-SDI-956 and CA-SDI-10573 may also contain the remains of house platforms. Abundant ceramic sherds were recovered from shallow cultural layers and surface scatters at the sites during archaeological surveys in the 1950s-70s (Wallace and Taylor, 1958; Wallace, 1962). Other artefacts include manos, metates, hammerstones, choppers, pestles, projectile points, debitage and shell beads. With the exception of CA-SDI-963, all sites are situated close to creeks where water would have been available, at least in the winter months. No direct evidence of ceramic production such as firing pits, tools, or unused ceramic raw materials have been found at any of the archaeological sites.

Sample	Site	Form	Lip Form	Recurved Rim	Decoration
1	CA-SDI-2336	Indeterminate	N/A	N/A	None
2	CA-SDI-2336	Indeterminate, flat	Rounded	No	None
3	CA-SDI-2336	Jar	Flattened, thickened to exterior	Slightly recurved	None
4	CA-SDI-2336	Jar	Flattened, thickened to exterior	No	None
5	CA-SDI-2336	Jar	Slightly flattened	No	None
6	CA-SDI-2336	Jar? (neckless or holemouth?)	Rounded	No	None
7	CA-SDI-2336	Jar	Flattened	No	None
8	CA-SDI-2336	Indeterminate	Rounded	No	None
9	CA-SDI-343	Bowl	Rounded	No	None
10	CA-SDI-343	Jar	Slightly flattened everted lip, thickened to exterior	Yes	None
11	CA-SDI-343	Bowl	Rounded	No	Burnished exterior, smoothed interior
12	CA-SDI-343	Jar (neckless or holemouth)	Rounded	No	None
13	CA-SDI-343	Indeterminate	Flattened, slightly thickened to exterior	No	None
14	CA-SDI-343	Jar	Flattened, everted	Slightly recurved	None
15	CA-SDI-343	Bowl	Slightly flattened, thickened to exterior	No	biconical drilled hole
16	CA-SDI-343	Jar (chimney neck)	Flattened	No	None
17	CA-SDI-343	Bowl	Slightly flattened, slightly thickened to interior	Indeterminate	None
18	CA-SDI-343	Indeterminate	Slightly flattened, thickened to exterior	No	Basket impression on interior surface? Possible red pigment on rim
19	CA-SDI-343	Bowl	Slightly flattened, slightly thickened to exterior	No	None
20	CA-SDI-343	Jar (chimney neck)	Slightly flattened everted lip, thickened to exterior	Slightly recurved	None
21	CA-SDI-343	Jar	Flattened, slightly to exterior	No	None
22	CA-SDI-343	Jar (chimney neck)	Slightly flattened everted lip, thickened to exterior	Yes	None
23	CA-SDI-343	Jar	Rounded	No	None
24	CA-SDI-343	Scoop?	Slightly flattened	No	None
25	CA-SDI-343	Bowl	Slightly flattened	No	incised rim
26	CA-SDI-343	Bowl (hemispherical)	Slightly flattened everted lip, thickened to exterior	Slightly recurved	None
27	CA-SDI-343	Jar	Slightly flattened, slightly thickened to exterior	Slightly recurved	None
28	CA-SDI-343	Jar	Rounded	No	None
29	CA-SDI-343	Disk or lid	Flattened, slightly thickened to ?	No	None
30	CA-SDI-343	Indeterminate	Flattened, everted	Yes	Beige slip?
31	CA-SDI-343	Jar	Rounded	No	None
32	CA-SDI-343	Bowl (straight-sided)	Flattened, thickened to exterior	Slightly recurved	None
33	CA-SDI-343	Jar	Flattened, everted. Slightly thickened to exterior.	Slightly recurved	none
34	CA-SDI-343	Jar	Rounded	No	Red paint? vertical band from lip. Light burnishing.
35	CA-SDI-343	Jar (neckless or holemouth)	Rounded	No	

Table 1. Details of the 70 Late Prehistoric sherds analysed in this study (continued in Table 2 below).

Sample	Site	Form	Lip Form	Recurved Rim	Decoration
36	CA-SDI-955	Jar (neckless or holemouth)	Rounded	No	None
37	CA-SDI-955	Jar (neckless or holemouth)	Rounded	no	molded basket or basket impressed anvil
38	CA-SDI-955	Bowl (hemispherical)	Flattened, slightly everted	No	Lightly burnished
39	CA-SDI-956	Bowl	Flattened	No	None
40	CA-SDI-956	Bowl (hemispherical)	Flattened	No	None
41	CA-SDI-956	Jar	Slightly flattened	No	None
42	CA-SDI-956	Bowl (hemispherical)	Flattened (thickened to interior and exterior)	No	incised rim
43	CA-SDI-956	Bowl	Flattened, thickened to interior and exterior	No	incised rim
44	CA-SDI-956	Bowl	Flattened	No	None
45	CA-SDI-956	Jar	Slightly flattened	Yes	None
46	CA-SDI-963	Bowl (straight-sided)	Flattened	No	None
47	CA-SDI-963	Bowl (hemispherical)	Everted lip, tapered to a point	No	None
48	CA-SDI-963	Jar (neckless or holemouth)	Flattened, slight upward inflection	No	None
49	CA-SDI-963	Bowl (straight-sided)	Rounded	No	None
50	CA-SDI-10571	Jar (chimney neck)	Flattened	No	None
51	CA-SDI-10571	Plate or lid	Rounded	No	None
52	CA-SDI-10571	Jar (chimney neck)	Slightly flattened	No	None
53	CA-SDI-10571	Jar (neckless or holemouth)	Rounded	No	None
54	CA-SDI-10571	Jar (necked)	Flattened	Yes	None
55	CA-SDI-10571	Bowl	Slightly flattened	No	None
56	CA-SDI-10573	Bowl	Flattened	No	anvil impression on interior?
57	CA-SDI-10573	Jar	Flattened, thickened to exterior	No	None
58	CA-SDI-10573	Bowl	Flattened, slightly thickened to interior	No	None
59	CA-SDI-10573	Bowl (hemispherical)	Flattened	No	None
60	CA-SDI-10573	Jar	Flattened	Yes	None
61	CA-SDI-10573	Bowl (straight-sided)	Flattened	No	None
62	CA-SDI-10573	Bowl	Rounded	No	basket impressed anvil on interior below rim
63	CA-SDI-10573	Bowl	Flattened	No	None
64	CA-SDI-10573	Jar (neckless or holemouth)	Rounded	No	Possible whitish slip
65	CA-SDI-10573	Bowl	Rounded	No	None
66	CA-SDI-10573	Bowl	Rounded	No	None
67	CA-SDI-10573	Bowl	Flattened, thickened to exterior	No	None
68	CA-SDI-10573	Bowl	Flattened, thickened to interior	No	None
69	CA-SDI-10573	Jar (neckless or holemouth)	Rounded, thickened to exterior	No	None
70	CA-SDI-10573	Bowl	Flattened	No	biconical drill hole

Table 2. Details of the 70 Late Prehistoric sherds analysed in this study (continued from Table 1 above).

A total of 70 rim sherds were selected from the ceramic assemblages of the seven sites (Tables 1 and 2). The samples tested comprised between 5-12% of each site assemblage, with 35 sherds selected from the northern and the southern ends of the study area. Most of the sherds had been originally collected from the site surface and are therefore of uncertain relative chronological assignment. In total, the studied sample consists mainly of bowls and jars, plus a lid, a plate and a scoop (Tables 1 and 2). Fragments of both hemispherical and straight-sided bowls as well as restricted chimney-neck jars and open neckless or 'holemouth' jars were present. Only 17 of the 70 sherds represented rim circumferences of 10% or more of the total aperture. Therefore the ability to identify reliable vessel size categories was limited and requires an expanded sampling program. Most of the vessels from which the sherds originated appear to have been undecorated, however, a few had incised rims and biconical drilled holes. Possible evidence for burnishing and painting was also found on a few sherds.

Figure 3. Late Prehistoric archaeological sites in the western Colorado Desert. a) Bow Willow Canyon with rocky granitic canyon walls and sandy alluvial soil, b) Prehistoric bedrock mortars at site CA-SDI-955 in Indian Valley.

Analytical Methods

Standard (30 μm) petrographic thin sections (Reedy, 2008, p. 1-3) were cut vertically through the vessel rim (Whitbread, 1996) of each of the selected sherds. These were analysed under the polarising light microscope using a modification of the holistic, descriptive approach pioneered by Whitbread (1989, 1995). This approach focuses on the nature of the clay matrix and voids as well as the more conspicuous aplastic inclusions. Using this method, it was possible to detect important microstructural and textural evidence in thin section for the techniques used to manufacture the ceramics. Such information is not readily detected by more quantitative petrographic methods such as point counting or the modal analysis of inclusions (Middleton *et al.*, 1985) that have been applied in the compositional analysis of southern California ceramics so far

(Pymale-Schneeberger, 1993; Griset, 1996; Gallucci, 2001, 2004; Hildebrand *et al.*, 2002).

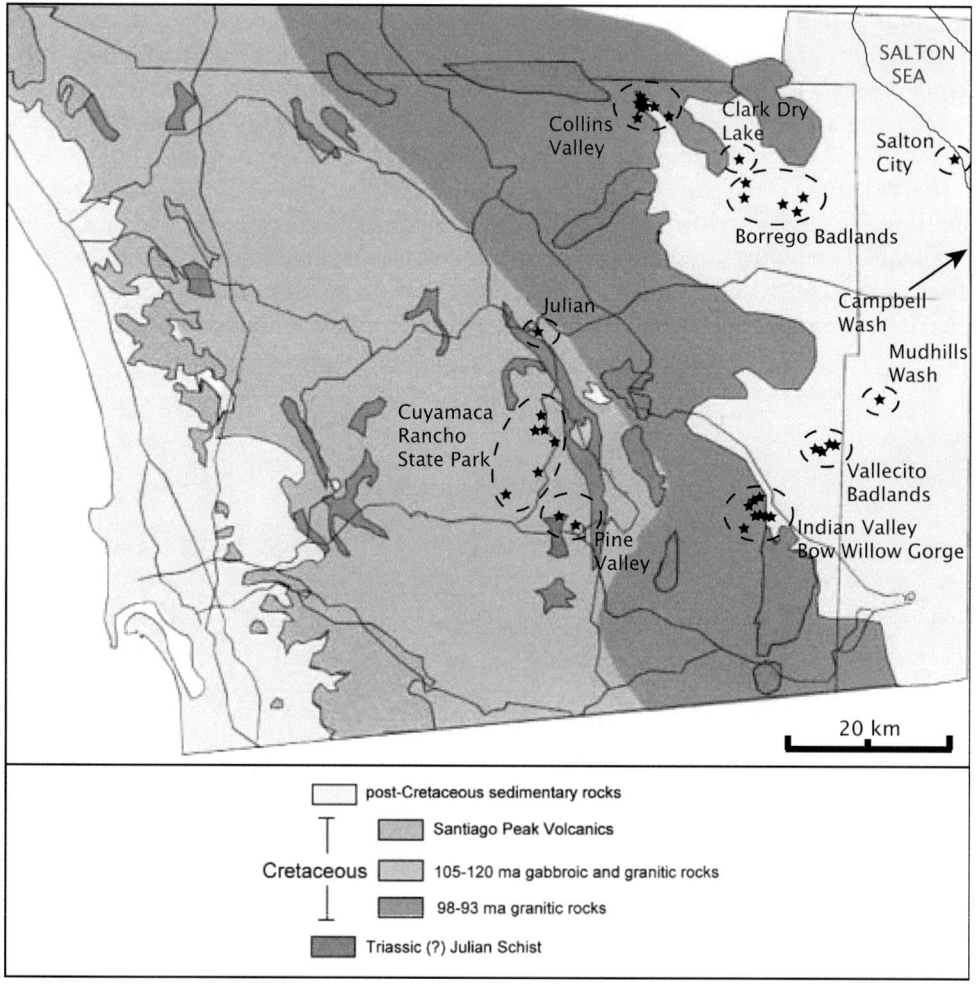

Figure 4. Generalised geology of southern California with the location of the geological field samples analysed in this study.

The ceramic thin sections were sorted into petrographic fabric groups, based on the overall composition of their inclusions, matrix and voids under the microscope, each representing a specific combination of raw materials and manufacturing techniques. Individual petrographic fabric groups were then characterised in detail by interpreting the type(s) of raw materials and the various steps involved in their manufacture, as well as their relationship to other classes. Compositional, microstructural and textural criteria were used to detect the presence of specific practices such as raw material processing, the intentional addition of different types of particulate matter or 'temper' and vessel forming techniques, as well as the atmosphere and degree of firing (Woods, 1984; Whitbread, 1986; Whitbread, 1995, p. 393-394; Whitbread, 1996; Rice, 1987, p. 409-411; Cuomo di Caprio and Vaughan, 1993; Roux and Courty, 1998; Reedy, 2008,

p. 146-148, 173-189).

In order to identify the possible sources of raw materials used to manufacture the Late Prehistoric ceramics, a program of geological field sampling and complementary analysis was undertaken. Geologically, the study area lies at the junction between the Mesozoic granitic and gabbroic igneous plutons of the Peninsular Range Mountains and the deep Cenozoic marine and non-marine sedimentary succession of the Salton Basin (Figure 4). All seven sites are situated within steep-sided northwest or southwest trending valleys cut into the eastern Peninsular Range Batholith. These valleys are characterised by light-coloured granitic rocks that weather into large boulders and form coarse sandy soil (Figure 3A).

Using geological maps (Strand, 1962; Rogers, 1965; Jennings, 1967) and field guides (Remeika and Lindsay, 1993; Clifford et al., 1997; Jefferson and Lindsay, 2006), detailed prospecting was carried out in the environs of the seven sites, as well as within the eroded sediments of the Salton Basin and the igneous terrain of the eastern Peninsular Range mountains (Figure 4; Table 3). Simple field tests on grain size and workability were used to identify and sample suitable clayey raw materials (Howard, 1982) that could have been used to produce ceramics. Samples of loose sandy sediment that may have represented suitable tempering material were also collected, as well as hard rock samples representative of the geology of each archaeological site.

Sample	Area	Type	Description	Sample	Area	Type	Description
1	Bow Willow	Sand	Recent alluvium	30	Collins Valley	Clay	Recent alluvium
2	Bow Willow	Rock	Granitic bedrock	31	Collins Valley	Sand	Recent alluvium
3	Bow Willow	Sand	Recent alluvium	32	Collins Valley	Rock	Granitic bedrock
4	Bow Willow	Rock	Metamorphic bedrock	33	Borrego Badlands	Clay	Pleistocene lacustrine Inspiration Wash Member
5	Bow Willow	Sand	Quaternary non-marine terrace	34	Borrego Badlands	Clay	Weathered intermediate igneous soil
6	Bow Willow	Sand	Weathered granitic soil	35	Cuyamaca Rancho State Park	Clay	Weathered intermediate igneous soil
7	Bow Willow	Rock	Granitic bedrock	36	Cuyamaca Rancho State Park	Clay	Weathered intermediate igneous soil
8	Vallecito Badlands	Clay	Recent alluvium	37	Cuyamaca Rancho State Park	Sand	Weathered intermediate igneous soil
9	Vallecito Badlands	Clay	Pliocene alluvium of Hueso Member	38	Clark Dry Lake	Clay	Recent lacustrine deposit
10	Vallecito Badlands	Sand	Pliocene alluvium of Hueso Member	39	Borrego Badlands	Clay	Pleistocene lacustrine Inspiration Wash Member
11	Vallecito Badlands	Clay	Pliocene alluvium of Hueso Member	40a	Borrego Badlands	Clay	Pliocene lacustrine Borrego Formation
12	Vallecito Badlands	Clay	Pliocene alluvium of Hueso Member	40b	Borrego Badlands	Clay	Pliocene lacustrine Borrego Formation
13	Vallecito Badlands	Clay	Pliocene laustrine Tapiado Member	41	Borrego Badlands	Clay	Pliocene lacustrine Borrego Formation
14	Vallecito Badlands	Rock	Pliocene laustrine Tapiado Member	42	Borrego Badlands	Clay	Pliocene lacustrine Borrego Formation
15	Vallecito Badlands	Clay	Pliocene alluvium of Diabo Formation	43	Borrego Badlands	Clay	Pliocene lacustrine Borrego Formation
16	Mudhills Wash	Clay	Pliocene marine Mudhills Member	44	Salton Sea	Clay	Recent lacustrine deposit
17	Mudhills Wash	Rock	Pliocene marine Mudhills Member	45	Campbell Wash	Clay	Quaternary lacustrine deposit
18	Indian Valley	Rock	Granitic bedrock	46	Campbell Wash	Clay	Quaternary lacustrine deposit
19	Indian Valley	Sand	Weathered granitic soil	47	Julian	Rock	Metamorphic bedrock
20	Indian Valley	Rock	Granitic bedrock	48a	Cuyamaca Rancho State Park	Clay	Weathered intermediate igneous soil
21	Indian Valley	Sand	Recent or Quaternary alluvium	48b	Cuyamaca Rancho State Park	Rock	Basic igneous bedrock
22	Indian Valley	Rock	Granitic bedrock	49a	Cuyamaca Rancho State Park	Clay	Weathered intermediate igneous soil
23	Collins Valley	Clay	Recent alluvium	49b	Cuyamaca Rancho State Park	Rock	Intermediate igneous bedrock
24	Collins Valley	Sand	Recent or Quaternary alluvium	50	Cuyamaca Rancho State Park	Rock	Granitic bedrock
25	Collins Valley	Rock	Metamorphic bedrock	51a	Cuyamaca Rancho State Park	Clay	Weathered intermediate igneous soil
26	Collins Valley	Clay	Recent alluvium	51b	Cuyamaca Rancho State Park	Rock	Intermediate igneous bedrock
27	Collins Valley	Sand	Sand dune	52a	Cuyamaca Rancho State Park	Clay	Weathered granitic soil
28a	Collins Valley	Rock	Metamorphic bedrock	52b	Cuyamaca Rancho State Park	Rock	Granitic bedrock
28b	Collins Valley	Rock	Metamorphic bedrock	53a	Pine Valley	Clay	Weathered intermediate igneous soil
28c	Collins Valley	Rock	Metamorphic bedrock	53b	Pine Valley	Rock	Intermediate igneous bedrock
29a	Collins Valley	Rock	Metamorphic bedrock	54	Pine Valley	Rock	Granitic bedrock
29b	Collins Valley	Rock	Metamorphic bedrock				

Table 3. Geological field samples collected and analysed in this study.

The collected clay samples were allowed to dry and were then crushed, re-hydrated and formed into test tiles or briquettes. These were fired in a laboratory kiln at 700°C under oxidising conditions and thin sectioned. Loose sandy sediment samples were set in resin before thin sectioning. All field samples were studied under the petrographic microscope and compared with the ceramic fabric classes in order to identify compositional matches that might be informative of the raw materials, technology and provenance of the archaeological samples.

Petrographic Fabric Classification

In thin section, the 70 Late Prehistoric ceramic samples exhibited a high degree of petrographic variability. Clear compositional similarities and differences between the samples in terms of mineralogy, petrography, texture and microstructure enabled a total of 18 different fabric groups to be identified, characterised by specific raw materials and manufacturing techniques (Table 4; Figures 5-7). These range from a large dominant group consisting of 29 out of the 70 analysed samples (Residual Granitic Fabric Group), to several groups composed of single unique sherds (e.g. Igneous Tempered Fabric Group, Fine Grog Tempered Fabric Group and Grog Tempered Calcareous Fabric Group).

Fabric Group	Samples	Sites
Residual Granitic Fabric Group	1, 2, 3, 4, 6, 8, 11, 14, 15, 17, 18, 21, 23, 25, 28, 37, 40, 43, 44, 48, 50, 51, 52, 53, 57, 58, 62, 66, 69, 70	CA-SDI-343, CA-SDI-2336 , CA-SDI-10571, CA-SDI-10573, CA-SDI-955, CA-SDI-956, CA-SDI-963
Well-Packed Alluvial Fabric Group	38, 41, 47, 55, 56, 59, 61, 67	CA-SDI-10571, CA-SDI-10573, CA-SDI-955, CA-SDI-956, CA-SDI-963
Grog Tempered Residual Granitic Fabric Group I	10, 20, 24, 27, 33, 34	CA-SDI-343
Residual Metamorphic Fabric Group	9, 19, 26, 29, 35	CA-SDI-343
Grog Tempered Fine Alluvial Fabric Group I	46, 49, 68	CA-SDI-963, CA-SDI-10573
Grog Tempered Fabric Group	30, 36, 42	CA-SDI-343, CA-SDI-955, CA-SDI-956
Sand and Grog Tempered Fabric Group	39, 54	CA-SDI-956, CA-SDI-10571
Sand and Grog Tempered Calcareous Fabric Group	63, 65	CA-SDI-10573
Gneiss Tempered Fabric Group	5, 7	CA-SDI-2336
Igneous Tempered Fabric Group	16	CA-SDI-343
Grog Tempered Residual Granitic Fabric Group II	60	CA-SDI-10573
Grog Tempered Residual Metamorphic Fabric Group	32	CA-SDI-343
Biotite-rich Residual Granitic Fabric Group	33	CA-SDI-343
Fine Biotite-Rich Grog Tempered II Fabric Group	31	CA-SDI-343
Grog Tempered Fine Alluvial Fabric Group II	12	CA-SDI-343
Igneous, Grog and Plant Tempered Fabric Group	64	CA-SDI-10573
Fine Grog Tempered Fabric Group	45	CA-SDI-956
Grog Tempered Calcareous Fabric Group	13	CA-SDI-343

Table 4. Petrographic fabric classification of the 70 Late Prehistoric sherds.

Significant petrographic variability was found to exist within the ceramic assemblages of each of the seven individual Late Prehistoric desert sites studied. A total of 11 different fabric groups, composed of geologically distinct raw materials and manufacturing techniques, were detected in the 27 sherds analysed from site CA-SDI-343. Even the three sherds analysed from the small ceramic assemblage of site CA-SDI-955 were compositionally distinct from one another in thin section. Some petrographic fabric groups (e.g. Sand and Grog Tempered Calcareous Fabric Group, Gneiss Tempered Fabric Group) were restricted to one or a few sites in the samples analysed, whereas others (e.g. Residual Granitic Fabric Group, Grog Tempered Fabric Group) had a more widespread distribution, occurring at sites in the north and south of the study area. Very little correspondence was detected in this study between the general form types of the sherds and their petrographic classifications. Common fabric classes included sherds from both bowls and jars with different lip forms and variously sized apertures. A larger sample size is needed to more rigorously test for associations between the fabric groups and vessel forms and sizes and their related functions.

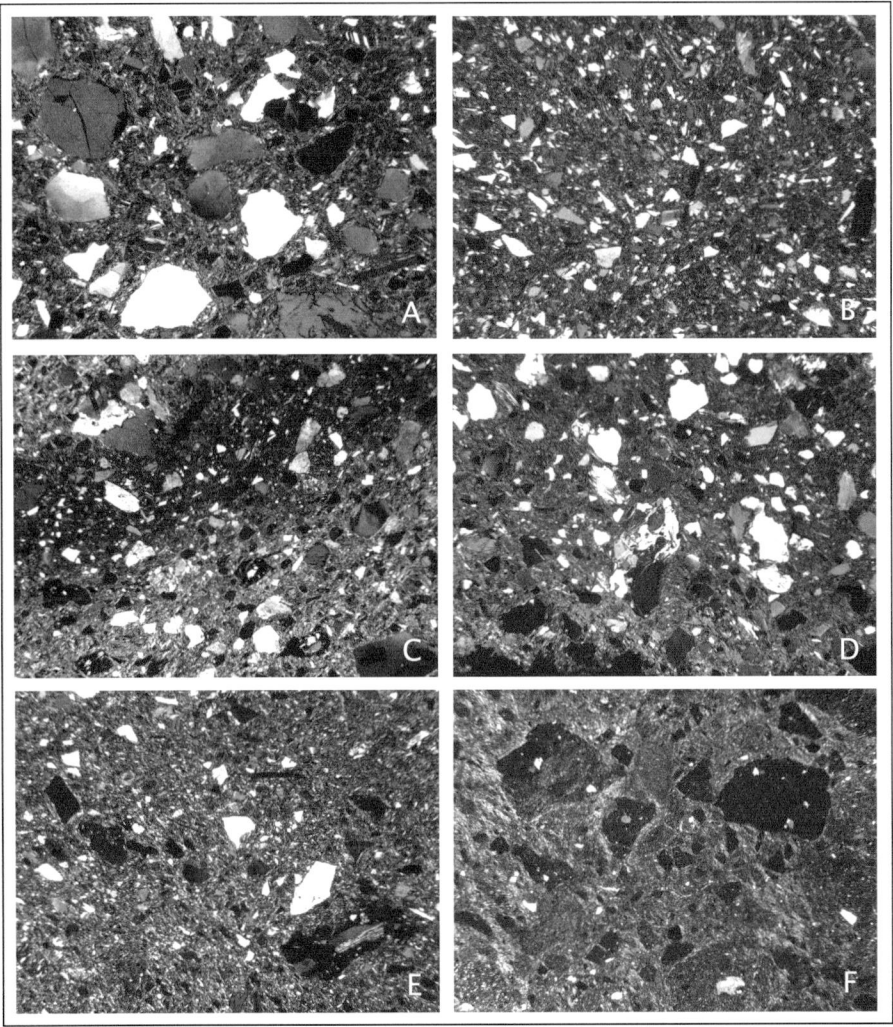

Figure 5. Photomicrographs of petrographic fabric groups detected in the Late Prehistoric sherds. a) Residual Granitic Fabric Group, sample 4, b) Well-Packed Alluvial Fabric Group, sample 41, c) Grog Tempered Residual Granitic Fabric Group I, sample 10, d) Residual Metamorphic Fabric Group, sample 19, e) Grog Tempered Fine Alluvial Fabric Group I, sample 46, f) Grog Tempered Fabric Group, sample 42. All images taken in crossed polars. Image width = 3.8 mm.

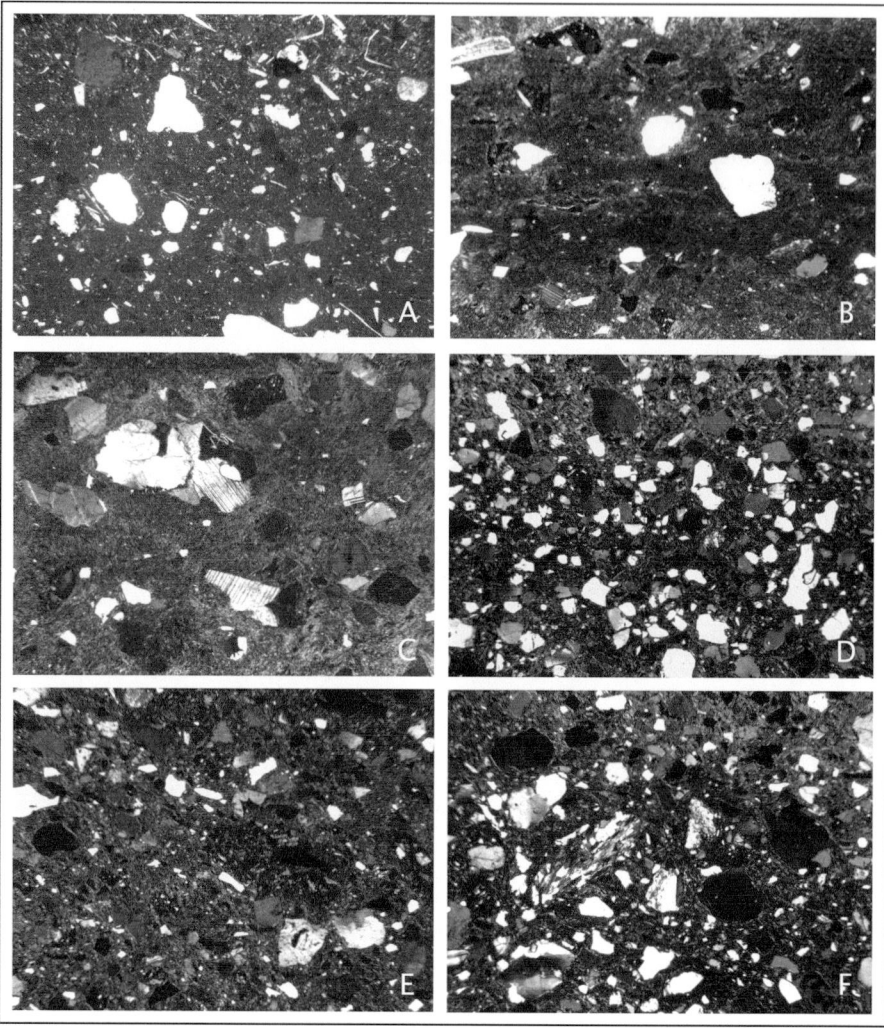

Figure 6. Photomicrographs of petrographic fabric groups detected in the Late Prehistoric sherds. a) Sand and Grog Tempered Fabric Group, sample 54, b) Sand and Grog Tempered Calcareous Fabric Group, sample 63, c) Gneiss Tempered Fabric Group, sample 7, d) Igneous Tempered Fabric Group, sample 16, e) Grog Tempered Residual Granitic Fabric Group II, sample 60, f) Grog Tempered Residual Metamorphic Fabric Group, sample 32. All images taken in crossed polars. Image width = 3.8 mm.

Ceramic Raw Materials

Based upon the mineralogy, petrography and texture of the 18 petrographic fabric groups and their comparison with the geological samples collected in the field, it has been possible to characterise the types of raw materials used for the production of the ceramics found at the seven Late Prehistoric desert sites. A surprising variety of different clay, particulate matter and hard rock appear to have been utilised by the indigenous potters of this region, although certain materials were used more commonly than others. By examining the availability of these raw material types using geological literature and the database of field samples, it has been possible in several cases to identify likely sources of the ceramic raw materials.

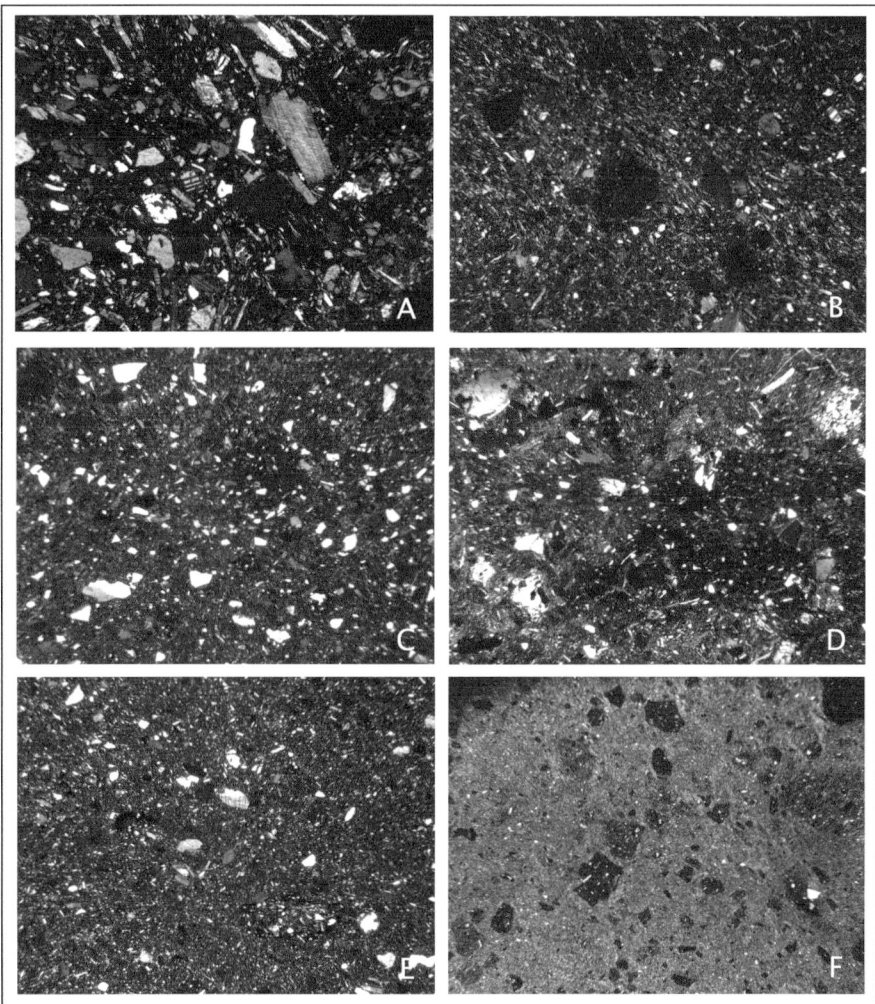

Figure 7. Photomicrographs of petrographic fabric groups detected in the Late Prehistoric sherds. a) Biotite-rich Residual Granitic Fabric Group, sample 22, b) Fine Biotite-Rich Grog Tempered II Fabric Group, sample 31, c) Grog Tempered Fine Alluvial Fabric Group II, sample 12, d) Igneous, Grog and Plant Tempered Fabric Group, sample 64, e) Fine Grog Tempered Fabric Group, sample 45, f) Grog Tempered Calcareous Fabric Group, sample 13. All images taken in crossed polars. Image width = 3.8 mm.

Coarse-grained, poorly sorted, residual clay, rich in quartz, plagioclase feldspar and biotite, deriving from the *in situ* weathering of granitic igneous rocks, appears to have been used for several fabric classes, such as the Residual Granitic Fabric Group (Figure 5A) and the Biotite-rich Residual Granitic Fabric Group (Figure 7A). The first of these accounts for 37 of the 70 sherds analysed and occurs at all seven sites. Field prospecting indicates that granitic igneous rocks such as granodiorite, quartz-diorite and tonalite occur in the three valleys in which the archaeological sites are situated (Geological samples 2, 7, 18, 20, 22 and 32). However, these weather slowly in the arid desert climate and break down into loose, coarse-grained, sandy soil, which is very poor in clay minerals. The same granitic rock types also occur in the eastern Peninsular Range mountains to the west of the study sites. Here in a wetter environment, the bedrock weathers chemically to form clay-rich soil (e.g. geological sample 52A) which is more likely to be the source of the ceramics of the of the Residual Granitic Fabric Group and the Biotite-rich Residual Granitic Fabric Group.

Variation in the mineralogy and texture of the large Residual Granitic Fabric Group exists in terms of the proportion of the mineral hornblende, which occurs in small amounts in ceramics from the northern sites, but is more or less absent in samples from the south. This pattern, which is also seen in the analysis of geological samples from the two ends of the study area, suggests that potters utilised several different granitic clay sources along the eastern edge of the Peninsular Range mountains. Several sherds from site CA-SDI-343 in the north belonging to the Residual Metamorphic Fabric Group (Figure 5D) and the Grog Tempered Residual Metamorphic Fabric Group (Figure 6F) are related to the main Residual Granitic Fabric Group, but also contain conspicuous sillimanite-bearing metamorphic rock fragments. The analysis of clay, hard rock and sand (Geological samples 27, 28a, 28b and 30) from Collins Valley has revealed a close match for this metamorphic material and may indicate that it is indicative of this area.

Several fabric groups, restricted to the southern sites, including the Well-Packed Alluvial Fabric Group (Figure 5B) and the Grog Tempered Fine Alluvial Fabric Group I (Figure 5E), are related mineralogically to the dominant Residual Granitic Fabric Group, but contain finer, more rounded and better sorted inclusions. This suggests that they were manufactured from immature alluvial clay deposits derived from the erosion of granitic rocks. Recent, locally derived, alluvial material rich in quartz, plagioclase feldspar and biotite exists near all archaeological sites studied (Geological samples 3, 19 and 21), but is generally very sandy. However, deposits of older Quaternary river terrace deposits that occur to the east of the southern sites contain clay-rich horizons (e.g. geological sample 5) that could represent the source of the Alluvial Fabric Group and the Grog Tempered Fine Alluvial Fabric Group I.

Very fine, sedimentary clay of either lacustrine or marine origin appear to have been used as a base-clay for a range of ceramics, such as those belonging to the Grog Tempered Fabric Group (Figure 5F), the Sand and Grog Tempered Calcareous Fabric Group (Figure 6B) and the Grog Tempered Calcareous Fabric Group (Figure 7F). These occur in low numbers at sites in the north and south within the ceramics analysed. Variation in the colour, texture and mineralogy of the clay matrices of these fabric groups, particularly their calcite content, suggests that they came from a range of different sources. Suitable clay beds outcrop within badlands that dissect the deep marine and non-marine sedimentary succession of the Salton Basin to the east of the studied sites. Field sampling in this area has revealed several possible matches such as the Early Pliocene marine Mudhills Member (Geological sample 16), which could be the source of the Grog Tempered Fabric Group and the Pleistocene non-marine Inspiration Wash Member (Geological sample 33) that may have been used to produce the ceramics of the Fine Biotite-Rich Grog Tempered II Fabric Group. However, confidently matching these fine sedimentary fabric groups to their sources of raw materials in thin section is difficult given their relative lack of inclusions.

In addition to the large number of different clay sources represented by the 18 petrographic fabric groups, an equally surprising range of natural particulate materials appear to have been used in the production of the ceramics analysed. Loose unconsolidated deposits of rounded, silt and sand-sized grains of quartz, feldspar, biotite and less commonly hornblende and muscovite were utilised for the ceramics belonging to the Sand and Grog Tempered Fabric Group (Figure 6A) and the Sand and Grog Tempered Calcareous Fabric Group (Figure 6B). Sandy alluvial material derived from the erosion of granitic rock is abundant in the vicinity of all sites (e.g. geological samples 3, 19 and 21) and sand dunes occur in some areas of the eastern desert (e.g. geological samples 27). Such deposits would have represented suitable sources of sandy raw materials for ceramic manufacture.

Other ceramics, such as those of the Igneous Tempered Fabric Group (Figure 6D) and the Gneiss Tempered Fabric Group (Figure 6C), appear to contain angular, crushed weathered rock. Hard igneous bedrock is abundant in the boulder-strewn valleys of the seven studied sites (Geological samples 2, 7, 18, 20, 22 and 32) and pinpointing the source of the material used in the Igneous Tempered Fabric Group is therefore difficult. However, a good match for crushed gneiss found in the Gneiss Tempered Fabric Group was found among the boulders of site CA-SDI-2336 (Geological samples 29A and 29B), close to where the sherds were found.

The geological interpretation of the 70 thin sections and their relationship to the database of field samples indicates that the potters who produced these ceramics made use of a wide range of different types of raw materials. This may suggest an intimate knowledge of the natural geodiversity of the region and perhaps an ability to adapt technologically to the availability of different types of raw materials during seasonal movements or long-term settlement shifts (Lyneis, 1988). Knowledge of the raw material sources used by southern California potters during the Late Prehistoric period is presently very poor (Schaefer, 1994b) and limited to a few historical accounts of favorable locations (Heizer and Treganza, 1972, p. 319, 333-334; Cline, 1984, p. 38; Hohenthal, 2001, p. 167). The compositional matches between the ceramics and geological field samples in this study are therefore significant in terms of determining where prehistoric people may have collected clay and other materials to produce pottery. The occurrence of certain common fabrics among the ceramics analysed, such as the Residual Granitic Fabric Group, seems to indicate that potters preferred certain raw material sources and used them repeatedly. This is supported by ethnographic accounts, which attest to the use of particular highly desirable quarry locations that were returned to on a regular basis (e.g. Wade, 1999, p. 3). However, whether potters in the past maintained personal clay resources for private use (Heizer and Treganza, 1972, p. 334; Hurd, et al., 1990) or considered raw materials to be public domain (Rogers, 1936, p. 4) cannot be determined based on current evidence.

Geological fieldwork in the dry rocky valleys in which the archaeological sites are situated did not reveal many local sources of clay-rich deposits that could have been used for ceramic manufacture. Instead, much of the raw materials used for the production of the ceramics appear to have come from elsewhere, either in the Peninsular Range mountains to the west, or in the Salton Basin to the east. Although ceramic resource procurement distances may have been significantly greater for mobile hunter-gatherers such as the Late Prehistoric period tribes of southern California (Rogers, 1936, p. 4; Heizer and Treganza, 1972, p. 334; Williams, 1989, p. 4) compared to those recorded for potters in more sedentary societies (Arnold, 1985, p. 32-60), it is likely that bulky raw materials would not have been transported over significant distances. Instead, ceramic production probably took place close to sources of clay, temper, water and fuel. With this in mind, much of the compositionally diverse ceramic assemblages recorded at the seven archaeological sites tested must be non-local in origin, having been made elsewhere and transported to the sites as finished pots. The absence of direct evidence for ceramic production at any of the studied sites supports this interpretation. Further, by examining the geographic patterning of the 18 petrographic fabric groups and the occurrence of comparable raw material sources, initial findings suggest that pottery vessels were transported over significant distances (>50 km) in numerous directions, within and beyond the desert. The patterns of movement revealed by the limited sample set in this study correlate well with historic accounts of ancient trail systems (Cline, 1979, p. 17, 18, 21, 1984, p. 13, 16-17)

Ceramic Technology

Based on the composition, microstructure and texture of the 70 Late Prehistoric sherds in thin section, it has been possible to identify many of the technological steps involved in their manufacture, including raw material processing, paste preparation, vessel forming techniques and the conditions of firing. These provide important evidence for the choices and behaviours of potters in the past, as well as their knowledge of raw material properties and their skill in the craft of pottery production.

The clayey raw materials of ceramics belonging to the Residual Granitic Fabric Group and the Biotite-rich Residual Granitic Fabric Group may have been 'cleaned' prior to use in order to remove very coarse mineral and rock inclusions. Whilst there is little direct evidence for this process in thin section, the coarse poorly-sorted nature of the residual weathered igneous raw materials that could have been used for these ceramics suggests that a degree of processing or cleaning must have been carried out before the clay was suitable for ceramic manufacture. This kind of activity has been reported in ethnographic studies of traditional southern California potters by Rogers (1936, p. 6), Wilken (1982) and Hohenthal (2001, p. 170).

The occurrence in the Grog Tempered Fabric Group and the Fine Biotite-Rich Grog Tempered II Fabric Group of fine argillaceous inclusions, with sharp to merging boundaries, that have an identical composition to the surrounding clay matrix (Figure 8A), may suggest that they were produced from dry, pulverised clay, which was wetted to form a paste. These inconspicuous inclusions appear to represent fine clay particles that were not sufficiently hydrated and therefore remained aplastic during the pottery manufacturing process. Clay used by traditional potters is often collected in dry state and a first step in the production of ceramics is to grind this into a fine powder. Observations of traditional pottery production in southern California attest to the crushing of chunks or 'clods' of clay in a mortar (Cline, 1984, p. 34) or on a flat rock (Rogers, 1936, p. 5) and subsequent grinding with a metate and mano (Rogers, 1936, p. 6; Bean and Lawton, 1965, p. 6; Wilken, 1982; Cline, 1984, p. 34; Hohenthal, 2001, p. 170). The evidence seen in thin section in this study suggests that such a practise was also carried out during the Late Prehistoric period.

The clay paste used to produce the ceramics of 13 out of the 18 fabric classes detected in this study appears to contain intentionally added particulate matter or temper (Figures 5-7). Temper was distinguished in thin section from naturally occurring aplastic inclusions using a combination of different criteria, including grain-size distribution, roundness, angularity and mineralogical composition (Rice, 1987, p. 409-411; Whitbread, 1995, p. 393). Both the number of fabric groups that contain temper and the range of different tempering agents that was added to the ceramics are surprising. In addition to sand and hard rock temper, noted above, other types of added particulate matter include crushed ceramic sherds or 'grog' (Figure 8B) and plant matter (Figure 8D). Grog temper is particularly common in the ceramics analysed, having been added to some ten different fabric classes including the Grog Tempered Fabric Group, the Grog Tempered Fine Alluvial Fabric Group I and the Grog Tempered Calcareous Fabric Group. In a few thin sections, grog inclusions were found to contain grog themselves (Figure 8C), suggesting that ceramics were repeatedly recycled and used as temper.

The act of tempering clay with particulate matter such as quartz (Rogers, 1936, p. 25), crushed rock (Kroeber, 1925, p. 722; Rogers, 1936, p. 25; Heizer and Treganza, 1972, p. 334), grog (Curtis, 1908, p. 27; Gifford, 1931, p. 42; Drucker, 1937, p. 22), ash (Wilken, 1982) and manure (Ferenga and Heredia, 1995, p. 4) has been recorded in ethnographic accounts of traditional southern California potters. Such actions may have been motivated by performance characteristics such as the workability of the clay paste or the strength of the vessel during firing (Heizer and Treganza, 1972, p. 334; Wilken, 1982; Campbell, 1999, p. 123). Indeed, within the ceramics analysed in this study, temper appears to be added mainly to those pots made from fine sedimentary clays with few naturally occurring inclusions, suggesting that the practise could be a response to the qualities of the available raw materials. However, several naturally coarse ceramic samples belonging to the Grog Tempered Fine Alluvial Fabric Group I, the Grog Tempered Residual Granitic Fabric Group II and the Grog Tempered Residual Metamorphic Fabric Group also contain grog inclusions (Figure 8E,F). The occurrence in the samples analysed of identical non-tempered versions of these two fabric groups (Residual Granitic Fabric Group and Residual Metamorphic Fabric Group) suggest that their residual base clays, which contained abundant naturally occurring non-plastic inclusions, did not need to be tempered for functional reasons. Equally, the motivation behind the addition of two or more different types of temper to ceramics of the Sand and Grog Tempered Fabric Group, the Calcareous Sand and Grog

Tempered Fabric Group and the Igneous, Grog and Plant Tempered Fabric Group are also not clear.

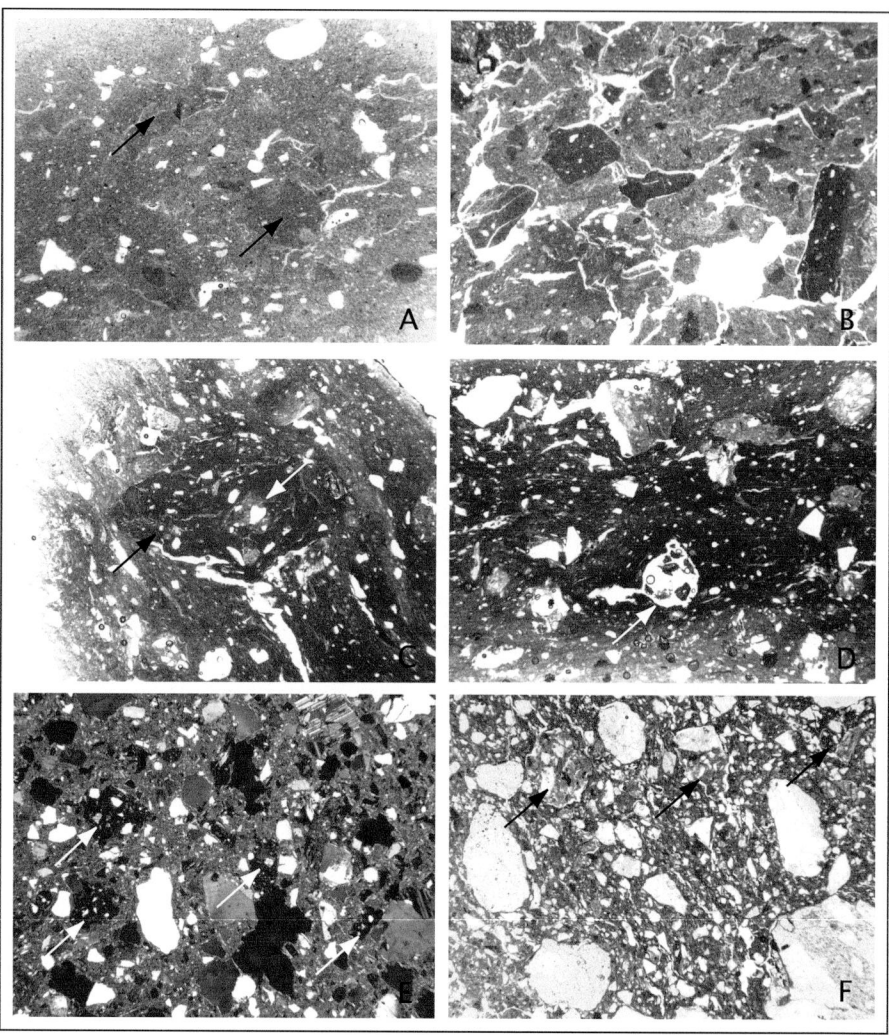

Figure 8. Photomicrographs of evidence for paste preparation techniques of the 70 Late Prehistoric sherds analysed in this study. a) Inconspicuous argillaceous inclusions in sample 65, that may be evidence of grinding dry clay, b) Grog inclusions in sample 36 indicating the addition of crushed ceramic temper, c) Grog inclusion (black arrow) with second generation grog (white arrow) in sample 64, indicating repeated recycling of ceramics, d) Void with charred organic matter in sample 64, possibly indicating addition of plant temper, e,f) Addition of grog to coarse clay rich in inclusions, samples 32 and 34. All images taken in plane polarised light, except e. Image width = 3.8 mm, except c = 2.4 mm.

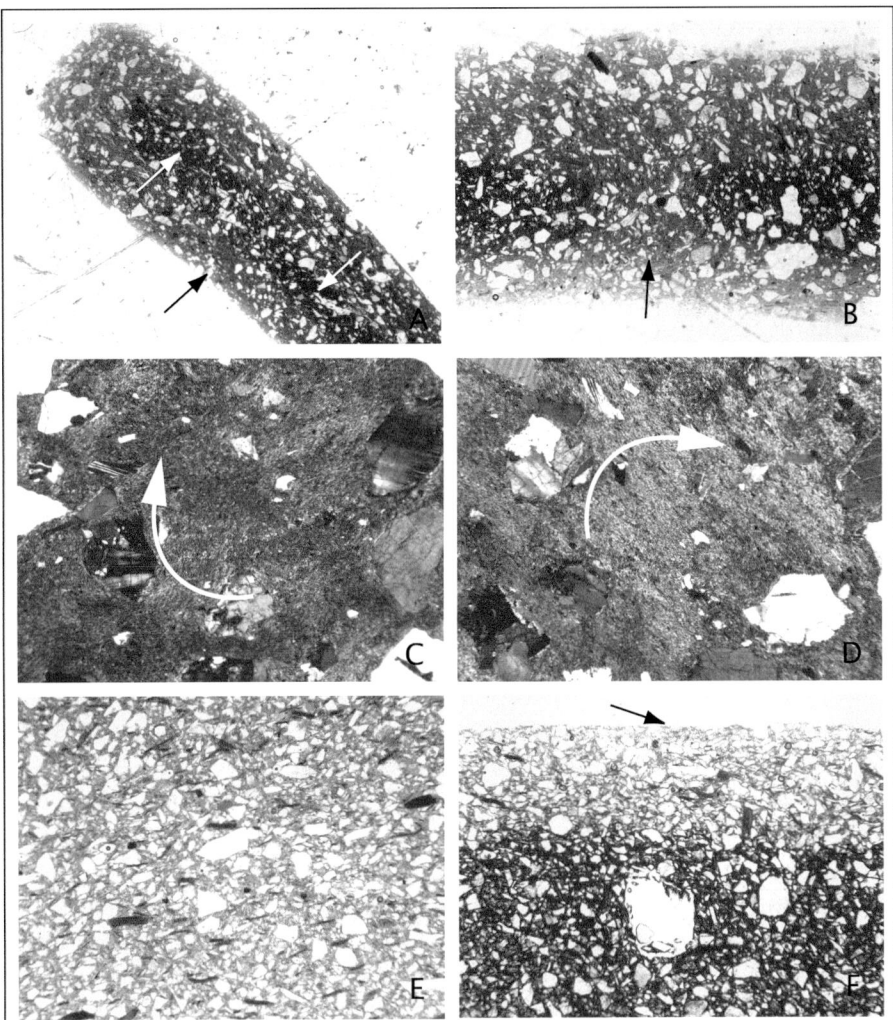

Figure 9. Photomicrographs of evidence for the forming and firing techniques of the 70 Late Prehistoric sherds analysed in this study. a) Relic coils (white arrows) and coil join (black arrow) in sample 58, suggesting construction of vessel by coiling, b) Join between adjacent coils indicated by degree of oxidation of core of vessel in sample 44, c,d) High optical activity of clay matrix during 45° rotation of sample 7 in crossed polars, indicating relatively low degree of firing, e) Light coloured clay matrix in sample 41 due to firing in oxidising atmosphere, f) Oxidised margin and reduced core in sample 55 due to low firing duration and/or high organic content of clay (arrow indicates vessel surface). Images a, c, d taken in crossed polars and b, e, f in plane polarised light. Image width = 3.8 mm, except c, d = 2.4 mm, a = 10.8 mm, b = 5.8 mm.

Microstructural evidence for the techniques used to form the Late Prehistoric desert ceramic vessels can be seen in many of the vertical thin sections analysed. The dominant forming method appears to have been the bonding of successive strands of clay or 'coiling'. Relic coils are highlighted in thin section by the concentric orientation of elongate inclusions and voids (Figure 9A), or by the presence of joins between successive coils (Figure 9B). Evidence for the practice of coiling was found in half of 18 fabric groups and was more evident in coarser grained ceramics, due to their larger, more abundant inclusions. However, relic coils and coil joins were also noted in some other, fine sedimentary ceramics such as the single sample belonging to the Grog Tempered Calcareous Fabric Group.

Coiling is suspected to have been the preferred method of vessel forming throughout the 'paddle and anvil' potting tradition of southern California (Griset, 1996, p. 9) and has been reported, albeit with minor variations in the implements used, in several ethnographic accounts of traditional pottery manufacture in this area (e.g. Rogers, 1936, p. 9; James, 1960, p. 63; Bean and Lawton, 1965, p. 6; Bean, 1978, p. 579; Kroeber and Hooper, 1978, p. 28; Wilken, 1982; Cline, 1984, p. 34; Hohenthal, 2001, p. 170). Coils, 'ropes' (Kroeber and Hooper, 1978, p. 28) or 'strings' (Cline, 1984, p. 34) were bonded together with the fingers and the vessel wall was thinned and beaten into shape using a wooden paddle and a clay anvil or river pebble. Evidence for the forming and shaping of ceramics with a paddle and anvil and the use of a basket mould was observed on the surfaces of a few of the sherds in hand specimen. In thin section, parallel alignment of inclusions and voids close to the exterior surface of some sherds may also be indicative of beating with an implement such as a paddle, or the drawing of the vessel walls with the fingers.

Based upon the hardness of the Late Prehistoric sherds in hand specimen and the nature of their clay matrices in thin section, most appear to have been moderately fired, at a temperature high enough to produce *terra cotta*, but almost certainly below 1000°C. Accurately determining the firing temperatures of ancient ceramics can be difficult (Rice, 1987, p. 426-435), however, a rough estimate of the degree of firing can be inferred from the optical state of the clay matrix in thin section (Peacock, 1971; Whitbread, 1995, p. 394; Rice, 1987, p. 431). Clay minerals, though too small to be seen under the polarising microscope, exhibit extinction in thin section, when rotated in crossed polars. This property is progressively lost in fired ceramics as the lattice structure of the clay minerals breaks down and they begin to vitrify, until at high temperatures the clay matrix becomes optically inactive. The clay matrices of the majority of the ceramics analysed in this study were either moderately or highly-optically active in thin section (Figure 9C,D). This suggests that they were not vitrified and were therefore fired at relatively moderate temperatures.

The colour and hue of the ceramics in hand specimen and in thin section suggests that they were fired under a range of different redox conditions, from highly oxidising (abundant oxygen) (Figure 9E) to strongly reducing (oxygen-starved) atmospheres. Samples characterised by a dark core and a lighter margin (Figure 9F), were likely to have been incompletely oxidised during firing, due to the presence of abundant organic matter in the clay paste and/or a short firing duration. The variation in firing atmosphere interpreted for the ceramics, as well as their moderate degree of firing is indicative of a relatively unsophisticated, poorly controlled, non-kiln firing technology (Rye 1981, p. 98; Rice, 1987, p. 109). Ethnographic studies of traditional ceramic production in southern California document the use of a simple firing pit (Rogers, 1936, p. 14; Cline, 1984, p. 38; Hohenthal, 2001, p. 171), in which vessels are placed and covered with fuel. Direct evidence for prehistoric firing technology is almost entirely absent in the region. Pottery may have been fired away from habitation or other activity areas (Rogers, 1936, p. 5; Griset, 1996, p. 288), making firing loci less likely to be found by archaeologists. In this respect, investigation of the prehistoric ceramics in this study provides vital evidence for the ways in which pottery was fired in the past.

Ceramic Technology, Tradition and Cultural Identity

Each of the different petrographic fabric groups that have been detected and characterised in this study represent specific combinations of raw materials and manufacturing technologies that were used to produce Late Prehistoric ceramic vessels. The particular sequence of steps involved in each case, from the selection and processing of clay and temper, through paste preparation, vessel forming, finishing and firing, defines what is referred to as the object's *chaîne opératoire* (Leroi-Gourhan, 1964) or 'technological style' (Lechtmann, 1977). This ontogenic sequence of how a synthetic artefact such as a ceramic vessel came to be is also a record of the potter who produced it, in terms of the conscious decisions that she or he made and the procedures carried out (Whitbread, 2001). These thoughts, choices and actions are likely to have been determined by a range of direct and indirect environmental, material, social, cultural and economic factors that structured the world in which the potter lived (Sillar and Tite, 2000). With this in mind, deciphering the technological styles of Late Prehistoric ceramic sherds can reveal a wealth of valuable archaeological information about the hunter-gatherer societies of southern California.

The petrographic and microstructural evidence for ceramic technology presented in this study and its comparison with ethnographic accounts of the craft provides evidence that the potter's choices and behaviours were guided by the specific material properties of the different types of resources utilised and acquired knowledge about how to manipulate these resources to produce a desirable product. For example, the addition of particulate matter to the fine sedimentary clay used in several fabric groups appears to have been an effort to compensate for the lack of naturally occurring inclusions in these raw materials, which otherwise may have resulted in clay that was too sticky to shape, or a vessel that was likely to crack during firing. Similarly, potters would have known through experience that insufficient reduction of clay raw materials would result in a failed product and therefore processed the clay by grinding and cleaning it of large impurities.

Given that potters, in investing time and effort into the ceramic production process, would have aimed to successfully create a functional object, it is likely that many of their technological decisions would have been based on practical and material constraints. However, within the ceramics analysed in this study, important evidence also exists to suggest that potters made choices and behaved in ways that were influenced by other than strictly functional factors that may have been of equal importance to them. Examples of such choices include the intentional addition of grog to ceramics made from coarse residual clay and the use of several different types of temper in a single clay paste. Ethnographic studies of traditional pottery manufacture in many parts of the world have highlighted the complex, socially-embedded nature of ceramic technology (Stark, 2003) and the reasons behind these behaviours may therefore be related to the potter's social milieu, including local and family traditions and beliefs, as well as a desire for individual expression.

Distinguishing among possible social and cultural influences on prehistoric technology may not be easy. In the case of the occurrence of grog in some coarse residual ceramics in this study, it is possible that the potter(s) that made these vessels added temper because their parents, ancestors or other artisans in their tribe or family group did the same. The potters may have been influenced by their own personal spiritual beliefs or may have been engaged in an established ritual process. Animistic beliefs and practises among southern Californian tribes have been documented (Patencio, 1934, p. 4; Rogers, 1936, p. 2, 5; Wilken, 1982; Bean et. al., 1991, p. 9; Hohenthal, 2001, p. 168). The addition of grog temper to clay may thus have been a symbolic action, representing renewal and remembrance of the deceased. Another possibility, suggested by the occurrence at a single site of both tempered and non-tempered varieties of the same residual clay paste, may be that the addition of grog was a matter of individual choice and expression. Recent ethnographic studies at the traditional potting village of Santa Catarina in Baja California have recorded significant variations in potters' methods that mark the distinctive style or hallmark of individual craftspeople (Wilken 1982). Potters in the past may likewise have carried out specific practises as a way of distinguishing their own products from those of others and expressing their own identity. Ethnohistoric records compiled by Rogers (1936, p. 22, 27) also indicate individual and group variation in the use of grog temper, though this practice appears to have been significantly less common in the early 20th century than in Late Prehistoric times based on the evidence presented in this study.

Moving beyond technical considerations to explore the social and cultural aspects of hunter-gatherer ceramic production is an exciting prospect. The results of this study combined with ethnohistoric and ethnographic reports suggest that petrographic analysis has the potential to help define different cultural or family-based ceramic traditions amoung the various ethno-linguistic groups that inhabited southern California. Despite the apparently homogeneous material culture complex that characterises most Late Prehistoric sites, important variability is likely to exist in the ways in which pots were made by different individuals, families, bands and tribes (Rogers, 1936, p. 2). In this study, a comparison of the ceramics from northern and southern sites within the Anza-Borrego Desert State Park revealed a high degree of variation within a small sample of sherds suggesting that a variety of techniques and raw material sources were used to produce the pottery vessels and that they were likely to have been transported across significant distances. Future comparative analyses of ceramics from a broader region will be needed to better understand the cultural patterning that exists in Late Prehistoric hunter-gatherer ceramic manufacture.

Summary

The compositional analysis of Late Prehistoric hunter-gatherer ceramics from the desert region of eastern San Diego County presented in this study has revealed a previously unexpected level of meaningful variability within these plain, undecorated artefacts. Detailed thin section ceramic petrography resulted in the definition of numerous distinct fabric classes or recipes, characterised by specific combinations of raw materials and technology. The compositional diversity of the ceramics and their correlation with geological field samples suggests that potters had an intimate knowledge of the geodiversity of the region and utilised a wide range of different naturally-occurring raw materials. Much of the pottery found at the seven sites analysed is likely to have been non-local in origin, having been made elsewhere in southern California and transported over significant distances in various directions, perhaps through seasonal movements or trade among hunter-gatherer social groups.

Detailed investigation of the steps involved in the manufacture of the various recipes has provided an important window into the nature of hunter-gatherer ceramic technology. Many of the aspects highlighted in this study could not have been detected by hand specimen studies or earlier geochemical analyses of southern California ceramics. A comparison of this data with ethnographic and historical accounts of the traditional craft as it existed in the 20th century in isolated locales suggests much continuity in practice, but also differences that may represent the impacts of European and Mexican settlers and modern tourism on indigenous cultural traditions (Griset, 1990; Wade, 2004).

By beginning to identify the choices and behaviours of potters in the past, we hope to establish a framework within which to evaluate the different environmental, social and cultural influences that underlie their particular ceramic traditions or styles. Evidence for technological choices based on the performance characteristics of materials was detected in many of the ceramics analysed. This indicates that potters possessed significant acquired skills and knowledge about different raw materials and their properties. In other cases, technological practices appear to have been guided by less readily apparent criteria that may have their origins in social or cultural practices and beliefs, or might be evidence of individual expression. Unraveling this deeper meaning within the plainware ceramics is a challenging undertaking, but holds significant potential for better understanding the role of ceramics and ceramic technology in the hunter-gatherer societies of southern California and elsewhere.

Acknowledgements

The research presented in this paper forms part of a larger initiative on the Late Prehistoric ceramics of southern California, funded by the Begole Archaeological Research Grant Program of the Colorado Desert Archaeology Society and the Anza-Borrego Foundation & Institute. The authors are grateful to staff and volunteers at the Begole Archaeological Research Center in Borrego Springs, California, including Robert Begole, Joan Schneider, Sue Wade and Bonnie Bade, for their generous support and help with the site collections, as well as State Park Geologist George Jefferson for his guidance and shared knowledge during the geological field sampling.

References

Arnold, D.E. 1985. *Ceramic Theory and Cultural Process*. Cambridge University Press.

Arnold, J.E., Walsh, M.R. and Hollimon, S.E. 2004. The Archaeology of California. *Journal of Archaeological Research* 12: 1–73.

Bean, L.J. and Lawton, H. 1965. *The Cahuilla Indians of Southern California*. Malki Museum Press, Canning, California.

Bean, L.J. 1978. Cahuilla. In: Heizer, R.F. (Ed.) *Handbook of North American Indians: Volume 8. California*. Smithsonian Institution, Washington: 575-587.

Campbell, P.D. 1999. Survival Skills of Native California. Gibbs-Smith, Salt Lake City.

Clifford, H., Bergen, F.W., Spear, S.G., Burns, D.M. and Tapper, G. 1997. *Geology of San Diego County: Legacy of the Land*. Sunbelt Publications, San Diego.

Cline, L. 1984. *Just Before Sunset*. LC Enterprises, Tombstone, Arizona.

Cline, L.C. 1979. *The ☐Kwaaymii: Reflections on a Lost Culture*. Occasional Paper 5, IVC☐ Museum Society, El Centro, California.

Cuomo di Caprio, N. and Vaughan, S.J. 1993. Differentiating Grog (Chamotte) from Natural Argillaceous Inclusions in Ceramic Thin Sections. *Archeomaterials*, 7: 21–40.

Curtis, E.S. 1908. *The North American Indian*. University Press, Cambridge, Massachusetts.

Drucker, P. 1937. Culture Element Distributions: V Southern California. *Anthropological Records*, 1: 1-52.

Eerkens, J.W. 2003. Residential Mobility and Pottery Use in the Western Great Basin. *Current Anthropology*: 728-738.

Eerkens, J.W., Neff, H. and Glascock, M.D. 2002. Ceramic Production among Small-Scale and Mobile Hunters and Gatherers: A Case Study from the Southwestern Great Basin. *Journal of Anthropological Archaeology*, 21, 200-229.

Ferenga, G.L. and Heredia, V.Y. 1995. *Pai Pai Ethnoarchaeology: Some Implications for California Archaeology*. Paper presented at the Society for California Archaeology southern data-sharing meetings, Los Angeles.

Gallucci. K.L. 2004. Ceramic Analysis at Wikalokal, San Diego County (CA-SDI-4787). *Proceedings of the Society for California Archaeology* 14: 119–123.

Gallucci, K.L. 2001. *From the Desert to the Mountains: Salton Brownware Pottery in the Mountains of San Diego*. Unpublished Master's thesis, San Diego State University.

Gifford, E.W. 1931. *The Kamia of Imperial Valley*. Bulletin of the Bureau of American Ethnology 97. Washington, D.C.

Griset, S. 1996. *Southern California Brown Ware*. Unpublished Ph.D. dissertation, University of California, Davis.

Griset, S. 1990. Historic Transformations of Tizon Brown Ware in Southern California. In: Mack. J.M. (Ed.) *Hunter-Gatherer Pottery in the Far West*. Nevada State Museum Anthropology Papers 23, Carson City, Nevada: 180-200.

Heizer, R.F. and Treganza, A.E. 1972 *Mines and Quarries of the Indians of California*. Ballena Press, Ramona, California.

Hildebrand J.A. Gross G.T. Schaefer J. and Neff H. 2002. Patayan ceramic variability: Using trace elements and petrographic analysis to study brown and buff wares in southern California. In: Glowacki D.M. and Neff H. (Eds.) Ceramic Production and Circulation in the Greater Southwest. Cotsen Institute of Archaeology Monograph 44. University of California, Los Angeles: 121–139.

Hohenthal, W.D. 2001. *Tipai Ethnographic Notes: A Baja California Indian Community at Mid Century*. Balena Press, Novato, California.

Howard H. 1982. Clay and the archaeologist. In: Freestone, I., Johns, C. and Potter, T. (Eds.) *Current Research in Ceramics*: Thin-Section Studies. British Museum Occasional Paper, 32: 145–158.

Hurd, G.S., Miller, G.E. and Koerper, H.C. 1990. An application of Neutron Activation analysis to the study of prehistoric Californian ceramics. In: Mack, J. (Ed.), *Hunter–Gatherer Pottery From the Far West*. Nevada State Museum Anthropological Papers 23, Carson City: 202–220.

James, H.C. 1960. *The Cahuilla Indians*. Westernlore Press, Los Angeles.

Jefferson, G. and Lindsay, L. 2006. *Fossil Treasures of the Anza-Borrego Desert*. Sunbelt, San Diego.

Jennings, C. W. 1967. *Geologic Map of California: Salton Sea Sheet. Scale 1:250,000*. State of California, Division of Mines and Geology. Sacramento.

Kroeber, A.L. 1925. *Handbook of the Indians of California*. Bureau of American Ethnology Bulletin 78. Washington, D.C.

Kroeber, A.L. and Hooper, L. 1978. *Studies in Cahuilla Culture*. Malki Museum Press, Banning, California.

Laylander, D. 1997. Last Days of Lake Cahuilla: The Elmore Site. *Pacific Coast Archaeological Society Quarterly* 33: 1–138.

Laylander, D. 1992 *Research Issues in San Diego Archaeology*. San Diego County Archaeological Society. Available on-line: http://home.earthlink.net/~researchissues

Lechtmann, H. 1977. Style in Technology: Some Early Thoughts. In: Lechtmann, H. and Merrill, T.S. (Eds.) *Material Culture: Style, Organization and Dynamics of Technology*. West Publishing Company, St Paul, Minnesota: 3-20.

Leroi-Gourhan, A. 1964. *Le geste et la parole I: Techniques et langage*. A. Michel, Paris.

Lyneis, M.M. 1988. Tizon Brown Ware and the problems raised by paddle-and-anvil pottery in the Mojave Desert. *Journal of California and Great Basin Anthropology*, 10: 146–155.

May, R.V. 1978. A Southern California Indigenous Ceramic Typology: A Contribution to Malcolm J. Rogers Research. *Journal of the Archaeological Survey Association of Southern California*, 2.

Middleton, A.P., Freestone, I.C. and Leese, M.N. 1985. Textural analysis of ceramic thin sections: Evaluation of grain sampling procedures, *Archaeometry*, 27: 64–74.

Patencio, F. 1943. *Stories and Legends of the Palm Springs Indians*. Times-Mirror Press, Los Angeles.

Peacock, D.P.S. 1971. Petrography of certain coarse pottery. In: Cuncliffe, B. (Ed.) *Excavations at Fishbourne, 1961-1969*. Research Report of the Society of Antiquaries, 27: 255-259.

Pymale-Schneeberger, S. 1993. Application of Quantifiable Methodologies in Ceramic Analysis:Petrographic and Geochemical Analysis of Ceramics from Riverside County, California. *Proceedings of the Society for California Archaeology*, 6: 257-276.

Reedy, C.L. 2008. *Thin-Section Petrography of Stone and Ceramic Materials*. Archetype, London.

Remeika, P. and Lindsay, L. 1993. *Geology of Anza-Borrego: Edge of Creation*. Kendall Hunt Publishing Company.

Rice, P.M. 1987. *Pottery Analysis: A Sourcebook*. University of Chicago Press.

Rogers, T.H. 1965. *Geologic Map of California: Santa Ana Sheet. Scale 1:250,000*. State of California, Division of Mines and Geology, Sacramento.

Rogers M.J. 1936. *Yuman Pottery Making*. San Diego Museum Papers 2. Museum of Man, San Diego.

Roux, V. and Courty, M.A. 1998. Identification of wheel-fashioning methods: technological analysis of 4th–3rd Millennium BC oriental ceramics. *Journal of Archaeological Science*, 25: 747–763.

Rye, O.S. 1981. *Pottery Technology: Principles and Reconstruction*. Taraxacum, Washington.

Sassaman, K.E. 2000. Agents of change in hunter-gatherer technology. In: Dobres, M-A. and Robb, J. (Eds) *Agency in Archaeology*. Routledge, London: 148-168.

Sassaman, K.E. 1993. *Early Pottery in the Southeast: Tradition and Innovation in Cooking Technology*. University of Alabama Press, Tuscaloosa, Alabama.

Schaefer, J. 1994a. The stuff of creation: Recent approaches to ceramics analysis in the Colorado Desert. In: Ezzo, J. (Ed.), *Recent Research Along the Lower Colorado River*, Statistical Research, Technical Series 51, Tucson, Arizona: 81-100.

Schafer, J. 1994b. The Challenge of Archaeological Research in the Colorado Desert: Recent Approaches and Discoveries. *Journal of California and Great Basin Anthropology*, 16: 60-80.

Schroeder, A.H. 1958. Lower Colorado Buff Ware: A Descriptive Revision. In: Colton, H. (Ed.) *Pottery Types of the Southwest*. Museum of Northern Arizona Ceramic Series 3D. Flagstaff, Arizona.

Sillar, B. and Tite, M.S. 2000. The Challenge of 'Technological Choices' for Materials Science Approaches in Archaeology. *Archaeometry*, 42: 2-20.

Strand, R.G. 1962. *Geologic Map of California: San Diego-El Centro Sheet. Scale 1:250,000*. State of California, Division of Mines and Geology. Sacramento.

Thompson, V.D., Stoner, W.D. and Rowe, H.D. 2008. Early Hunter-Gatherer Pottery along the Atlantic Coast of the Southeastern United States: A Ceramic Compositional Study. *Journal of Island & Coastal Archaeology*, 3: 191-213.

Van Camp, G.R. 1979. *Kumeyaay Pottery: Paddle-and-Anvil Techniques of Southern California*. Ballena Press, Socorro, New Mexico.

Wade, S.A. 2004. *Kumeyaay and Paipai pottery as evidence of cultural adaptation and persistence in Alta and Baja California*. Unpublished masters thesis, San Diego State University.

Wade, S.A. 1999. *Analysis of Ceramics from SDI-10156 A and B: Topomai*. Unpublished Report on file at Begole Archaeological Research Center, Borrego Springs, California.

Wallace, W.J. 1962. *Archaeological explorations in the southern section of Anza-Borrego State Park, California*. Archaeological Report 5, State of Calfornia Parks and Recreation, Sacramento California.

Wallace, W.J. and Taylor, E.S. 1958. An archaeological reconnaissance in Bow Willow Canyon, Anza-Borrego Desert State Park. *The Masterkey*, 32: 155–166.

Waters, M.R. 1982. The Lowland Patayan Ceramic Typology. In: McGuire, R.H. and Schiffer, M.B. (Eds.) *Hohokam and Patayan: Prehistory of Southwestern Arizona*. Academic Press, New York: 537-570.

Waters, M.R. 1983. Late Holocene Lacustrine Chronology and Archaeology of Ancient Lake Cahuilla, California. *Quarternary Research*, 19: 373-387.

Whitbread, I.K. 2001. Ceramic Petrology, Clay Geochemistry and Ceramic Production: From Technology to the Mind of the Potter. In: Brothwell, D. R. and Pollard, A. M. (Eds.) *Handbook of Archaeological Sciences*, Wiley, New York: 449–458.

Whitbread, I.K. 1996. Detection and Interpretation of Preferred Orientation in Ceramic Thin Sections, In: *Proceeding of the 2nd Symposium of the Hellenic Archaeometrical Society (26–28 March 1993)*: 413-425

Whitbread, I.K. 1995 *Greek Transport Amphorae: A Petrological and Archaeological Study*. Fitch Laboratory Occasional Paper 4. British School at Athens.

Whitbread, I.K. 1989. A proposal for the systematic description of thin sections towards the study of ancient ceramic technology. In: Maniatis, Y. (Ed.) *Archaeometry: Proceedings of the 25th International Symposium*. Elsevier, Amsterdam: 127–138.

Whitbread, I.K. 1986. The characterization of argillaceous inclusions in ceramic thin sections. *Archaeometry*, 28: 79-88.

Wilken, M. 1982. The Paipai Potters of Baja California: A Living Tradition. *The Masterkey*, 60:18-26.

Williams, S.L. 1989. *Micro-analysis of Sherds from CA-SDI-10780*. Unpublished Report, RECON Archaeology, California.

Woods, A. 1984, Use of tangential thin sections on experimental and archaeological material, *Bulletin of the Experimental Firing Group*, 3: 108-114.

ACKNOWLEDGEMENTS

This book would have not been possible without the help of numerous individuals, who kindly provided their expertise during its production or assisted with the organisation of the Sheffield meeting on *Petrography of Archaeological Materials* from which is was inspired.

Former and current committee members of the Ceramic Petrology Group (CPG), including Caroline Cartwright, Ian Freestone, Louise Joyner, Andrew Middleton, Susan Pringle, Alan Vince and Ian Whitbread provided important moral support in getting the Sheffield meeting off the ground and kindly permitted the use of valuable CPG funds. Hosting the meeting at the Department of Archaeology would have not been possible without the help of a competent team of PhD and MSc student volunteers consisting of Noemi Müller, Elisa Alonzo Lopez, Barbara Borgers, Noirin Hurley, Derek Pitman, Angelos Sotiropoulos and David Broughton. Several other members of the Sheffield Department prepared delicious food that was enjoyed by all at the meeting reception. Invited keynote speakers Evangelia Kiriatzi and Lara Maritan did an excellent job in opening the conference and Peter Day provided an equally fitting conclusion by summarising the proceedings of the weekend. Lastly, all delegates that attended the Sheffield meeting are thanked for their part in making it a vibrant and stimulating experience.

The process of peer reviewing the papers in this volume benefited greatly from the expert opinion of numerous individuals, many of who are also authors of chapters in the book. In addition, Michael Smith, Chris Doherty, Ken Dorning, Linda Howie, Michela Spataro, Ben Jervis, Alice Hunt, Peter Hommel, Noemi Müller, Laura Campbell, Elisa Alonzo Lopez and Barbara Borgers also provided valuable comments on specific chapters. All contributors are thanked for their patience and cooperation in adhering to the specific instructions for authors and presentation guidelines. Hopefully the end product was worth the extra time and effort. Finally, Ian Whitbread's efforts in composing a most appropriate foreward under very difficult circumstances are greatly appreciated.

The year or so that it has taken to put this volume together was made more bearable by the patience, support and understanding of Mrs. Jessica Quinn and Miss Karolina Quinn. Thank you.

Dr. Patrick Sean Quinn
Department of Archaeology
University of Sheffield, UK